ADVANCES IN CHEMICAL ENGINEERING

Volume 3

Advances in

CHEMICAL ENGINEERING

Edited by

THOMAS B. DREW

*Department of Chemical Engineering
Columbia University
New York, New York*

JOHN W. HOOPES, JR.

*Atlas Chemical Industries, Inc.
Wilmington, Delaware*

THEODORE VERMEULEN

*Department of Chemical Engineering
University of California
Berkeley, California*

VOLUME 3

Academic Press · New York and London · 1962

Copyright 1962 by Academic Press Inc.

ALL RIGHTS RESERVED

NO PART OF THIS BOOK MAY BE REPRODUCED IN ANY FORM
BY PHOTOSTAT, MICROFILM, OR ANY OTHER MEANS,
WITHOUT WRITTEN PERMISSION FROM THE PUBLISHERS

ACADEMIC PRESS INC.
111 Fifth Avenue
New York 3, N. Y.

United Kingdom Edition
Published by
ACADEMIC PRESS INC. (London) Ltd.
Berkeley Square House
London W. 1

Library of Congress Catalog Card Number: 56-6600

PRINTED IN THE UNITED STATES OF AMERICA

CONTRIBUTORS TO VOLUME 3

JOHN BEEK, *Shell Development Co., Emeryville, California*

ALAN FERGUSON, *Stanford Research Institute, Menlo Park, California*

C. S. GROVE, JR., *Department of Chemical Engineering and Metallurgy, Syracuse University, Syracuse, New York*

DANIEL HYMAN, *Stamford Research Laboratories, American Cyanamid Company, Stamford, Connecticut*

ROBERT V. JELINEK, *Department of Chemical Engineering and Metallurgy, Syracuse University, Syracuse, New York*

RUSSELL C. PHILLIPS, *Stanford Research Institute, Menlo Park, California*

HERBERT M. SCHOEN, *Department of Chemical Engineering and Metallurgy, Syracuse University, Syracuse, New York*

DOUGLASS J. WILDE, *Department of Chemical Engineering, The University of Texas, Austin, Texas*

PREFACE

The range of papers in this third volume of *Advances in Chemical Engineering* exemplifies again the vitality and range of chemical engineering. The unit operations are represented in papers on crystallization and on mixing; the first draws heavily upon modern chemistry and physics, the second upon fluid mechanics. Process techniques of the future are foreseen in the article on high temperature technology. Finally, mathematical methods are applied to apparatus design, in the specific case of catalytic reactors, and to management policy and practice, in the field of optimization of plant operations.

A sustained and steadily intensified need exists in the chemical engineering profession for authoritative and comprehensive reviews of the current status of the surging tide of research. To meet this need, an increased frequency of publication of *Advances* is foreseen, probably an annual schedule. While many of the reviews presented in this and earlier volumes are invited contributions, the Editors will be (and have been) most receptive to suggestions regarding appropriate topics and authors, and to independent proposals to prepare a review of the desired type.

New York, New York THOMAS BRADFORD DREW
Wilmington, Delaware JOHN WALKER HOOPES, JR.
Berkeley, California THEODORE VERMEULEN
February, 1962

CONTENTS

Contributors to Volume 3 v

Preface vii

Crystallization from Solution

C. S. Grove, Jr., Robert V. Jelinek, and Herbert M. Schoen

I. Introduction 1
II. Solubility Principles 4
III. Nucleation 13
IV. Crystal Growth 22
V. Crystal Size Distribution 31
VI. Crystallization Equipment 47
Nomenclature 56
References 58

High Temperature Technology

F. Alan Ferguson and Russell C. Phillips

I. Introduction 61
II. Temperature Definitions 62
III. Temperature Measurements 70
IV. Means for Attaining High Temperatures 83
V. Trends in High Temperature Technology 107
Nomenclature 113
References 115

Mixing and Agitation

Daniel Hyman

I. Introduction 120
II. General Characteristics of Mixing Processes and Agitated Vessels . . 121
III. One-Liquid-Phase Systems 133
IV. Gas-Liquid Systems 157
V. Liquid-Liquid Systems 167
VI. Solid-Liquid Systems 176
VII. Heat Transfer 183
VIII. Scale-Up of Heterogeneous Systems 187
IX. Experimentation with Agitated Systems 190
Nomenclature 196
References 198

Design of Packed Catalytic Reactors
John Beek

I. Introduction	204
II. Reduction of Chemical and Rate Equations to an Independent Set	205
III. Equations Describing Simultaneous Reaction and Transport Processes	211
IV. Estimation of Transport Properties	224
V. Numerical Solution of Equations	235
VI. Stability of Packed Tubular Reactors	257
VII. Scale Models of Packed Tubular Reactors	259
Nomenclature	268
References	269

Optimization Methods
Douglass J. Wilde

I. Introduction	273
II. Search Problems	277
III. Interaction Problems	292
IV. Feasibility Problems	314
Acknowledgment	331
References	331

AUTHOR INDEX	333
SUBJECT INDEX	340

CRYSTALLIZATION FROM SOLUTION

C. S. Grove, Jr., Robert V. Jelinek, and Herbert M. Schoen*

Department of Chemical Engineering and Metallurgy
Syracuse University, Syracuse, New York

I. Introduction	1
A. Crystal Types	2
B. Crystallographic Systems	3
II. Solubility Principles	4
A. Solubility Relationships	4
B. Supersaturation	11
III. Nucleation	13
A. Basic Concepts	13
B. Energy Relationships	15
C. Nucleation Kinetics	17
IV. Crystal Growth	22
A. Geometric Characteristics	23
B. Growth Mechanism and Rate	24
C. Dissolution	30
V. Crystal Size Distribution	31
A. Crystal Size in Relation to Nucleation and Growth	31
B. Multicrystal Growth in Seeded Solutions	32
C. Growth in Spontaneously Seeded Solutions	37
D. Crystal Size Distribution for a Continuous Process	42
VI. Crystallization Equipment	47
A. History	47
B. Classification of Equipment	48
C. Operation and Design	49
VII. Addendum	55
Nomenclature	57
References	58

I. Introduction

Crystallization problems are of great interest in modern science and technology. Modern solid state physics is closely connected with the study and use of large uniform crystals. A complex and deep study of crystallization processes is required in order to grow these crystals. Crystal studies are also important in biology, particularly in virology. Improvements in contemporary metallurgy and the chemical, ceramic

* Present address: Radiation Applications, Inc., Long Island City, New York.

and other industries depend to a large extent on a thorough knowledge of crystallization processes.

It is apparent that crystallization is important as a unit operation because of the large number of solid materials which are marketed in the crystalline form. A crystal growing from an impure solution is itself much purer than the solution. This affords a practical method of obtaining solid chemicals in a pure form suitable for handling, storing, and marketing. The most important criteria of a crystallization process are product yield and purity. There often are other requirements as to shape, size, and range of sizes of the crystals, which markedly affect the design, production, and operating characteristics of crystallization equipment.

A crystal may be defined as a homogeneous particle of solid which is formed by the solidification, under favorable conditions, of a chemical element, a compound, or an isomorphous mixture, whose boundary surfaces are planes symmetrically arranged at definite angles to one another in a definite geometrical form. Many chemical substances solidify into various such forms, and each substance thus crystallized takes a specific characteristic form. A crystalline substance can be defined as one in which the internal atomic or molecular arrangement is regular and periodic in three dimensions over intervals which are large compared with the unit of periodicity. Crystallization, then, is the process or unit operation of producing crystals or crystalline substances.

Crystallization can occur from melts, solutions, or vapors. Since crystallization from aqueous solutions is most pertinent to chemical engineering, this aspect of the general topic is stressed in the following presentation. Historical and descriptive material are minimized. The fundamental principles underlying solubility, nucleation, and crystal growth are presented first, followed by a brief discussion of their application in modern practice, so that the reader may be apprised of recent significant advances in the design and operation of crystallization equipment.

A. Crystal Types

Crystalline solids need not be single crystals. Usually they are composed of an aggregate of crystals which can be distinguished as separate entities under the microscope. They can also be composed of crystallites, that is, of crystals in which the pattern repeats itself a few times in each direction. In this case, they cannot be resolved by ordinary microscopic examination. Crystallite size is extremely variable and not easily measurable. However, sometimes crystallites are so nearly parallel to each other that the solid as a whole can be called a single crystal (L4).

Crystalline substances may be classified into five major types (S11). They vary in the kind and strength of the bond between the constituent atoms or ions, and in their electrical, magnetic, and mechanical properties. These types are metal crystals, ionic crystals, valence crystals, semiconductor crystals, and molecular crystals.

Metal crystals are formed from the atoms of the electropositive elements. In alloys, a definite orientation of the atoms of each constituent may or may not exist. Metal crystals are characterized by a very strong interatomic bond and by excellent electrical and thermal conductivities, because of high electron mobility.

Ionic crystals are formed by combinations of highly electropositive and highly electronegative elements, such as ordinary salts. Ions rather than atoms occupy positions in the space lattice and are held together by coulomb electrostatic forces. Ionic crystals obey valence rules and are good ionic conductors of electricity when molten.

Valence crystals are formed by combinations of atoms of the lighter elements in the middle column of the periodic table, such as diamond and silicon carbide. These crystals conform to valence rules, and the interatomic bonds are due to the sharing of electron pairs. Valence crystals are characterized by very great hardness, poor cleavage, and poor electrical conductivity.

Semiconductor crystals add impurities easily. This modifies the properties of the pure crystals so that they do not obey normal valence rules. Zinc oxide is an example of this type of crystal. Pure crystals of this type are characterized by deficient space lattices.

Molecular crystals are formed by inactive atoms, such as the rare gases, and by molecules with completed shells, such as organic compounds. Their weak bonds consist essentially of weak residual forces of the van der Waals type. Molecular crystals are relatively soft and some can be evaporated molecularly.

B. Crystallographic Systems

Crystalline compounds exist in a great many crystal forms. The accepted method for the crystallographic classification of crystals is based on the angles between the crystal faces. In this classification system, the types of crystal forms are not related to the relative sizes of the crystal faces, since the relative development of the faces is characteristic of the specific material. The cubic system, for example, is characterized by the fact that the faces of a cubic crystal can be referred to three equal and mutually perpendicular axes. The actual macrocrystal may be a cube, a needle, a plate, or an aggregate of imperfect crystals.

As long as the faces are perpendicular to each other, the crystals fall crystallographically into the cubic system.

In 1930, Hessel showed that thirty-two different classes of crystal symmetry or point groups were possible. He based his conclusions on geometrical considerations and on the occurrence of different elements of symmetry passing through a single point, taken in conjunction with the law of rational indices (G4). These thirty-two classes fall conveniently into seven crystallographic systems. Each system is characterized by axial angles and ratios; six of these systems are represented by three crystallographic axes. One, the hexagonal system, is conveniently represented by a set of four axes.

These seven crystallographic systems are illustrated in Fig. 1 and may be described as follows:

(a) The triclinic or anorthic system. Three axes not at right angles, all axes of unequal length.

(b) The monoclinic system. Three axes of which two form a right angle with the third axis inclined, all axes of unequal length.

(c) The orthorhombic or rhombic system. Three axes at right angles, all axes of unequal length.

(d) The rhombohedral or trigonal system. Three axes equally inclined, all angles equal but not at right angles, all axes of equal length.

(e) The tetragonal system. Three axes at right angles, two axes of equal length, but not equal to the third axis.

(f) The hexagonal system. Three equal axes, coplanar at 60°, and a fourth axis of different length at right angles to the other three.

(g) The cubic or regular system. Three axes at right angles, all axes of equal length.

II. Solubility Principles

An understanding of crystallization processes begins with a knowledge of fundamental solubility relationships. The important principles of solubility and "supersolubility" are briefly discussed in the following paragraphs. Several complete treatises have been written on these subjects and cover the theory and data in depth (C2, R3).

A. SOLUBILITY RELATIONSHIPS

When a liquid (solvent) and a solid (solute) are brought into contact, the attractive forces of the liquid tend to break apart the surface of the solid and to disperse its ions or molecules into the liquid in the form of discrete mobile units. This process is termed "solution" or, preferably, "dissolution." The extent to which the solute dissolves is de-

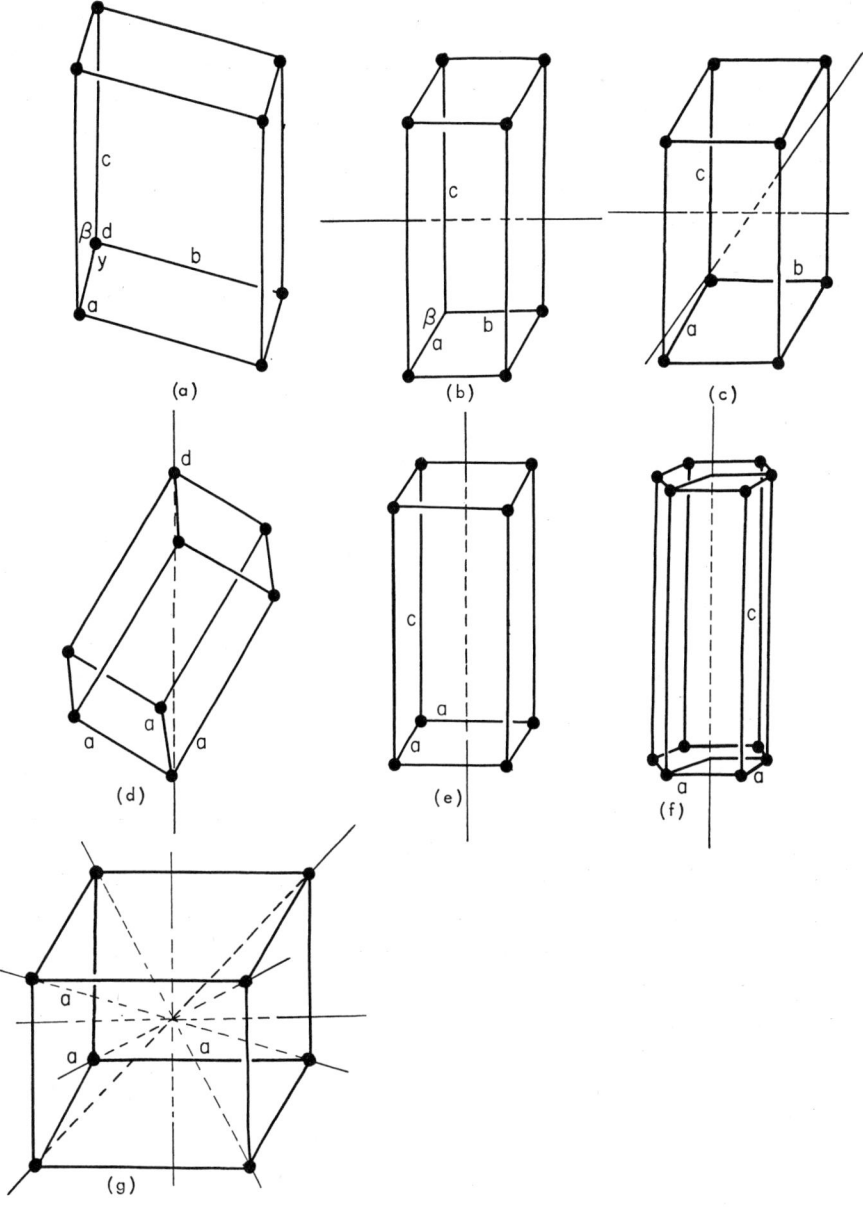

FIG. 1. Unit cells of the seven systems. (a) triclinic, (b) monoclinic, (c) orthorhombic, (d) rhombohedral, (e) tetragonal, (f) hexagonal, (g) cubic.

fined as its "solubility." More specifically the term "solubility" is used to denote the equilibrium-limit value of solution concentration, at which point the solution becomes "saturated." Under certain conditions (e.g., subcooling of an unseeded saturated solution) this equilibrium-limit value of solution concentration may be exceeded, leading to "supersaturation." This is discussed in more detail later.

The process of dissolution may be pictured as resulting from the weakening of bonds within the solid by the solvent. Thus a salt is soluble in water because its ions have a greater attraction for the water than for each other. Interionic forces may be calculated from the following equation (S12):

$$F = \frac{\epsilon_1 \epsilon_2}{d^2 \psi} \tag{1}$$

where

F = attractive force between ions
ϵ_1, ϵ_2 = charges on the ions
d = distance between the ions
ψ = dielectric constant of the solvent

Water is a good solvent for ionic crystals because it is a strong dipole and has a high dielectric constant. These characteristics tend to neutralize the attractive forces between the ions.

From thermodynamic considerations, it is possible to predict the extent of the solubility or, more specifically, the composition of the saturated solution at equilibrium. When solid A is dissolved in a solvent to give a saturated solution, there is a continual exchange of particles (ions or molecules) between the solid and the saturated solution. At equilibrium, the chemical potential of pure solid A must equal the chemical potential of dissolved A. This is equivalent to saying that the fugacity of the solid must equal the fugacity of the same substance in solution. If $f_{A(s)}$ denotes the fugacity of the pure solid and $f_{A(l)}$ denotes that of the corresponding pure liquid at the same temperature, then, assuming Raoult's law in the solution, the following relationship holds at saturation:

$$f_{A(s)} = f_{A(l)} y_A \tag{2}$$

Since the system under consideration is one in which a solid material is being dissolved in an appropriate solvent, it is evident that pure liquid A cannot be the soluble modification of A under the conditions assumed. Therefore, $f_{A(l)}$ must be fugacity of the super-cooled liquid. This in turn is the fugacity of the vapor which would be in equilibrium with the super-cooled liquid. Such vapor would be supersaturated in relation to the solid so that it would have a spontaneous tendency to form crystals. The vapor pressure of A in solution must be reduced to the point where the above equation holds. This provides the basis for calculating the composition of the saturated solution (W2).

This discussion can be illustrated by Fig. 2, in which the fugacities (approximately vapor pressures) of the supercooled liquid, solid, and solution are plotted against the temperature. The system will be in equilibrium where the fugacity curve for the solid meets that of the

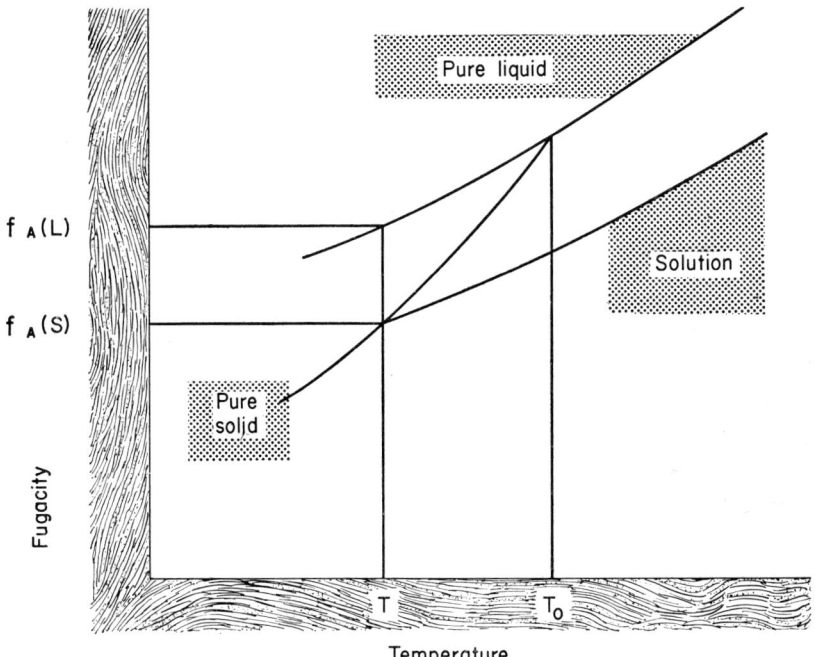

Fig. 2. Fugacity relationships at saturation. After Wall (W2).

solution. The solubility at any temperature will be determined by that amount of solvent which lowers the fugacity of the pure supercooled liquid solute to that of the solid solute at the given temperature. It is obvious that as the temperature is raised, the solubility of the solid increases.

The concentration of solute in a saturated solution (the composition of the saturated solution) is known as the solubility of the solute in the solvent. It is obvious that the solubility is dependent upon the nature of the solute, the nature of the solvent and the temperature. Pressure has a slight effect on solubility, but, for most practical cases, this effect can be neglected. It can be seen that for each given solute-solvent system, a value can be obtained for each given temperature. The curve obtained by plotting the saturation values of a given solute in a given solvent versus the temperature is known as a solubility curve or, less frequently, as a saturation curve.

Various materials exhibit characteristic and rather different solubility relationships, as evidenced by Fig. 3 and Table I.

TABLE I
SOLUBILITIES OF TYPICAL SALTS[a]

Temperature (°C.)	$CuSO_4$	Na_2SO_4	NaCl	$K_2Al_2(SO_4)_4$
0	14.3	5.0	35.7	3.0
10	17.4	9.0	35.8	4.0
20	20.7	19.4	36.0	5.9
30	25.0	40.8	36.3	8.4
40	28.5	48.8	36.6	11.7
50	33.3	48.7	37.0	17.0
60	40.0	45.3	37.3	24.8
70	—	—	37.8	40.0
80	55.0	43.7	38.4	71.0
90	—	—	39.0	109.0
100	75.4	42.5	39.8	—

[a] All solubilities are expressed as gm. of anhydrous per 100 gm. of water (P2, H9).

Although the composition of an ideal solution can be predicted theoretically, few solutions are ideal, and fugacities and activity coefficients are seldom available for real systems. Hence, in general, too little is known of the direct relationships between solubilities and the specific properties of the solute and the solvent to permit prediction of solubility curves. The characteristics of each system must be determined experimentally. In many cases, it is not even possible to predict the effect of temperature on the solubility values of a given solute-solvent system.

Ordinarily, the solubility of a given solute in a given solvent at a given temperature is not affected by the particle size of the residual undissolved solute. However, when submicroscopic particles are involved, the tendency of a solid to dissolve is affected by particle size in a manner analogous to the effect of particle size on vapor pressure. This is shown by the action of crystals which are in contact with their saturated solutions. While the relative amounts of solute and solvent must remain unchanged in such a solution, a dynamic exchange exists wherein the small crystals, possessing a greater tendency to dissolve than the large ones, go into solution. At the same time, the large crystals increase in size by a corresponding amount of solute. This is an important phenomenon in commercial crystallization. Equally important, sharp edges and corners of irregular solute masses also exhibit a tendency to dissolve and redeposit the plane surfaces. Thus irregular solutes tend to grow into regular shapes or into homogeneous crystals.

The dependence of solubility on the size of crystal grains present in the solution has been studied by many workers. Some of these have

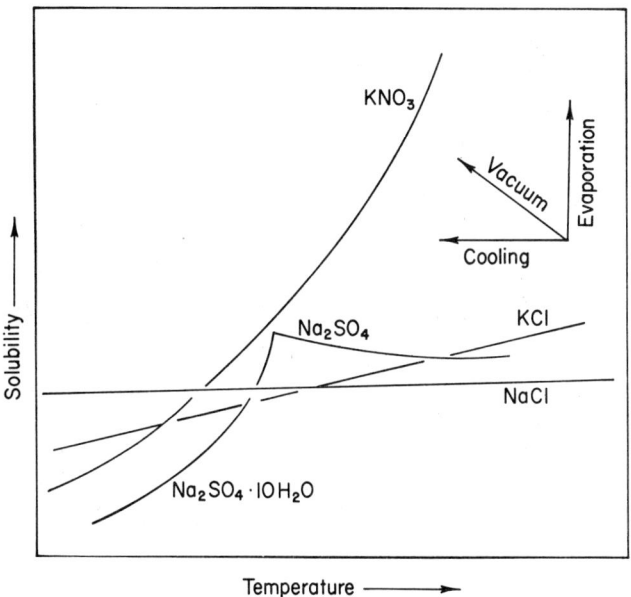

Fig. 3. Solubility curves and modes of crystallizer operation.

presented mathematical derivations to account for this dependence. These are discussed in detail by Buckley (B8), together with some numerical examples. The following equation (F4) applies:

$$\frac{RT}{M} \ln \left(\frac{C_{s(r)}}{C_{s(\infty)}} \right) = \frac{2\sigma}{\rho r} \qquad (3)$$

where

$C_{s(r)}$ = solubility of particles of radius = r
$C_{s(\infty)}$ = normal solubility (at $r = \infty$)
M = solute molecular weight
ρ = solute density
σ = surface tension, solid/liquid

When ions are involved, the left side of Eq. (3) should be multiplied by the term $(1 - \alpha + m\alpha)$, where m is the number of ions per mole and α is the degree of ionization (B8).

Mutual solubility relationships are of great importance in many industrial processes. The principles of fractional crystallization are used

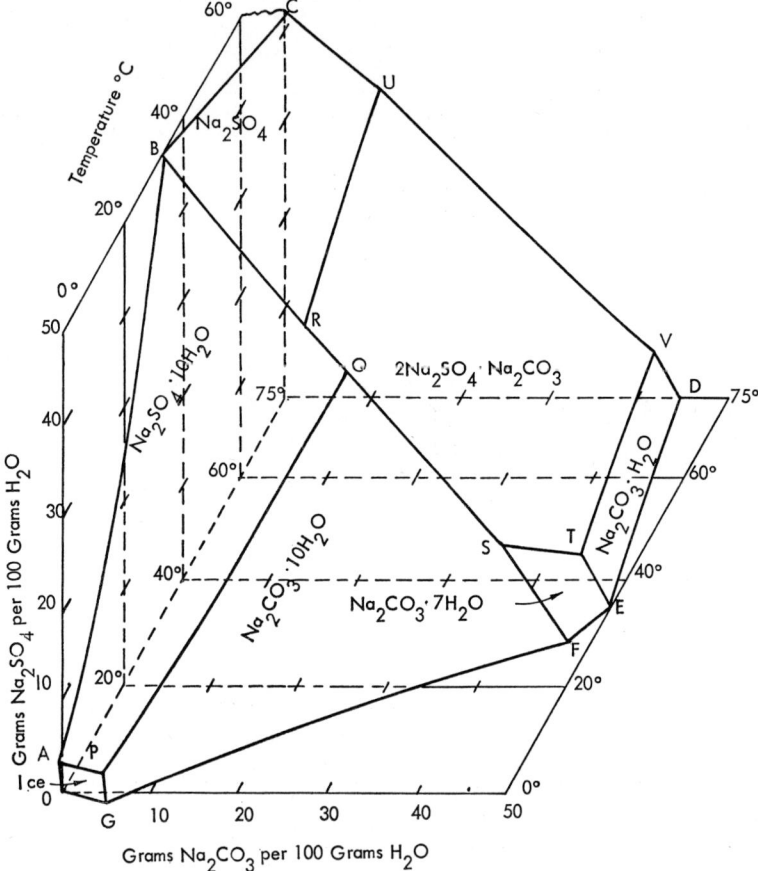

Fig. 4. Solubility of sodium carbonate–sodium sulfate in water. After Hougen and Watson (H11).

in the purification and separation of valuable soluble substances from less valuable or less desirable substances. Successive fractional crystallizations are often required in order to obtain the necessary high purity in the final product. Solubility data for many multicomponent systems are available in the literature. The data for a three-component system, consisting of two solutes and one solvent, can be presented in an isometric, three-coordinate diagram. The solubility data in such a diagram determine a series of surfaces which, with the axial planes, form an irregular-shaped envelope in space. The surfaces represent conditions under which equilibria are achieved for the solutes and the saturated solutions. A plot of this type is shown in Fig. 4 for sodium carbonate — sodium

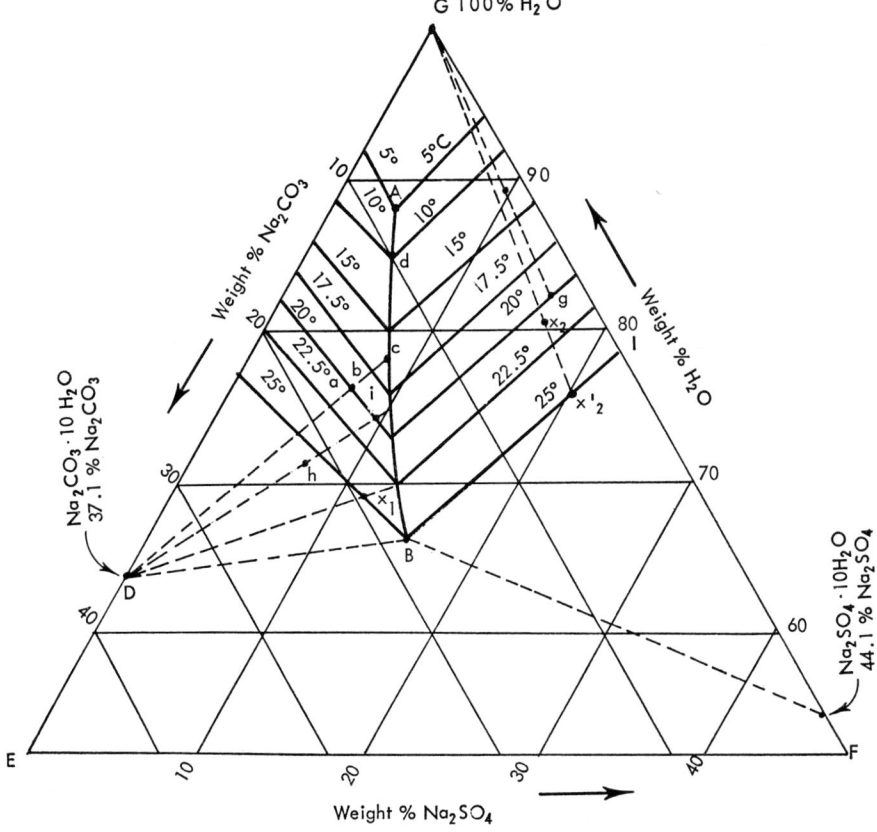

Fig. 5. Solubility of the system $Na_2CO_3 - Na_2SO_4$-H_2O at low temperatures. After Hougen and Watson (H11).

sulfate — water. For calculation purposes, however, data are more useful if they are plotted as a family of isothermal solubility curves on a triangular diagram. The same system is shown in this way in Fig. 5 (H11).

B. Supersaturation

The early investigators in crystallization learned that crystal growth does not take place in a solution which is just saturated. Crystallization is similar to any other mass transfer process in that a driving force is necessary. In crystallization, this driving force is a concentration gradient from the solution to the face of the growing solute crystal. This force arises from the fact that solubility can be exceeded; it is expressed as the difference between the supersaturated concentration and the solubility value.

The terms "supersaturation" and "supersolubility" were apparently first suggested by Ostwald to explain some of the growth phenomena

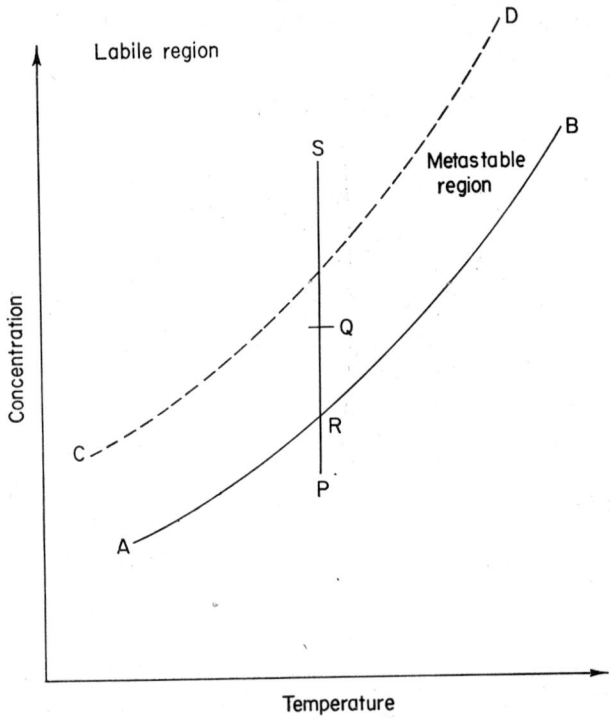

FIG. 6. Miers supersolubility concept.

which he observed in crystallization (O2). He also introduced the concepts of "metastable" and "labile" to clarify these phenomena. In a series of investigations Miers and his co-workers developed evidence which substantiated Ostwald's theory (M8). The "supersolubility" value is the lowest solution concentration at which crystal nuclei will form spontaneously in a supersaturated solution at a given temperature. Miers postulated that for each solute-solvent system a characteristic concentration-temperature relationship exists which defines a "supersolubility" curve. As illustrated in Fig. 6, this curve (*CD*) is approximately parallel to the normal solubility curve (AB) at a somewhat higher concentration. According to Miers, "self-nucleation" (spontaneous formation of crystal nuclei) can occur only in the "labile" region above the supersolubility curve (CD). But crystal growth can occur in both the "metastable" (between curves AB and CD) and labile regions.

In Fig. 6, point P represents a solution that is unsaturated; at this concentration neither nucleation nor crystal growth will occur. Point S represents a labile solution which will nucleate spontaneously, with concentration falling to point R as nucleation and crystal growth occur. Point Q represents a metastable concentration at which growth will take place if crystal seeds are present or added. Although the supersolubility limit is affected by external factors, discussed later, the Miers concept is useful as an empirical representation of nucleation behavior and in treating crystallization from unseeded solutions.

Crystallization of the solute from a solution can be brought about in three different ways. (1) The composition of the solution may be changed by removing the solvent through evaporation until supersaturation and consequent crystallization occur. (2) A decrease in temperature produces a condition of reduced solubility with consequent supersaturation and crystal growth. (3) A change in the nature of the system may induce crystallization. For example, the addition of alcohol to aqueous solutions of certain inorganic salts will often change solubility relationships and lead to crystallization. The first two of these methods, either separately or in combination, are used in commercial crystallization processes, as illustrated by the arrows in Fig. 3.

Many experimental results have been reported which show that materials other than the solute and solvent present in the solution may inhibit crystal growth and induce a greater degree of supersolubility than usual. For example, many years ago, Marc (M6) showed that the presence of even small amounts of the dye Ponceau 2R extends the supersaturation limit of potassium chlorate solution. The general effects of impurities and additives on nucleation and growth are discussed further below and in detail by Buckley (B8).

III. Nucleation

A. Basic Concepts

Crystallization is a two step process, requiring first nucleation and then crystal growth. In practice, as discussed below in Section V, the two steps occur concurrently, but their explanation is simplified by first considering them separately. Supersaturation is necessary for both nucleation and growth, but its effect is different in the two steps.

In the crystallization of melts, where relatively large degrees of supersaturation are attainable, nucleation and growth phenomena are more easily separated and studied experimentally than in crystallization from solution, which is characterized by rather narrow metastable regions. However, the basic concepts of nucleation are the same in both types of processes. In fact, much of the experimental verification of nucleation theory has come from studies of condensation and precipitation from the vapor phase. The highly publicized rainmaking experiments of several years ago made significant contributions (S6).

Nucleation is the generation within a metastable mother phase of the smallest particles of a more stable phase capable of growing spontaneously. Like other kinetic processes, nucleation involves the combination and activation of smaller unstable particles, called embryos. The critical rate-determining embryo becomes a nucleus, which differs from an equal number of ordinary molecules in solution by possessing sufficient excess surface energy to form a new phase (L1). This energy may be calculated via a general formula of Gibbs (G3).

Critical sizes characteristic of different substances vary from less than ten to several hundred molecules. The earliest measurements were made by Ostwald (O1), who found 10^{-10} g. to be the limiting mass of sodium chlorate in initiating crystallization from water solution. Volmer and Flood (V8) have shown that the critical nucleus in water vapor condensation is about 80 molecules. Extensive studies of barium sulfate precipitation (C4, L2) indicate that in this system the critical nucleus is no more than the unit cell, consisting of four sulfate and four barium ions.

Ostwald may be regarded as the father of modern nucleation theory. In addition to his significant experimental work, he contributed order and clarity to the confused mass of data accumulated in the 18th and 19th centuries by Fahrenheit, Lowitz, Gay Lussac, and many others. He popularized the basic concepts of supersaturated states which today are generally accepted and interpreted the fundamental contributions of Gibbs to thermodynamic stability and metastable states. LaMer (L1) summarizes a number of Ostwald's teachings; these are discussed in greater detail by Volmer (V7).

Nucleation may be either induced or spontaneous. Complete spontaneity is rarely, if ever, achieved experimentally, since extreme supersaturation reaching the labile zone is necessary and complete elimination of all catalytic influences is difficult. The principal factors (M3) which aid nucleation are: (a) mechanical impact, such as may be produced by vigorous stirring or sharp collisions between crystals; (b) the catalytic effects of existing crystals, foreign nuclei, or container walls; and (c) local variations in concentration, which can be produced by uneven agitation, temperature gradients, or surface evaporation. In practice, nuclei may be added in the form of small seed crystals, particularly at the start of an operation. Measures to remove excess nuclei are often used to control crystal size distribution.

B. ENERGY RELATIONSHIPS

Metastability is difficult to treat by the conventional methods of thermodynamics. The essential requirement of a metastable condition

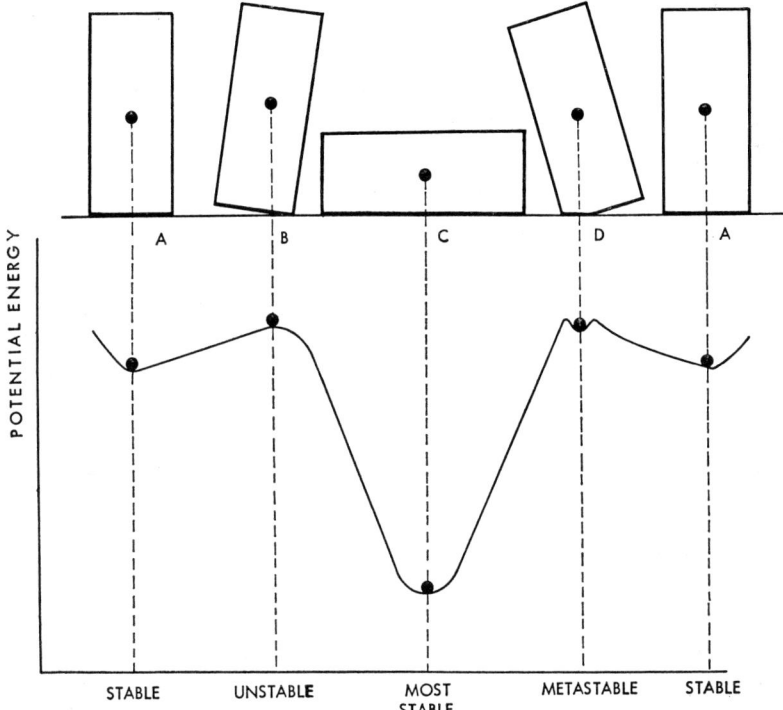

FIG. 7. Stability and metastability illustrated by brick on table. After LaMer (L1).

is that an energy maximum must be exceeded before spontaneous transition to a stable lower-energy state can occur. In other words, "activation energy" is required to disturb metastability.

LaMer's analogy (L1) of the brick on a table is a good illustration of metastability. This analogy is reproduced in Fig. 7. A and C represent the two stable states in this system; in either of these positions, if the brick is subjected to a small finite displacement, it will return readily when the disturbing influence is removed. From an unstable (labile) state at B, the disturbed brick never returns automatically to its position of maximum energy; instead it proceeds to a more stable state of lower energy (A or C). The metastable state, D, is illustrated by beveling a corner edge so that a more stable position than B is possible. In state D the brick can survive limited displacements whose magnitude is determined by the height of the energy barriers surrounding the small depression in the corresponding energy curve.

The key to calculating the activation energy associated with nucleation from supersaturated (metastable) solutions is Gibbs' formula (G3) for the work of forming a new phase within a homogeneous fluid:

$$w = \frac{\sigma s}{3} \qquad (4)$$

where

σ = interfacial tension
s = surface area of new phase particle

Starting with this equation and considering the generalized nucleation process

Homogeneous phase $A \longrightarrow$ droplets of phase B dispersed in A

LaMer (L1) derived the following equation for the free energy change accompanying nucleation:

$$\Delta G = 4\pi\sigma \left(r^2 - \frac{2}{3}\frac{r^3}{r_c} \right) \qquad (5)$$

which is plotted in Fig. 8. The maximum in this curve corresponds to the critical droplet radius (r_c) that must be attained for spontaneous growth to proceed. This is a metastable condition, since a particle of size r_c has an equal chance of growing or decreasing in size, both processes resulting in lower free energy. Particles larger than r_c can only increase in size. An important conclusion from Fig. 8 or Eq. (5) is that a minimum-sized particle is required before growth can proceed spontaneously.

Using the Kelvin-Gibbs equation for the effect of radius of curvature on vapor pressure of a spherical droplet (T3, L1)

$$\ln\left(\frac{p}{p_\infty}\right) = \frac{2\sigma V_B}{kT} \times \frac{1}{r} \qquad (6)$$

Frenkel (F2) derived an important equation for the work of nucleogenesis in terms of p/p_∞, the degree of supersaturation in the vapor phase

$$\mathcal{W} = \frac{16}{3} \frac{\pi \sigma^3 V_B^2}{[kT \ln (p/p_\infty)]^2} \tag{7}$$

where

V_B = volume of one molecule in the liquid phase.

While this equation may not be directly applicable to the formation of solid nuclei in solution, it is reasonable to believe that an analogous re-

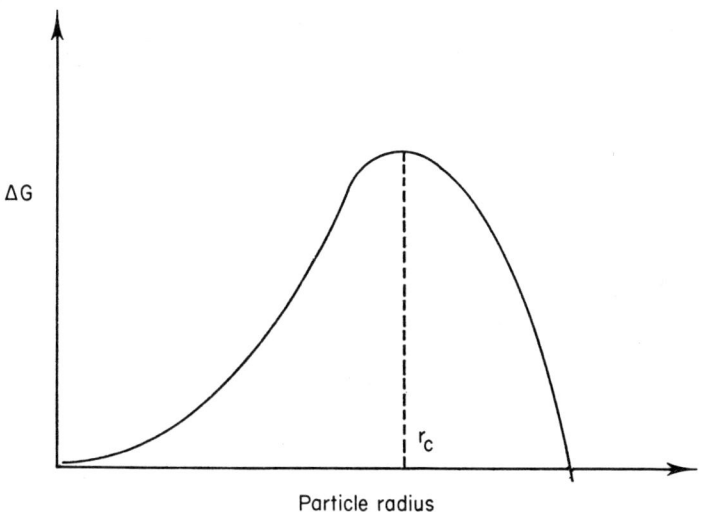

Fig. 8. Free energy of nucleation.

lationship exists involving a supersaturation ratio in terms of solution concentration.

C. Nucleation Kinetics

The effect of solution concentration on nucleation rate is shown qualitatively in Fig. 9. At low levels of supersaturation, the rate is essentially zero; but, as concentration is increased, a fairly well defined critical supersaturation is reached (point 1), beyond which nucleation rate rises steeply (curve 1–2). Point 1 may be regarded as the threshold of the labile region. Data from a series of such curves at different temperatures establish the locus of points at which nucleation starts, i.e., the Miers "supersolubility" curve discussed in Section II.

Although the existence of a supersolubility curve has been demonstrated as characteristic of most crystallizing systems, its position depends on a number of external factors. These include the presence of solid particles (either seeds or foreign nuclei), agitation, mechanical shock, holding-time, vessel surface area, and vibrations. For example, Van Hook and Frulla (V3) have shown that strong sonic irradiation in the frequency range 8–16 kc. initiates sucrose nucleation from solutions only 10% supersaturated, under conditions requiring at least 30% supersaturation for spontaneous nucleation. Extensive studies of the formation of ice crystals in air conducted by the General Electric Company (S6) have established that self-nucleation of water vapor requires supercooling to −39°C. Seeding with silver iodide, which closely resembles ice in its crystal structure, raises the nucleation temperature to about −5°C. Atmospheric dusts and pollens are capable of initiating nucleation at

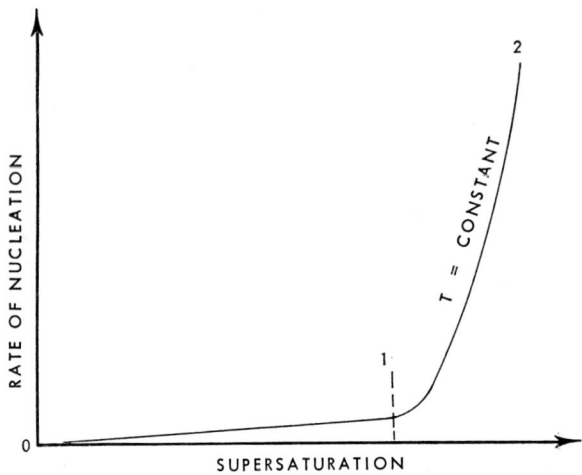

FIG. 9. Effect of supersaturation on nucleation rate.

−20 to −30°C. Negative effects of impurities have also been reported, such as in the studies of silver chloride crystallization from aqueous solution by Davies (D1), who observed that introduction of surface-contaminated seed crystals retarded nucleation. Many other interesting catalytic phenomena associated with nucleation are described in the literature (B3, B5, H5, M11, N4, T2, T6).

Temperature has also been studied as a variable in determining nucleation rate, particularly in crystallization from melts or from solutions maintained at constant supersaturation. Figure 10 shows a typical result, obtained with a piperine melt by Tammann (T1) and interpreted by Coulson and Richardson (C5). Here, as temperature is lowered, the degree of supersaturation and driving force causing nucleation are increased, causing a steep rise in the curve. A maximum is attained, how-

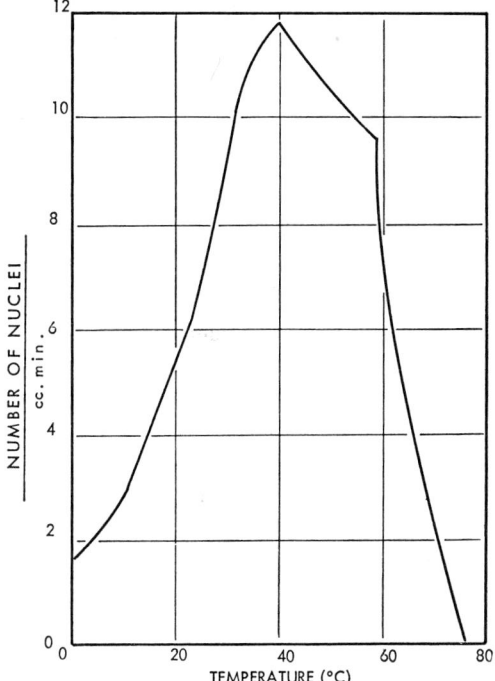

FIG. 10. Nucleation of piperine from melt. After Coulson and Richardson (C5).

ever, because further lowering of temperature adversely affects the kinetics of the process.

Quantitative treatment of nucleation kinetics usually begins with the assumption that the process follows the familiar general rate relationship

$$\text{Rate} = \nu e^{-\Delta G\ddagger/kT} \qquad (8)$$

where

ν = a frequency factor
$\Delta G\ddagger$ = free energy of activation

This expression can be applied directly to rate data for a particular system, and values of ν and $\Delta G\ddagger$ can be obtained experimentally. This has been done by various investigators.

The exponential term in Eq. (8) can be approached directly on the basis of the previous energy considerations, but the frequency factor is more difficult to predict theoretically since this requires postulation of a nucleation mechanism. If $\Delta G\ddagger$ is taken as equivalent to the work of forming a new phase, then Eq. (7) may be substituted directly into Eq. (8) indicating that the exponent is inversely proportional to [ln

$(p/p_\infty)]^2$. As pointed out by LaMer (L1), this term dominates the rate expression, making possible the prediction of critical supersaturation ratios within 10%, despite a hundredfold error in estimating the frequency factor. Close agreement between theory and experiment in the condensation of various vapors is demonstrated by the data of Volmer and Flood (V8) and discussed by Pound (P3).

This treatment may be extended to nucleation of solids from solution by substituting a supersaturation ratio in concentration units (C/C_s) into Eq. (7). The work of Preckshot and Brown (P4) with potassium chloride solutions represents a typical application.

In attempting to predict the frequency factor, one approach would be to assume that all of the atoms necessary to form the nucleus collide simultaneously. This seems rather unlikely, however, in view of the large numbers of atoms required in most cases. A more probable mechanism is one that postulates a stepwise series of collisions between embryos and single atoms, until a reasonable number of embryos grow to critical size and become nuclei.

The latter theory was first suggested by Becker and Doering (B2), who applied a quasi-equilibrium treatment and developed the following equation for the nucleation rate of condensing vapors:

$$n_0 = 9.5 \times 10^{25} \left(\frac{p_\infty}{T}\right)^2 \left(\frac{p}{p_\infty}\right)^2 \sqrt{\frac{\sigma M}{\rho}} \exp -\left[17.49 \left(\frac{M}{\rho}\right)^2 \left(\frac{\sigma}{T}\right)^3 \left(\ln \frac{p}{p_\infty}\right)^{-2}\right] \quad (9)$$

where

M = atomic weight
ρ = liquid density
n_0 = droplet formation rate (number/sec./cc.)

This expression is of the same general form as Eq. (8) and the exponential term corresponds to that derived above from Eq. (7). According to Becker and Doering, then, the frequency factor is proportional to the square of the supersaturation ratio. As pointed out by Pound (P3), the Becker-Doering equation closely predicts critical supersaturation values in the condensation of various liquids. For example, for water vapor at 275°K., the calculated ratio $p/p_\infty = 4.2$ agrees exactly with experimental results.

Since a critical size is associated with nucleation, the process is not instantaneous, but rather requires an induction period (or "waiting time"), which is a function of the supersaturation ratio. Time-lag data have been reported by several investigators (A1, C4, L2, P4, V10), showing generally that waiting time decreases sharply as supersaturation increases. Preckshot and Brown (P4) found a linear relationship between log (waiting time) and log^{-2} (supersaturation ratio) in the nucleation of potassium chloride solutions. In analyzing time-lag data for barium sulfate precipitation, LaMer (L1) shows a linear plot of log concentration (in terms of mean ionic mobility) versus log time. Since the slope of this plot is six, he concludes the reaction is seventh-order and suggests that the rate-determining step is the addition of a seventh ion to a cluster of three Ba^{++} and three SO_4^{--} ions. Van Hook (V2) presents a detailed discussion of time lag and other nucleation phenomena reported for sugar solutions.

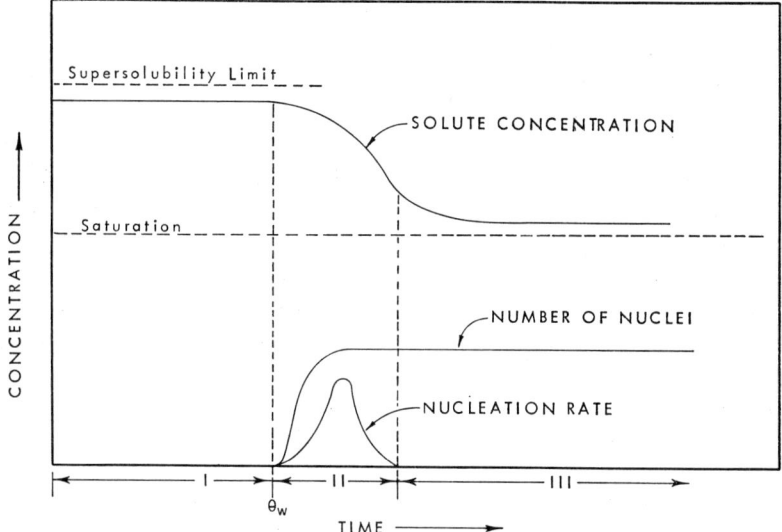

FIG. 11. Schematic representation of a nucleating system.

Figure 11 is a schematic summary of the nucleation process. Region I on the plot represents the induction period, region II the period of rapid nucleation and partial relief of supersolubility, and region III the period of steady growth on a constant number of nuclei.

Much of the treatment of nucleation catalysis in the literature is still qualitative, especially analysis of influences such as mechanical shock and ultrasonic vibrations. Heterogeneous nucleation (i.e., catalytic effects of foreign nuclei), however, has received some theoretical attention, building on the early work of Volmer (V6). Considering vapor condensation on a catalyst surface, he proposed using an interfacial contact angle ϕ as a mathematical parameter, defined by the equation

$$\cos \phi = \frac{\sigma_{CV} - \sigma_{CL}}{\sigma_{LV}} \tag{10}$$

where

σ_{CV} = catalyst-vapor interfacial tension
σ_{CL} = catalyst-liquid interfacial tension
σ_{LV} = liquid-vapor interfacial tension

The effective work of forming a nucleus is then expressed by Eq. (7), multiplied by a factor $0.25\ (2 + \cos \phi)(1 - \cos \phi)^2$.

This treatment was extended successfully to nucleation of crystals from solutions and melts by Turnbull and Vonnegut (T6), who discuss its applicability to representative literature data. In general, potent nucleation catalysts have low index planes with an atomic configuration similar to that of certain low index planes of the crystallizing substance. Telkes (T2) suggested that correspondence of lattice constants within 15% is necessary and Preckshot and Brown (P4) demonstrated positive catalytic effects with a series of materials corresponding within 10%. Turnbull and Vonnegut (T6) proposed using the reciprocal of disregistry as an index of catalytic potency. Later, Newkirk and Turnbull (N2) showed that for some systems the logarithm of the critical supersaturation ratio is proportional to the square of the disregistry.

While observations such as these contribute to the necessary accumulation of knowledge in this area, their empirical nature is evidence that a comprehensive quantitative understanding of nucleation catalysis is still in the process of development. Many seemingly isolated phenomena are still treated separately, and rather formidable experimental difficulties retard their rationalization and correlation. This situation parallels that in the area of chemical catalysis, where theory seems to be somewhat more advanced, but practice is still often empirical.

IV. Crystal Growth

The over-all process of crystal growth in a seeded solution is analogous to other mass transfer situations encountered in chemical engineering and may be treated as a diffusional step in series with a surface reaction step. Solution supersaturation provides the driving force required for each step, as portrayed schematically in Fig. 12. First, solute molecules or ions diffuse through the solution to the growing crystal. Second, upon reaching the surface, the molecules or ions must be accepted and incorporated into the crystal lattice.

A general knowledge of the mechanism of the surface step is necessary to the understanding of crystallization, but its discussion here will not

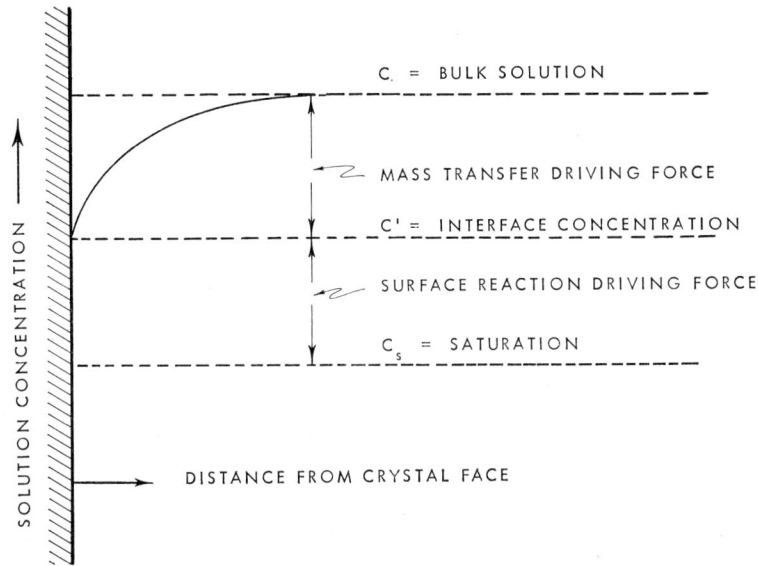

FIG. 12. Schematic concentration profile.

be detailed, in view of the extensive recent literature on this subject (B8, B9, F4, W3).

A. Geometric Characteristics

Section I above introduced the subject of crystal geometry and listed the seven classes of crystal systems, based on interfacial angles and relative lengths of crystal axes. As a crystal grows, these angles remain constant, so that each face is always parallel to its original position. This is known as the principle of parallel displacement of faces (M2) and leads to the concept of invariant crystals, illustrated in Fig. 13. Each of the geometrically similar polygons in this figure depicts the crystal outline at a different time.

In terms of translation velocity, defined as the rate at which a face moves in a direction perpendicular to its original position, the smaller faces grow faster than the larger ones. Under constant external conditions, however, the relative translation velocities of the several faces remain proportional, so that an invariant crystal retains geometrical similarity as it grows or dissolves. This is the basis of McCabe's "ΔL law," which states that all geometrically similar crystals of the same material suspended in the same solution grow at the same rate, regardless of initial size (M1). That is to say, if ΔL is the increase in linear dimension of one crystal, it is also equal to the increase in the corresponding dimension of each of the other crystals, provided that all crystals in the suspension are treated alike. This important condition is not always attained in commercial practice, as discussed in Section V.

The term "crystal habit" is often used to describe the relative sizes of the faces of a crystal. Crystal habit is readily modified by conditions of nucleation and growth, and it is rather difficult to prepare crystals with all faces of the same form equally developed (M2). Small amounts of soluble impurities, especially dyes, which may be adsorbed selectively on the different faces of a crystal, cause these faces to be suppressed in favor of others. This can alter the external geometry of a crystal completely, except for its interfacial angles. Many examples of crystal habit modification are reported in the literature (B8), and in some commercial

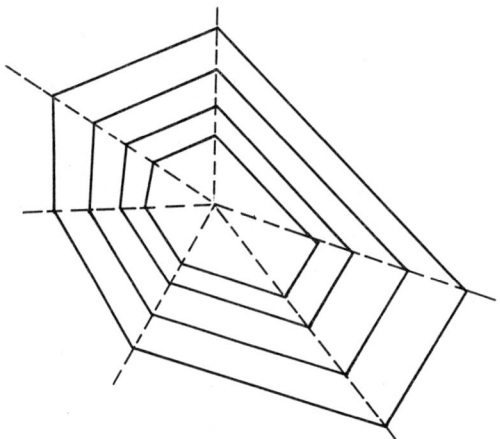

Fig. 13. Invariant crystal.

operations habit-modifying additives are used to control crystal size and shape (G2). For example, various dyes of the mono-azo, bis-azo, pyrazolone and amino-triphenylmethane types are effective in modifying the growth of potassium and sodium nitrates. Tartrazine and Amaranth produce thin, friable ammonium sulfate crystals, reducing caking tendencies in storage (W4).

B. Growth Mechanism and Rate

1. *Crystal Surface Development*

Figure 14 is a simplified schematic representation of a crystal surface at some arbitrary point in its development. It shows an exposed portion of a complete layer and an incomplete layer of atoms or molecules; for simplicity these are drawn as cubes. Consistent with the requirements of statistical thermodynamics, the configuration is dynamic and tends toward a minimum total free energy as an equilibrium condition. The numbered atoms increase in free energy in the order 1, 3, 2, 5, and 4. Position 6 is the most likely site for a new atom to occupy and is known as a "Kossel site" (K2, S13).

When the crystal surface is in contact with a supersaturated solution, the rate of deposition of molecules or ions exceeds the rate of dissolution. The Kossel sites are filled progressively until the started surface layer is completed. In order for growth to continue, a new layer must then be

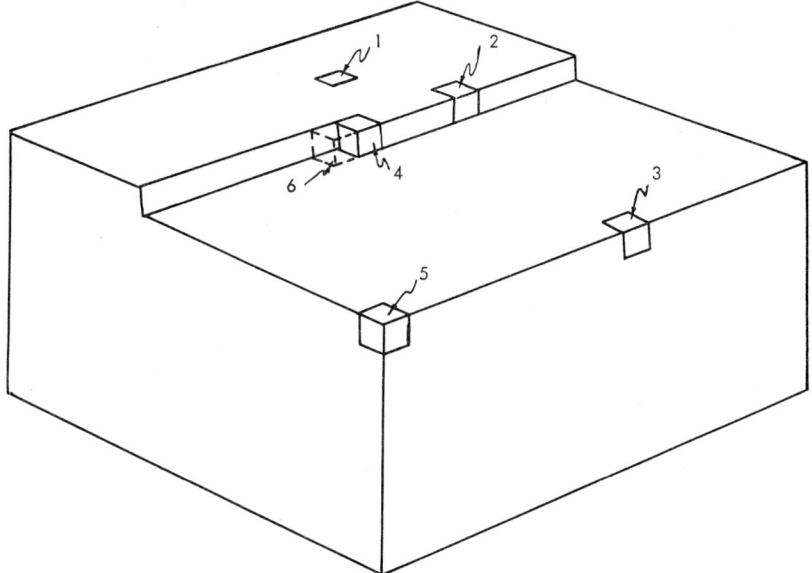

Fig. 14. Simplified representation of crystal surface.

initiated. Applying the criteria of nucleation theory, a supersaturation level of 25–50% appears necessary to form a stable nucleus for a new layer (F4). As early as 1931, however, Volmer and Schultze (V9) succeeded in growing crystals at only 1% supersaturation. This wide discrepancy between theory and experiment remained unexplained until 1949, when Frank proposed the "screw dislocation" mechanism for forming a non-self-extinguishing step on the growing surfaces (F1).

Illustrated in Fig. 15, Frank's model suggests a crystal imperfection of the type that would result if a cut were made part way through the crystal and the two sides skewed a distance of one layer at the edge of the crystal. Growth normal to the step occurs by filling of the Kossel sites. As the step advances across the crystal face, a spiral growth pattern results, and the dislocation is continually regenerated as each surface layer is completed. The screw dislocation theory has been widely verified by experiment (B9, F4, R2, V5). Some observations suggest that multiple dislocations and multilayered crystal steps occur.

There is still considerable speculation on the cause of screw dislocations. Some may come from irregularities in foreign nuclei upon which many crystals form; others may be started by inclusions of dirt or foreign atoms in the crystal. A strong mechanical disturbance may cause a grow-

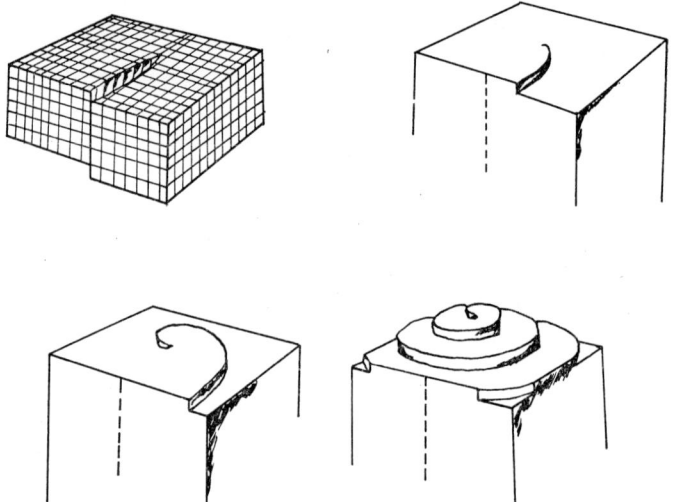

FIG. 15. Spiral growth resulting from screw dislocation. After Fullman (F4).

ing crystal to buckle and shear; indeed, growth has been induced in thin crystals by pressing a fine needle point against them (F4).

2. Theories of Crystal Growth

Early investigators (N3, N4) assumed that crystal growth was controlled by solute diffusion and that the solution contacting the crystal surface was saturated. Berthoud (B4) was the first to suggest that the rate of the surface reaction must be taken into account. Valeton (V1) later added further support to this theory, and the following equations were developed for rate of crystal growth:

$$\frac{dm}{d\theta} = \frac{DA}{\delta}(C - C') = k'A(C' - C_s) \tag{11}$$

Eliminating C', this becomes

$$\frac{dm}{d\theta} = \frac{A(C - C_s)}{1/k' + \delta/D} \tag{12}$$

where

D	= diffusivity of solute in solution
m	= mass deposited in time θ
A	= crystal surface area
δ	= thickness of laminar layer
k'	= rate constant for surface process
C, C', C_s	= solute concentrations in bulk solution, at interface and at saturation, respectively

The Berthoud-Valeton model conforms to Fig. 12 and treats the surface reaction as first-order. Although the latter is not always the case, this general approach is fundamentally valid, and differences between recent investigators lie primarily in correlation techniques, rather than in basic theory.

The relative contributions of surface reaction and diffusion to the over-all growth rate vary with the crystallizing system and with temperature. For example, in unagitated crystallization of sucrose the ratio of diffusion rate to surface reaction rate is 25.7 at 20°C., 2.1 at 40°C. and 0.03 at 70°C. (V2). In agitated copper sulfate-water and magnesium sulfate-water systems, both the surface reaction and the diffusional step must be taken into account (H7). The copper sulfate surface reaction appears to be second-order and considerably faster than mass transfer in the solution, whereas the magnesium sulfate surface reaction is first-order and proceeds at nearly the same rates as mass transfer. Barium sulfate crystal growth appears to be controlled by a fourth-order surface reaction at solution concentrations below 0.4 mM and by diffusion at higher concentrations (N5).

In general, growth rate depends principally on temperature when the surface reaction predominates (i.e., is the slower, rate-determining step) and mainly on degree of agitation when diffusion controls. Diffusional mass transfer rates can be predicted satisfactorily from generalized correlations, but surface phenomena are more complex. In addition to rate processes attending the crystal surface development mechanisms discussed above, the rates of intermediate steps such as ion dehydration also come into play. There is strong evidence that in some systems an adsorbed monolayer of hydrated ions forms a rate-determining barrier at the crystal surface (D1). Interdependence of the various factors involved is not yet understood well enough to permit reliable prediction of mass transfer rates at the interface. Consequently, experimental rate data at projected operating conditions are usually required for crystallizer design.

3. Growth Coefficients

A paper by McCabe and Stevens (M4) illustrates an empirical approach to the correlation of growth rate coefficients. These authors reasoned that mass transfer to the interface consists of two parallel processes, a diffusion effect independent of velocity and a flow effect linear in velocity. They expressed this by a rate term $(r_0 + \beta u)$. This was related to r_g, the over-all growth rate coefficient, by the conventional expression for rate processes in series,

$$\frac{1}{r_g} = \frac{1}{r_i} + \frac{1}{r_0 + \beta u} \tag{13}$$

where

r_i = the rate of the interface reaction
u = solution velocity

McCabe and Stevens applied this equation successfully to data on the growth of copper sulfate crystals suspended in a forced-convection U-tube apparatus. Their calculation method is useful in designing and operating crystallizers, within the limitations of its rather simplified empirical basis.

Hixson and Knox employed a more fundamental theoretical approach in their analysis of the effect of agitation on growth rate of single crystals (H7). They started with the basic equation for mass transfer of solute across a plane parallel to an interface and at a fixed distance from it:

$$w_a = -\frac{D_m}{1-y}\left(\frac{dy}{dx}\right) \tag{14}$$

where

w_a = molal rate of transfer of solute
D_m = diffusivity constant
y = mole fraction of solute in solution
x = distance in direction of mass transfer

Next, they defined a mass transfer coefficient, F_d, by means of the relation

$$w_a = F_d \int_{y_v}^{y_i} \frac{dy}{1-y} = F_d \ln\left(\frac{1-y_i}{1-y_v}\right) \tag{15}$$

where subscript v refers to conditions in the bulk solution and subscript i to conditions in the solution at the liquid-crystal interface. They proposed that F_d be correlated with solution properties and flow rate in the familiar Chilton-Colburn manner (C3).

$$\frac{F_\text{d} d_\text{e}}{D_\text{m}} = c \left(\frac{d_e u \rho}{\mu}\right)^{0.6} \left(\frac{\mu}{M_\text{m} D_\text{m}}\right)^{0.3} \tag{16}$$

where

- d_e = equivalent diameter of a sphere having the same surface area as the crystal.
- u = linear velocity of solution past the crystal
- ρ = solution density
- μ = solution viscosity
- M_m = mean molecular weight of solution
- c = experimental constant characteristic of system

Equation (16) was found quite applicable to copper sulfate and magnesium sulfate systems, with c equal to 0.29 in the former and 0.48 in the latter.

The surface reaction was assumed to follow an equation of the form

$$w_\text{a} = F_\text{R}(y_\text{i} - y_\text{s})^n \tag{17}$$

where

- y_s = mole fraction at equilibrium (saturation)
- F_R = rate coefficient of the surface reaction

The exponent n was reported as 2 for copper sulfate and 1 for magnesium sulfate crystallization. F_R is independent of solution velocity, but varies exponentially with the reciprocal of temperature.

Hixson and Knox also suggest an over-all mass transfer coefficient, F_0, defined by the relationship

$$w_\text{a} = F_0 \ln\left(\frac{1 - y_\text{s}}{1 - y_\text{v}}\right) = F_\text{d} \ln\left(\frac{1 - y_\text{i}}{1 - y_\text{v}}\right) = F_\text{R}(y_\text{i} - y_\text{s})^n \tag{18}$$

from which

$$\frac{1}{F_0} = \frac{1}{F_\text{d}} + \frac{\ln(1 - y_\text{s}/1 - y_\text{i})}{F_\text{R}(y_\text{i} - y_\text{s})^n} \tag{19}$$

In a sense, Eqs. (13) and (19) are analogous. However, the latter is clearly on a more sound theoretical basis and amenable to wider application through the use of Eq. (16).

F_R may vary considerably from face to face in the same crystal. Consequently, mean values of F_R or F_0 are of interest in the design of crystallization apparatus, whereas experimental studies with single crystals may measure mass transfer rates at specifically oriented surfaces. McCabe and Smith (M3) suggest a procedure for estimating the contribution of individual surface rate coefficients to mean values and a method for calculating the increase in linear crystal dimensions from the latter.

C. Dissolution

Dissolution, the opposite of crystallization, finds widespread use in chemical processing and related industries. Both gravity dissolvers and mechanically agitated apparatus are used, with the latter accounting for about 25% of all agitation equipment presently in use (H2). Rates of dissolution are usually diffusion-controlled and more rapid than corresponding crystallization velocities at comparable solution-concentration driving forces (V5). Although the diffusion terms in relations such as Eqs. (15) and (17) can be expected to apply, dissolution is not mathematically the reverse of crystallization, except in cases where the surface reaction is very rapid.

A number of experimental and theoretical studies of mass transfer in solution processes have been published. Since this literature is fairly well known, it will be mentioned briefly, but not analyzed in detail. Most of the earlier work in agitation employed dissolution rates as performance criteria (H6, H8, W5). Experimental studies of dissolution itself have employed suspended solute plates (B7, W1), single crystals (M12, P5), revolving crystals (D2), and packed beds (G1, L3, M5, V4). Recently, several theoretical analyses of literature data have appeared (E1, H1, R3). A number of Russian investigators have also studied dissolution (N6, Z1); they prefer to correlate data in terms of individual variables rather than the dimensionless groups customary in English and American literature.

Most of these studies have several limitations which restrict their general extension to dissolution processes of commercial interest. These limitations arise from use of sparingly soluble salts, of rather narrow temperature ranges, and of regularly-shaped particles. In an attempt to overcome some of these difficulties and to obtain information directly applicable to the design of gravity dissolvers, Heath (H2) investigated the dissolution of sodium chloride and copper sulfate pentahydrate crystals. He studied both falling particles and packed beds over a range of Reynolds numbers from 1.5 to 2000 and Schmidt numbers from 250 to 3000. He found that the familiar Chilton-Colburn mass transfer factor (C3) applies to falling particle data in the form

$$J_d = \left(\frac{k_a}{u_m}\right)\left(\frac{\mu}{\rho D}\right)^{2/3} \tag{20}$$

He related this factor to Reynolds number (based on terminal velocity) by the equation

$$J_d = 0.01 + \left(\frac{d_s u_m \rho}{\mu}\right)^{-0.6} \tag{21}$$

In Heath's equations,

k_a = dissolution rate coefficient
u_m = maximum superficial velocity of falling particle
d_s = equivalent spherical particle diameter (based on mean volume)
D = diffusivity of solute in solution

Assuming a tetrahedral packing model, Heath developed expressions for an effective Reynolds number and a modified J_d factor which extended Eqs. (20) and (21) to his packed bed data. He also showed that his information on natural crystals of fairly soluble materials correlated well with the literature data previously cited.

Results such as these suggest that dissolution may be treated mathematically via generalized solution mass transfer correlations, with surface reaction having a negligible effect on determining the over-all rate. However, more systems should be tested before sweeping conclusions are drawn.

V. Crystal Size Distribution

A. Crystal Size in Relation to Nucleation and Growth

In Sections III and IV, the principles of nucleation and growth were discussed separately. Now the crystallization process as a whole will be considered. In any practical application of crystallization, a stable solid phase must first be formed from the metastable liquid phase, and then additional molecules are deposited on the nucleus to form the macroscopic crystalline solid. Since nucleation and growth are taking place simultaneously, the theoretical principles discussed earlier are difficult to apply quantitatively to crystallization practice. Consequently, empirical expressions are still generally used in the design of equipment and prediction of its operation.

In the production of a crystalline material to be sold as a consumer product, crystal size, shape, and uniformity are often quite important. Sometimes a crystal product can be manufactured most economically by a particular process, but the product may not have the desired sales appeal. In such instances more costly manufacturing techniques may have to be employed in order to achieve the desired crystal characteristics. Factors such as these must be considered in economic calculations on a crystallization process. Often, as in the case of table salt, size and uniformity of size are important for practical reasons. Fines are undesirable, since they may be lost as an objectionable "dust," while coarse crystals may not fit through small salt shaker openings. Uniform size also aids in reducing caking. Screening (H4) may be necessary to obtain the desired size range.

Even when size and size distribution *per se* impose no restrictions on the process, the filtration, washing, or centrifugation steps which usually follow crystallization may call for special crystal characteristics. Since fine crystals have a large specific surface area, excessive loss of product during washing may be encountered. Time and cost of filtration or centrifugation are highly dependent on crystal size distribution.

The relative effects of nucleation and growth during crystallization are shown schematically in Fig. 16. Rates of both nucleation and growth increase with increasing supersaturation; nucleation rate increases very rapidly once the labile region is penetrated. The width of the metastable region is not clearly defined but, rather, is dependent on a number of factors such as cooling rate, degree of agitation, and the size and quantity of seeds present (T5). In Fig. 16, the solid curves represent nucleation and growth rates under quiescent conditions while the broken curves show the effect of agitation.

The ratio of growth rate to nucleation rate is a measure of the crystal size obtained from a given process; the larger this ratio, the coarser the product. It is also clear that production rate falls off with decreasing supersaturation. The operation of a crystallizer is a compromise between these two factors. If size is of little importance, high supersaturations will result in high production rates and small crystals. Conversely, coarse product is obtained at lower supersaturation, but at the expense of production rate. Crystallizer operation will be discussed in greater detail in Section VI.

B. Multicrystal Growth in Seeded Solutions

1. *Presentation of Crystal Size Distributions*

In addition to crystal quality and purity, crystal size distribution is one of the most important measures of crystallizer operation. Crystal size distributions are most readily obtained using standard sieving techniques (T7). In addition to the usual statistical and sampling errors, other errors are encountered in sieving (H3); these arise from: (a) effects of sieve loading and sieving time; (b) effects of random orientation of particles; (c) variation in the sieve aperture; (d) errors due to random sampling; and (e) effects of type of equipment and operation. Although these errors cannot be completely eliminated, they can be minimized by the use of reasonably small samples (50–100 g.) and standardized techniques. Several tests at different loading and sieving times should serve to locate the optima for the particular crystals to be analyzed.

Data obtained from a sieving operation are commonly presented as the weight per cent of crystals associated with each crystal size range. From these data a cumulative plot, showing the total weight per cent of crystals, finer or coarser than a given size, may be constructed.

Since the cumulative size distribution is often S-shaped, various types

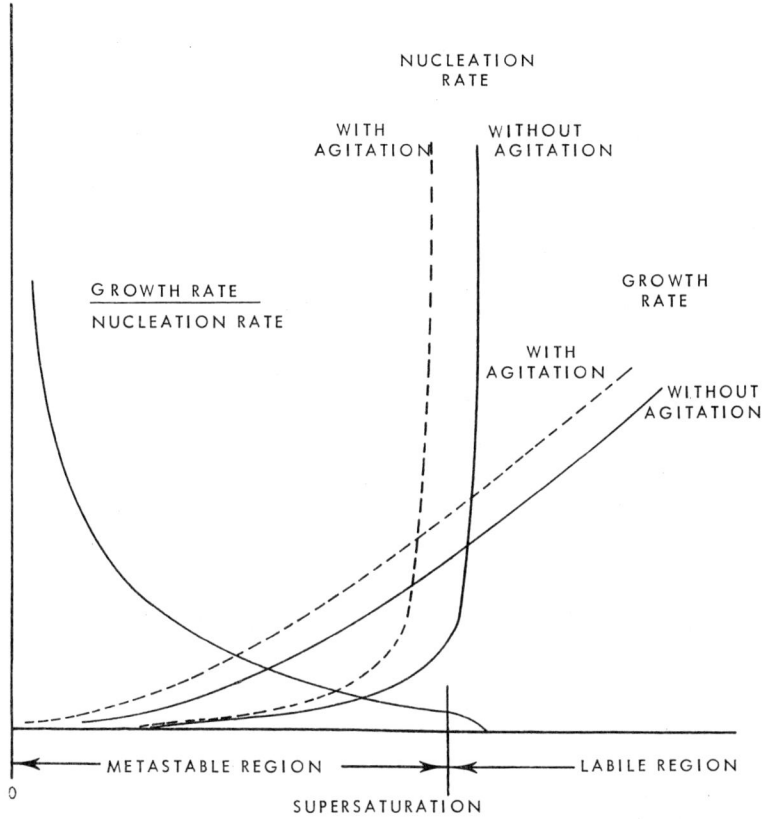

FIG. 16. Nucleation and growth rate dependence on supersaturation (schematic). Autenrieth *et al.* (A2).

of coordinates may be used to straighten out the plot. Logarithmic-probability and normal-probability coordinates are often used for this purpose. The mesh openings of standard sieves are in the ratio of the square root or fourth root of two, making it convenient to plot the data directly on paper ruled in these ratios.

A typical differential weight distribution curve is shown in Fig. 17. The abscissa has the dimensions of crystal size l and the ordinate units of weight. The weight and length are related

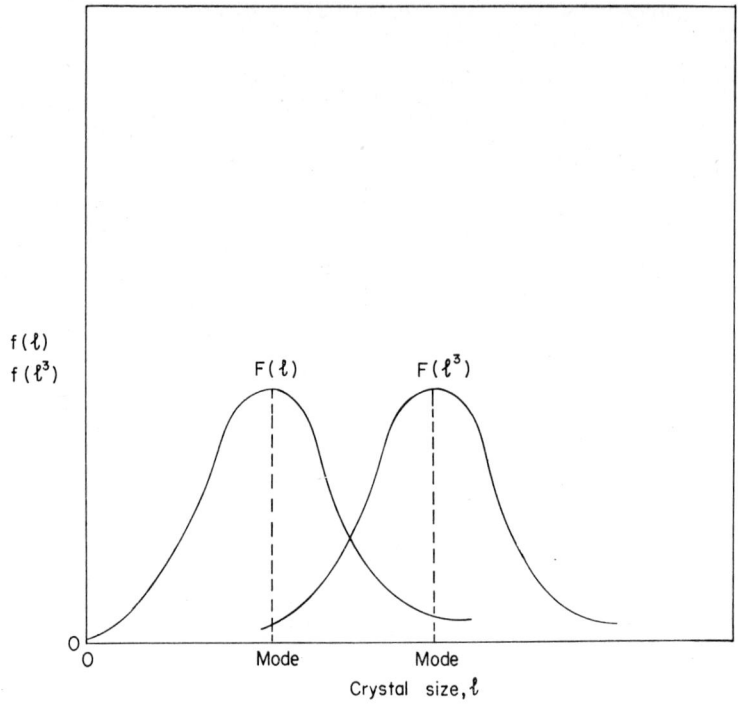

Fig. 17. Differential number and weight distributions (schematic).

$$W = f(l)^3 = \rho\alpha_1 l^3 \tag{22}$$

where α_1 is the shape factor relating length and volume. The differential number distribution $f(l)$, although difficult to obtain by direct measurement, can be determined from the weight distribution. In a like manner, the differential surface area distribution may also be calculated:

$$A = f(l^2) = \alpha_2 l^2 \tag{23}$$

where α_2 is the shape factor relating length and surface. Cumulative distributions may be expressed as follows:

$$W_c = \int_0^{l_{max}} \rho\alpha_1 l^3 \, dl \tag{24}$$

Table II summarizes several distribution functions.

2. *Growth from Uniform Seeds*

One of the earliest quantitative attempts to describe and predict the size distribution from a continuous seeded crystallization was reported

TABLE II
Particle Size Distribution Functions

Function	Equation form	
	Differential	Cumulative
Weight (W)	$f(l^3)$	$\int_0^{l_{max}} f(l^3)\, dl$
Area (A)	$f(l^2)$	$\int_0^{l_{max}} f(l^2)\, dl$
Number (N)	$f(l)$	$\int_0^{l_{max}} f(l)\, dl$

by Montillon and Badger (M10). They proposed the following empirical equation for the crystallization of magnesium and sodium sulfates:

$$\frac{\Delta W}{\Delta A} = ae^{b\theta} \qquad (25)$$

where

W = increase in crystal weight
A = increase in crystal surface area
a, b = constants, dependent on operating conditions
θ = time

Their experimental results led to the conclusion that larger crystals grow at a faster rate than smaller ones. Subsequent studies have shown that growth rate is independent of crystal size if all crystals are treated in a like manner (M1). Selective crystal treatment could exist, for example, in a tubular crystallizer with a slow moving spiral flight used to agitate the solution and move the crystals through the length of the crystallizer. In such a unit, it is quite feasible that larger crystals are picked up from the bottom and allowed to fall freely through the solution, while the smaller ones are relatively stationary with respect to the solution. This difference in crystal motion, relative to the solution reaction, would result in different growth rates in a diffusion controlled crystallization.

In order to study crystal growth in seeded solution, Grove (G6) developed a rotating drum crystallizer. The drum, a perforated cylinder, was submerged in a supersaturated solution and contained the growing seeds. The perforations were large enough to pass fines formed due to spontaneous nucleation or attrition, and these were removed continuously from the system, permitting growth of a fixed number of seeds to be carried out.

Using this type of equipment, Palermo (P1) investigated the growth of potassium alum crystals. The following equation was derived and verified experimentally:

$$\frac{W_p^{1/3} - W_s^{1/3}}{N^{1/3}} = K \int_0^\theta \Delta C \, d\theta \qquad (26)$$

where

N = number of crystals
K = growth constant
ΔC = supersaturation = $C - C_s$
W_p = weight of product crystals
W_s = weight of seed crystals

For an isothermal run, the growth constant K may be evaluated by determining the initial and final weights of the seeds, the number of crystals, and the variation of supersaturation with time. Values of the growth constant obtained at several different temperatures may be used with Eq. (26) to predict nonisothermal operation. Palermo's work is in agreement with McCabe's (M1) earlier work, for it is in essence an analysis limited to a single crystal size, rather than a distribution of sizes.

The growth of potassium chloride from an aqueous solution of sodium and potassium chlorides was studied by Davion (D3). Using uniform-size seeds, growing at a constant level of supersaturation, the crystallization growth constant K', was determined by the expression below.

$$W_p^{1/3} - W_s^{1/3} = K' \alpha \, \Delta C \, \frac{\theta}{3} \qquad (27)$$

This relation is equivalent to Eq. (26) at a constant supersaturation; then $\alpha = 3N^{1/3}$ and $K = K'$. Davion found that product size distribution closely approximated the normal distribution and that the growth constant was independent of seed size. In certain ranges of supersaturation, growth rate apparently decreased with increasing supersaturation. This could be attributed to agglomeration observed during these runs, which would in effect decrease the surface area available for growth. Factors such as attrition and agglomeration may introduce considerable errors when attempting to predict crystal size distributions in practice.

3. *Growth from Nonuniform Seeds*

The McCabe "ΔL law" (M1) is probably the best known method for predicting product crystal size distribution.

$$W_p = \int_0^{W_s} \left(1 + \frac{\Delta l}{l_s}\right)^3 dW_s \qquad (28)$$

Where W_p is the weight of product obtained from W_s grams of seeds, l_s and W_s are the coordinates of the cumulative screen analysis curve for the seeds. If all the seed crystals are geometrically similar, then the true size, L, of a crystal is related to the sieve opening that will just pass the crystal, then $L = \alpha l$ and $\Delta L = \alpha \Delta l$, and Eq. (28) may be written in terms of L and ΔL.

For a given seed distribution, the product distribution can be calculated if the yield is known; conversely, the yield can be determined for any desired increase in seed size. The calculations are somewhat tedious, since the integration is carried out by graphical trial and error. A nomograph (H11) showing l_s, Δl and $(1 + \Delta l/l_s)^3$ aids somewhat in these calculations.

Jang's work (J1) with a rotating drum crystallizer pointed out the limitations of an expression such as Eq. (26) for evaluating a growth constant when two widely different seed sizes were used in the same run. The growth constant for the larger crystals fell below the expected value, while the constant for the small crystals was higher than expected.

Although it is accepted that within a given crystallizer the linear growth rate is independent of size *per se*, the growth rates may differ due to effects such as the dependence of the free settling velocity on size. If all crystals, regardless of their size, were treated in exactly the same manner, the equations presented in this section could be used to predict product size for any seed size distribution.

C. Growth in Spontaneously Seeded Solutions

The difficulties surrounding study of spontaneously seeded growth may be illustrated by considering the precipitation of sparingly soluble salts. In this case, nuclei may form in one or more of several ways: (a) Nuclei (or embryos capable of forming nuclei) may exist before mixing in stable solutions of the two precipitating ions. (b) They may form spontaneously, at the instant of mixing. (c) They may originate from a homogeneous chemical process during the early portions of the experiment. (d) Impurities or catalytic surfaces may promote nucleation.

In studying barium sulfate precipitation, Nielsen (N4) found that the number of crystals formed depends upon the method of cleaning the glass vessel. Steam-washing the glass vessel reduced the number of barium sulfate nuclei from 2000 to 100 per mm³. He concluded that for this system nucleation is primarily catalytic rather than homogeneous.

In another paper (N5), Nielsen reports further kinetic studies of barium sulfate crystal growth, followed turbidimetrically and conductimetrically. When the concentrations of Ba^{++} and SO_4^{--} were larger than about 0.4 mM he found that the crystal surface reaction was first-order and the over-all process was diffusion-controlled:

$$k_d \theta = \int_0^z z^{-1/3}(1 - z)^{-1} dz \tag{29}$$

where z = the fraction of barium sulfate having deposited on the crystals at time θ.

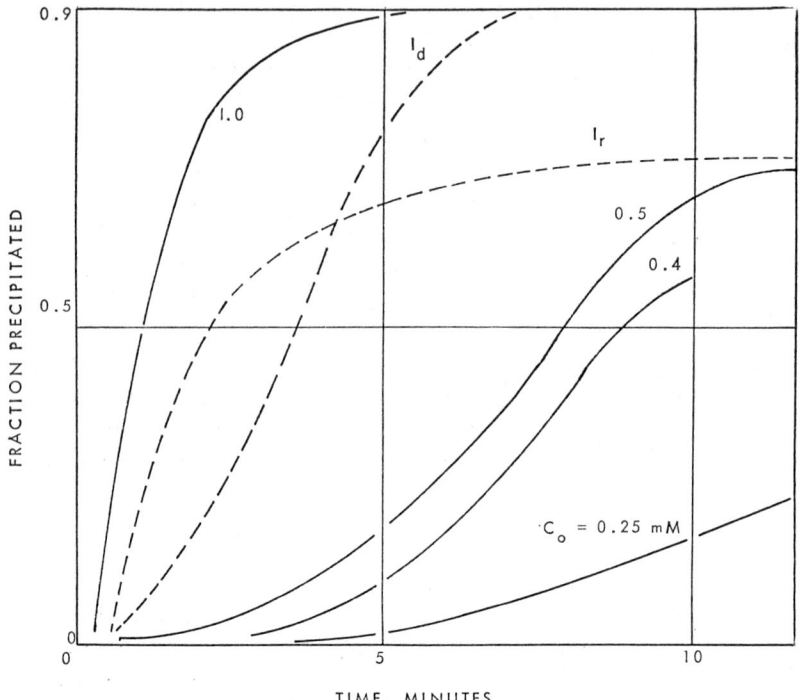

Fig. 18. Fraction of barium sulfate precipitated as a function of time. After Nielsen (N5).

At concentrations below 0.4 mM, a fourth-order reaction at the crystal surface controls:

$$k_r \theta = \int_0^z z^{-2/3}(1 - z)^{-4} dz \tag{30}$$

Figure 18 shows the fraction of BaSO$_4$ present which has deposited on the crystals as a function of time. C_0 is the initial concentration, I_d and I_r are the theoretical curves in arbitrary time units corresponding to Eqs. (29) and (30). It can be seen that the experimental curves (solid lines) resemble I_d at small times (high concentrations) and I_r at large times (low concentrations).

In Fig. 19, the dependence of linear growth rate on concentration is shown. Below 0.4 mM (log $C = -0.4$) the rate increases proportional to the fourth power of the concentration. In a batch operation, since the concentration changes continually, the order of the reaction may also change, depending on the controlling mechanism. In view of this, and the

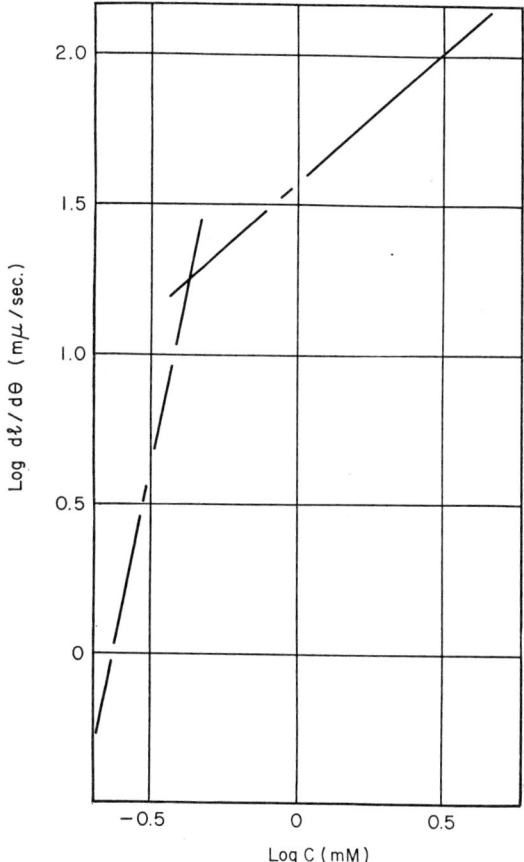

Fig. 19. Dependence of linear growth rate on concentration. After Nielsen (N5).

fact that nucleation also takes place during at least part of the operation, prediction of the final crystal size distribution is at best a difficult task.

In the crystallization of calcium sulfate dihydrate, Schierholtz (S7) found that the growth process appears to follow a first-order rate law for a large part of the reaction (see Fig. 20). The solid line shows the experimental results obtained during a typical run. The time interval preceding θ_0 is taken as the induction period. Deviations from linearity at the ends of the curve indicate more complex mechanisms at the start and end of the run. The decay of supersaturation with time was found to depend greatly on pH as well as various additives.

Pertinent to this discussion of spontaneously seeded batch operations is an investigation by Schlichtkrull (S8) in which the seeded and unseeded cases are compared. In these experiments with insulin crystals,

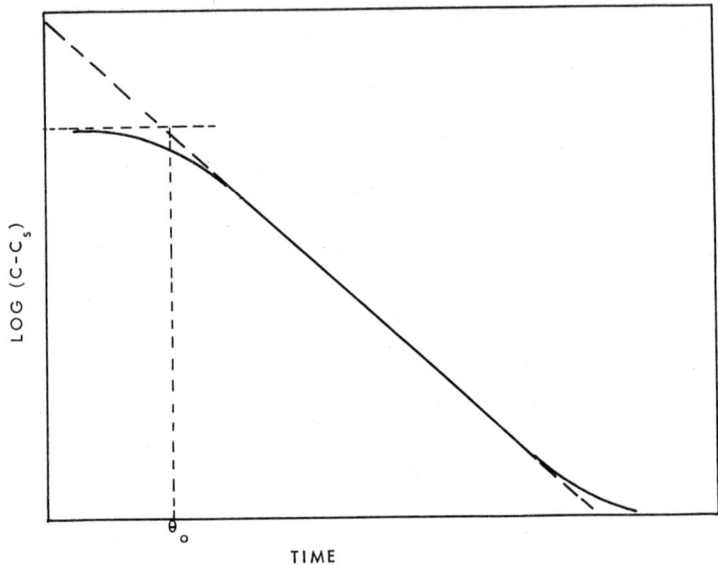

Fig. 20. Concentration versus time crystallization of calcium sulfate dihydrate. After Schierholtz (S7).

uniform-size seeds were used. Their weight, number, and size are related by

$$W_s = \alpha N l_s^3 \tag{31}$$

The proportionality constant (α) includes the crystal density as well as a shape factor. The total weight W_p of product crystals is related to the final seed size by the following equation:

$$\beta W_p = \alpha N l^3 \tag{32}$$

where β is the fraction of the product contributed by the grown seeds, and $(1 - \beta)$ the fraction contributed by spontaneously formed crystals. Combining Eqs. (31) and (32), and setting $\gamma = W_s/W_p$,

$$l = l_s \sqrt[3]{\beta/\gamma} \tag{33}$$

or

$$\beta = \gamma (l/l_s)^3 \tag{34}$$

The fraction β of uniform crystals (i.e., grown seeds) depends on the crystallization conditions as well as the seed size l_s, and γ. If the crystallization is carried out so that essentially no spontaneous nucleation takes place, $\beta = 1.0$, and the maximum crystal size, from Eq. (33), becomes

$$l_{\max} = l_s \sqrt[3]{1/\gamma} \tag{35}$$

Cumulative product size distribution for two identical runs, with and without seeding, is shown in Fig. 21. The results obtained by Schlichtkrull from three seeded runs are shown in Fig. 22. In curve (a) it is assumed

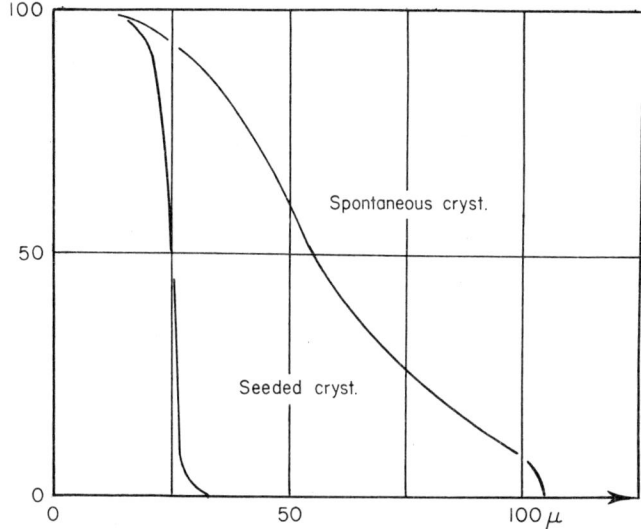

FIG. 21. Cumulative distribution of insulin crystals. After Schlichtkrull (S8).

that spontaneous nucleation is negligible ($\beta = 1.0$) where γ is relatively large (0.12). Taking the median product size l_{50} as the product size, and β equal to unity, from Eq. (33) the seed size l_s is found to be equal to 3.0 microns (μ).

Using Eq. (34) for cases (b) and (c), β is found to be 0.9 and 0.7, respectively. These values of β are found to be the ordinates corresponding to the deflections of the curves in Fig. 22. These points presumably divide the distributions into two parts, one containing spontaneously formed crystals and the other grown-up seeds. Both the spread and the median size of the product depend upon the extent of seeding. Table III summarizes the relationships corresponding to the curves of Fig. 22.

TABLE III
Variation in Amount of Seeds[a]

Curve	γ	l_{50} (microns)	$\beta = \gamma(l/l_s)^3$
(a)	12×10^{-2}	6	1.0^b
(b)	21×10^{-4}	23	0.9^b
(c)	30×10^{-5}	39	0.7^b

[a] After Schlichtkrull (S8).
[b] $l_s = 3$ microns.

D. Crystal Size Distribution for a Continuous Process

Although several types of continuous crystallizers are commonly used in industry, their design and operation has long been more of an art than

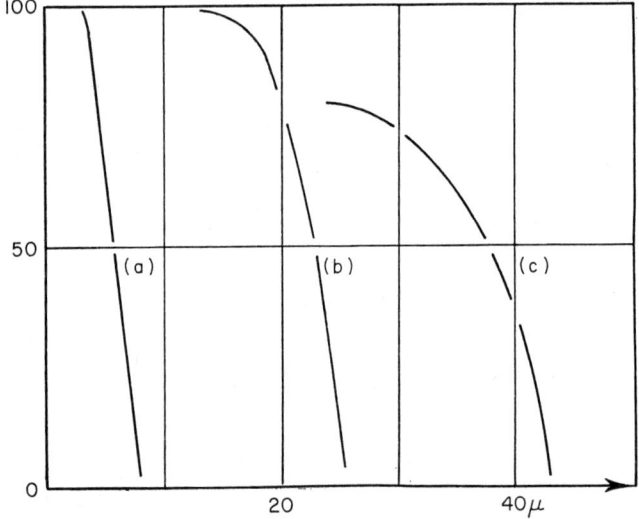

Fig. 22. Cumulative distribution in seeded runs. After Schlichtkrull (58).

a science. Recently, however, attention has been directed at applying proven principles governing crystal growth toward predicting and controlling crystal size distribution in continuous processes, notably by Saeman and his associates (M9, S1, S4, S5).

A continuous crystallization process ultimately reaches a steady state, in which the rates of nucleation and growth are constant with time. For a given set of operating conditions, crystal size distribution depends considerably upon the degree to which product classification is practiced. Figures 23 and 24 illustrate schematically the possible extremes between

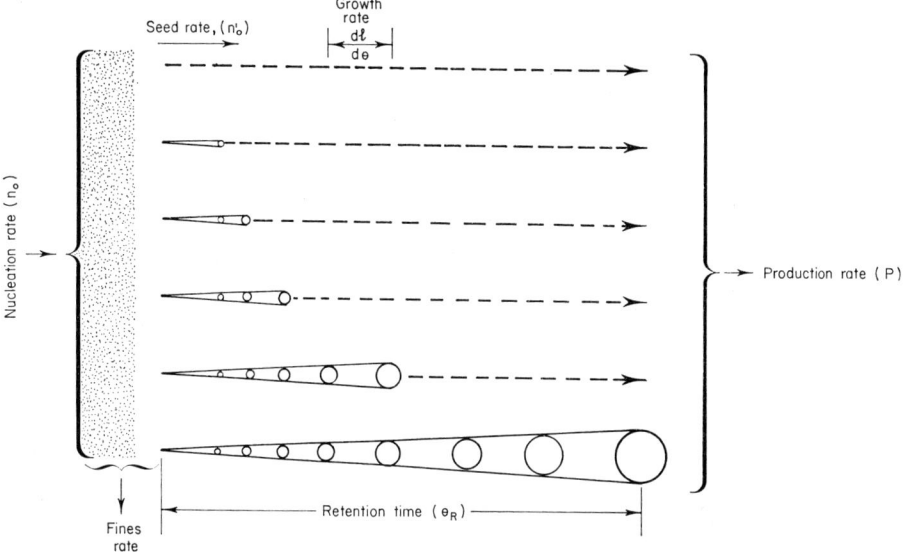

Fig. 23. Idealized crystal size distribution for mixed product removal. After Saeman (S1).

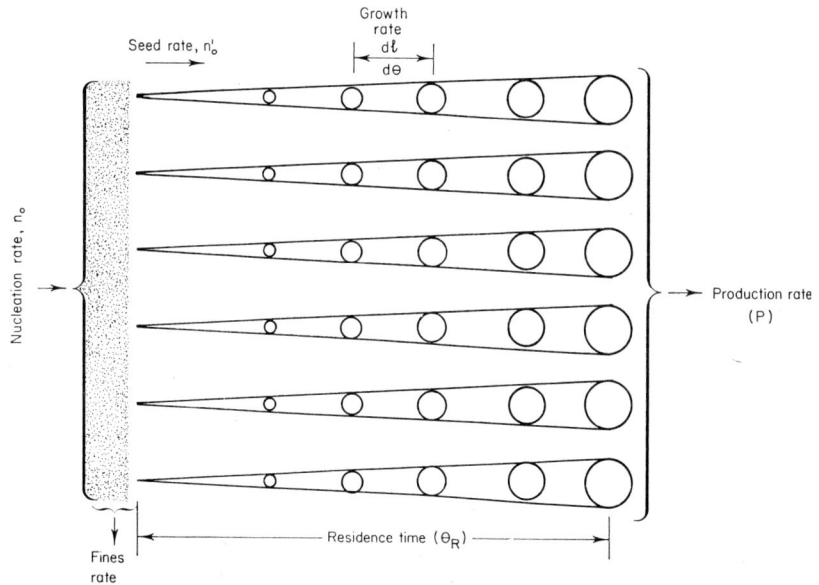

Fig. 24. Idealized crystal size distribution for classified product removal. After Saeman (S1).

fully mixed and completely classified suspensions. The use of mixed suspensions allows operation at high product rates, but produces a range of crystal sizes. When large uniform crystals are desired, classification is advantageous and a lower product rate is necessary.

1. *Perfectly Mixed Crystallizer*

In his analysis of this system, Saeman (S1) made the following basic assumptions: (a) The suspension is completely mixed. (b) The system is operating at steady state. (c) The numerical rate of withdrawing product plus fines equals the nucleation rate. (d) The weight rate of withdrawing product plus fines equals the crystallization rate. (e) The shape factor is constant (linear crystal size proportional to cube root of volumetric size). (f) The linear rate of crystal growth is constant and proportional to the supersaturation.

He derived the following equation for the differential weight distribution function:

$$W(l) = \frac{dW}{dl} = \frac{\alpha_1 n_0' \rho \theta_R}{L} l^3 e^{-l/L} \tag{36}$$

where

- α_1 = shape factor, relating volume and size of crystal
- ρ = crystal density
- n_0' = suspension seed rate (number/unit volume/unit time), which equals the nucleation rate if no fines are withdrawn separately
- θ_R = residence (or "draw-down") time = W/P
- W = cumulative weight of crystals in suspension
- P = production rate (weight/unit time)
- L = crystal size at time θ_R

The use of relative size l/L permits analysis of crystal growth without detailed knowledge of the mathematical relation between growth rate and supersaturation. Integrating Eq. (36) from the lower limit $W = 0$ at $l = 0$, the expression for cumulative weight of crystals up to size l is:

$$W = 6(\alpha_1 n_0' \rho L^3) - e^{-l/L}\left[6 + 6\left(\frac{l}{L}\right) + 3\left(\frac{l}{L}\right)^2 + \left(\frac{l}{L}\right)^3\right] \tag{37}$$

These two equations are plotted in Fig. 25, from which it is seen that in a perfectly mixed crystallizer the dominant-size fraction (mode) appears at $l/L = 3$. Saeman illustrated his analysis with data from large-scale vacuum crystallization of ammonium nitrate (M9, S1).

In earlier treatment of this type of system, Bransom *et al.* (B6) studied continuous crystallization of cyclonite from concentrated nitric acid in a small, well-agitated laboratory vessel. They developed the

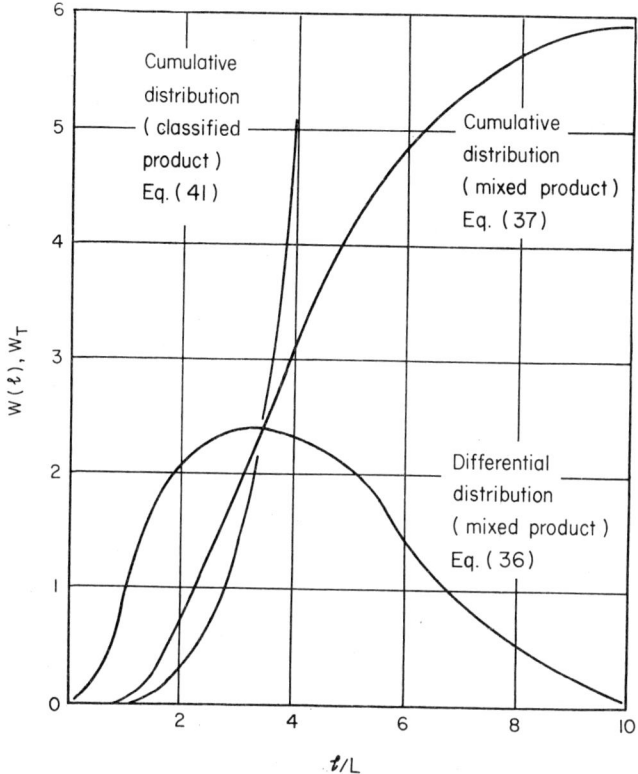

Fig. 25. Relative weight distribution curves for classified and mixed product removal. After Saeman (S1).

following expressions for crystal size distributions on a number basis, $N(l)$, and weight basis, $W(l)$, respectively:

$$N(l) = \frac{n_0}{f(s)} \exp\left(-\frac{vl}{Vf(s)}\right) \tag{38}$$

$$W(l) = \alpha\rho l^3 N(l) = \frac{\alpha n_0 \rho l^3}{f(s)} \exp\left(-\frac{vl}{Vf(s)}\right) \tag{39}$$

where

n_0 = number of crystals formed/unit volume/unit time
v = volumetric product rate
V = crystallizer volume
$f(s) = dl/d\theta$ = linear crystal growth rate = function of supersolubility

In logarithmic form, Eq. (38) becomes

$$\log N(l) = \log\left(\frac{n_0}{f(s)}\right) - \frac{vl}{V f(s)} \tag{38a}$$

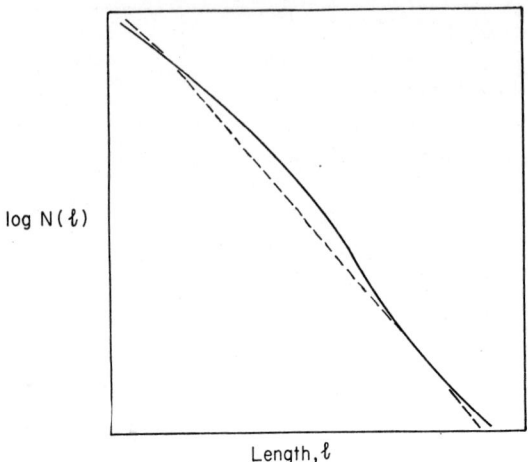

Fig. 26. Number distribution. After Bransom et al. (B6).

A plot of log $N(l)$ against l should be linear, and from its slope and intercept, respectively, $f(s)$ and n_0 may be calculated. Figs. 26 and 27 are typical plots of Bransom's data. In each case, the solid line shows actual results and the dotted line represents theory—Eqs. (38) and (39). It is also possible to obtain $f(s)$ from a plot such as Fig. 27, since it may be shown that the maximum in the weight distribution curve occurs at

$$l_{\text{mode}} = (3V/v)f(s) \tag{40}$$

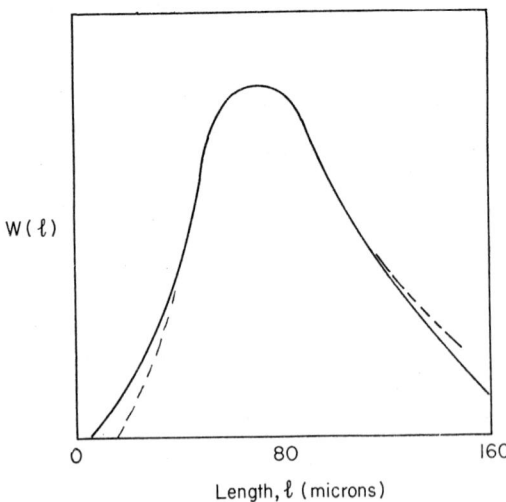

Fig. 27. Weight distribution. After Bransom et al. (B6).

Except for nomenclature, Eqs. (39) and (36) are identical, since $L = v/Vf(s)$ for steady operation, where $f(s)$ is constant.

2. Classified Product Removal

A well-mixed continuous crystallizer may be operated with a classified product take-off; only crystals which have attained the desired size are withdrawn, as depicted in Fig. 24. This may be achieved in a separate vessel or by incorporating classifying action into the circulation system of the crystallizer (M2). Saeman's discussion and mathematical analysis (S1) include this type of operation. His basic assumptions listed above still apply, with the additional condition that product crystals are withdrawn at full size only.

Saeman showed that for completely classified product removal,

$$W = \tfrac{1}{4}\alpha n_0'\theta_R L^3(l^4/L) \tag{41}$$

and that $l = 4L$. Hence,

$$W = 64\alpha n_0'\rho\theta_R L^3 \tag{42}$$

Eq. (41) is included in Fig. 25 to show the contrast between classified and mixed-product operation.

If the two types of operation are compared on the basis of equal growth rate, production rate, and suspension weight, then residence time (θ_R) and reference size (L) will also be equal. The two cases can be equated only by an adjustment in seed rate to satisfy the relation

$$(n_0') \text{ classified} = 0.094(n_0') \text{ mixed} \tag{43}$$

Thus, the required seed rate for mixed product is about eleven times the number required for classified product. On a weight basis, however, this is not very significant, since Saeman's equations show that 77% of the mixed product will contain the same number of crystals that is found in 100% of the classified product. Seed-rate control is achieved in practice by segregation and removal of excess nuclei (fines) as quickly as possible. Several examples of actual operation are described by Saeman (S1).

VI. Crystallization Equipment

A. History

A detailed historical development of crystallization equipment is not the purpose of this presentation, but a few general remarks are appropriate. Crystallization has been practiced for centuries and its beginnings, like those of other classical unit operations, are lost in antiquity (C1, S10). The earliest crystallizers were shallow open tanks, operated batchwise and cooled by surface evaporation. Simple, inexpensive equipment of this kind is still used in various parts of the world where atmospheric conditions are favorable. Gradually, paralleling the evolution of other chemical process equipment, refinements such as agitation, controlled cooling, continuous operation, and product classification were introduced. Griffiths (G5) reviews the early state of the art. Present-day equipment and operating procedures are described in standard textbooks on unit operations (B1, C5, M2, M3). Garrett and Rosenbaum (G2) discuss selection, design, and current costs of crystallizers.

Despite the long history and extensive use of crystallization and the rather exacting product specifications often encountered, equipment design is still largely on an empirical basis. Systematic application of crystal growth principles, developed largely in laboratory and small-scale studies, and of mass-transfer principles, stemming from experience with other unit operations, is just beginning. Much still remains to be done to place crystallization practice on a fully scientific basis.

B. Classification of Equipment

1. *General*

Since crystallization has developed largely as an art, orderly classification of equipment is difficult. Several systems have been proposed by various authors, but disagreement and confusion persists in the literature. The broadest classification is based on the nature of the phases involved: (a) crystallization from solution, (b) crystallization from the melt, and (c) crystallization from the vapor phase.

The first of these is of greatest general interest to chemical engineers. The second is used in metallurgy, in the preparation of solids with special properties, such as semiconductors, and in the growth of large, perfect, single crystals. The third has various specialized applications, such as the vapor plating of metals. Since this whole presentation has been restricted to crystallization from the solution, the following discussion considers only equipment in this category, specifically, apparatus for effecting multicrystal growth from solutions.

2. *Multicrystal Growth from Solutions*

Even in this specific area, there are still a number of possible methods for classifying crystallizers. Operating classifications such as continuous versus batch and agitated versus unagitated are obvious but much too general.

Classification of crystallizers according to the shape of the solubility curve in the crystallizing system is quite meaningful and has been discussed by several authors (M7, S9, T3).

(a) Solubility increases rapidly with increasing temperature (e.g., KNO_3–H_2O).

(b) Solubility increases moderately with increasing temperature (e.g., KCl–H_2O).

(c) Solubility increases slightly with increasing temperature (e.g., $NaCl$–H_2O).

(d) Solubility decreases with increasing temperature (e.g., Na_2SO_4–H_2O).

The most suitable means of achieving supersaturation in each case broadly defines the type of crystallizer to be used. In case (a) cooling, (b) vacuum, and in (c) and (d) evaporative, are generally employed. These three operating paths are depicted schematically by the arrows in Fig. 3. For example, it is clear that for sodium chloride a large amount of cooling would provide the same yield as a small amount of evaporation.

Newman and Bennett (N3) classify crystallizers according to three methods of operation: unagitated tank, circulating liquor, and circulating magma.

Although unagitated tank crystallizers are quite commonly used, they are greatly limited with respect to control of desired crystal product characteristics. The only variables are the initial and final temperatures and initial concentration, which determine the product yield. Rate of cooling may be controlled to some extent by the use of insulation or a tempered water jacket. The only liquor circulation which takes place is due to natural convection currents. As crystals grow and settle to the bottom of the vessel, agglomeration and occlusion of mother liquor may take place. Serious difficulties arise from the inability to control crystal product size and quality.

In circulating-liquor crystallizers, such as the Krystal-Oslo unit, supersaturation is created in what is essentially clear liquor. This clear liquor is then transported to the crystal bed where the supersaturation is released.

The circulating-magma design describes equipment in which crystals are intentionally transported with the liquor to the region where supersaturation is being created. Swenson-Walker, Wulff-Bock, and double-pipe crystallizers are examples of this type. Forced-circulation design is common.

C. Operation and Design

Since systematic application of scientific principles is just beginning in this area, the literature relating to crystallizer design is seriously limited. The contributions of Saeman and his associates (M9, S1–5) are significant; these and other recent papers are discussed below.

1. Circulating Magma

A recent development in circulating-magma crystallizer design was reported by Newman and Bennett (N3). This design consists of a closed vessel with a vertical draft tube surrounding a propeller located near the bottom (see Fig. 28). The outer annulus of the vessel serves as a settling area and permits regulation of magma density and control of nuclei re-

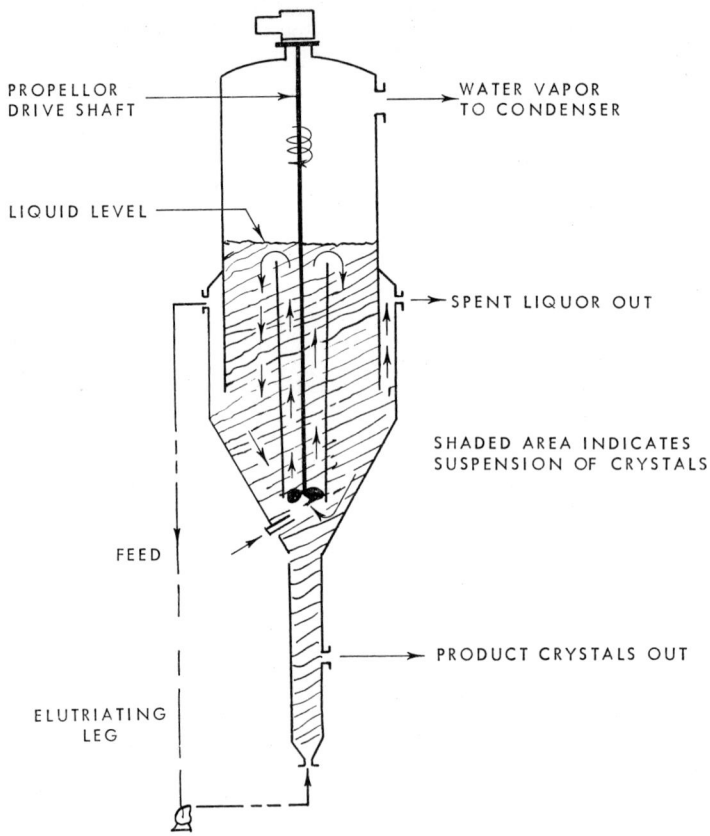

FIG. 28. Vacuum crystallizer. Draft tube and baffle type. After Newman and Bennett (N3).

moval. Internal circulation is achieved against a head of only a few inches of water, thereby requiring less horsepower than external circulation loops.

Newman and Bennett applied dimensional analysis in order to compare forced-circulation and draft-tube crystallizers. The first variables considered were the linear dimensions defining the crystallizer geometry. Since only geometrically similar crystallizers were considered, one characteristic dimension was sufficient to define the unit. Secondly, the kinematic and dynamic characteristics of flow, such as power consumed, velocity and gravity were considered. In the forced-circulation unit the inlet velocity of the fluid supplies the energy needed to keep the fluid in motion, opposed by the viscous and gravitational forces. In the draft-tube unit, the motion is supplied by the propeller of diameter d, rotating at N rev./min. The last important variable is the submergence S required to suppress vapor formation in the superheated liquid. The two modes of operation and the applicable dimensionless groups are summarized in Fig. 29.

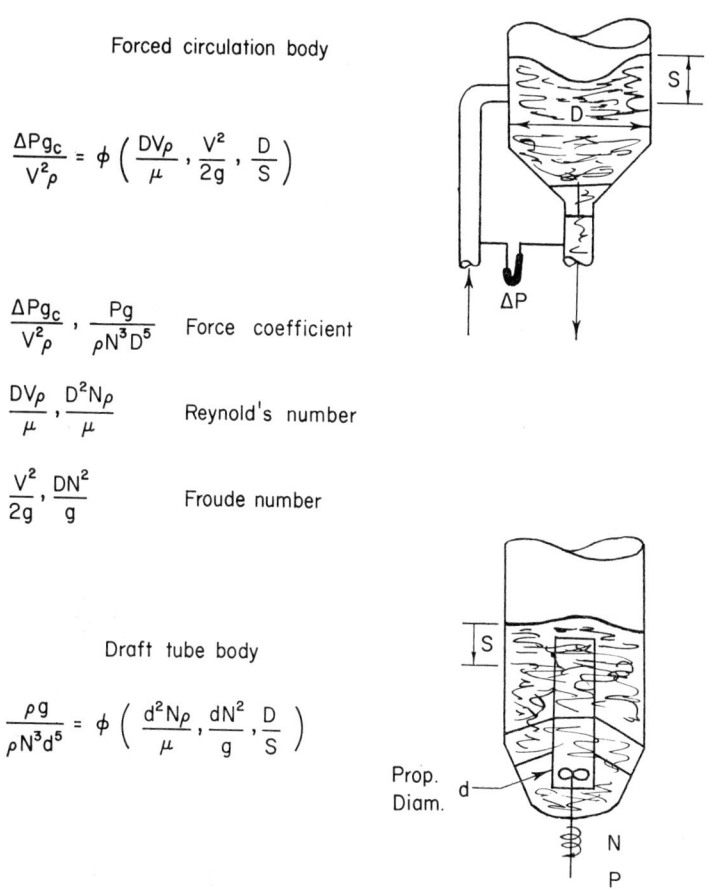

Forced circulation body

$$\frac{\Delta P g_c}{V^2 \rho} = \phi \left(\frac{DV\rho}{\mu}, \frac{V^2}{2g}, \frac{D}{S} \right)$$

$\dfrac{\Delta P g_c}{V^2 \rho}, \dfrac{Pg}{\rho N^3 D^5}$ Force coefficient

$\dfrac{DV\rho}{\mu}, \dfrac{D^2 N \rho}{\mu}$ Reynold's number

$\dfrac{V^2}{2g}, \dfrac{DN^2}{g}$ Froude number

Draft tube body

$$\frac{Pg}{\rho N^3 d^5} = \phi \left(\frac{d^2 N \rho}{\mu}, \frac{dN^2}{g}, \frac{D}{S} \right)$$

FIG. 29. Crystallizer scale-up factors. After Newman and Bennett (N3).

In scaling up a crystallizer for producing a crystal product of equivalent characteristics, the following must be kept constant:

(a) Driving force (supersaturation)
(b) Seed surface characteristics (size, shape, and concentration)
(c) Other environmental factors (temperature, viscosity, etc.)
(d) Hydraulic factors:
  (1) Power number
  (2) Reynolds number
  (3) Froude number
  (4) Submergence ratio, d/S
  (5) Geometric similarity

If these conditions are met, it can be expected that the nucleation and growth rates will be the same, and, consequently, that the same product characteristics can be obtained.

2. *Fines Removal*

Ideally, a crystallizer should operate with an inherent nucleation rate which will yield the desired product size distribution, without the necessity of a fines removal system. Since it is not always possible to control nucleation at its source, it is often necessary to control crystal size by segregating, removing, and disposing of excess nuclei. The high supersaturations which result in high growth rates also favor increased nucleation and, consequently, a smaller-size product. Nuclei may be produced in several ways: spontaneously, by attrition (crystal-on-crystal, agitators, pumps, etc.), or by seeding (recycle of crushed product).

Saeman (S1, S4) clearly explained the importance of correctly adjusting the suspension seed rate. If, for example, a four-fold increase in the linear size of the product is desired, the seed rate must be lowered by a factor of 4^3, or 64-fold. If spontanous nucleation cannot be controlled, then, for every product crystal, 63 nuclei must be removed and destroyed. If the weight of excess nuclei is to be kept at a reasonable level, say 1% of the product weight, they must be removed when their size is 1/20 of the product size $[63 \times (1/20)^3 = 0.8\%]$. The liquor must therefore be circulated through the fines trap a minimum of 20 times during the life cycle of the product crystal. Bypassing within the crystallizer will call for even higher circulation rates through the fines trap. Since the average product size is completely determined by the ratio of growth rate to seed rate, complete product size control could be achieved only with a 100% effective seed-rate control system.

According to Saeman, the following are the key crystallizer design criteria: (a) The actual residence time of crystals in the suspension; (b)

Optimum seed control size in relation to product size and inherent nucleation rate of the system; (c) Fines trap capacity to assure effective control of seed size chosen; (d) Phasing of seed rate adjustments in relation to corrections or changes in product size; and (e) Sensitivity of measurement and control procedures to assure effective and steady size control of product.

Surface fines removal from the Oslo-Krystal design is shown in Fig. 30. In this arrangement, fines segregate on the top of the classified suspension. This is an effective method of removing fines, since they are

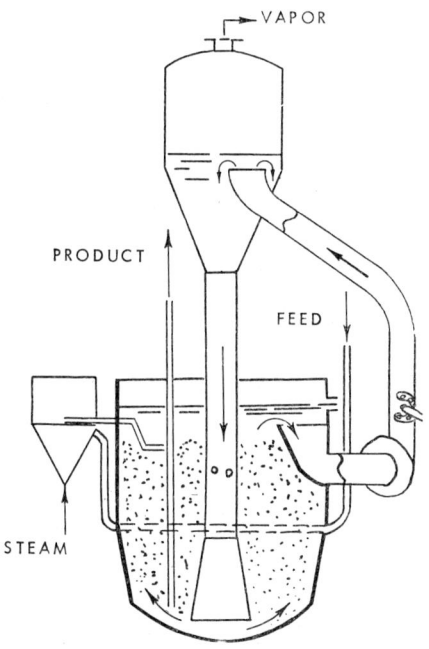

FIG. 30. Surface fines removal from classified suspension.

located in the region of lowest residual supersaturation. The production capacity is limited, however, because low superficial velocities are required to segregate the fines at reasonable small size.

An example of surface fines removal from a mixed suspension is shown in Fig. 31. In this system, there is considerable bypassing of solution; but solution velocity through the fines trap can be adjusted independently of the suspension turnover rate, which permits fines to be removed at a very small size.

A self-regulating fines removal system is shown in Fig. 32. The upper edge of the baffle is provided with adjustable openings to permit control of the flow of solution back into the suspension. Flow is induced by the density difference in the main suspension body and the annular area containing fines. For each flow rate, the fines come to steady state and may be withdrawn in a concentrated form into the dissolver.

3. Cyclic Nature of Crystallizer Operation

Although a crystallizer may be operating continuously at steady state with respect to a small portion of the circulating liquid, over-all the unit

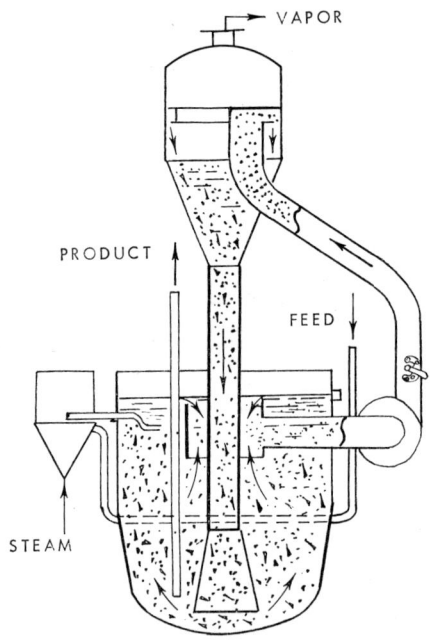

Fig. 31. Surface fines removal from mixed suspension. After Saeman (S2).

behaves in a cyclic manner. Typical crystallizer cycles may be compared on a plot of supersaturation versus time, as in Fig. 33.

In a circulating-liquor crystallizer, as the liquor flashes in the vapor head, it becomes supersaturated (path a–b). Ideally, the liquor will remain at a constant supersaturation (b–c) as it is transported to the seed bed. Spontaneous nucleation, however, will tend to lower the supersaturation (b–d). The remaining supersaturation is released on the crystals in the seed bed (d–e). As hot feed is mixed with the recirculating liquor, the mixture becomes unsaturated (e–a). The cycle recommences as the liquor is transferred to the vapor head.

If recirculation rate is increased, the length of the cycle decreases and consequently the residual supersaturation at the end of the cycle is higher. If the amount of flashing is assumed constant, the supersaturation level in the vapor head is higher, tending towards increased spontaneous nucleation.

The supersaturation profile for unagitated batch operation is also

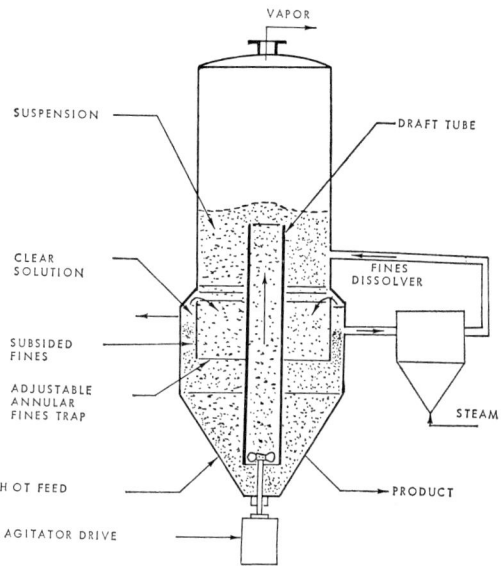

FIG. 32. Vacuum crystallizer with submerged self-regulating fines trap. After Saeman (S3).

shown in Fig. 33 for comparative purposes. At the start of the batch, the liquor may or may not be supersaturated, depending upon its composition and temperature.

In a circulating-magma crystallizer the supersaturation created in the vapor head does not rise to point b, for it is being deposited on the crystals as it is formed by evaporation. Higher recirculation rates and/or magma densities permit operation at lower actual supersaturations, thereby decreasing spontaneous nucleation, but increasing the number of nuclei formed by attrition.

VII. Addendum

Since completion of this manuscript, the crystallization literature has grown substantially in volume. However, as pointed out in recent reviews (G7, G8, S15, S16), this has stressed the technology of crystallization, and no major theoretical advances have occurred. Two new textbooks have appeared (M13, V3a), more extensive and detailed but otherwise rather similar to this chapter in their coverage of crystallization theory and practice.

A significant study of particle integration rate has been made by Cartier et al. (C7). By means of direct optical microscopy, they observed the growth of single crystals of citric and itaconic acids from water solution and succeeded in correlating growth rates with supersaturation and temperature, according to equations proposed by Amelinckx (A3).

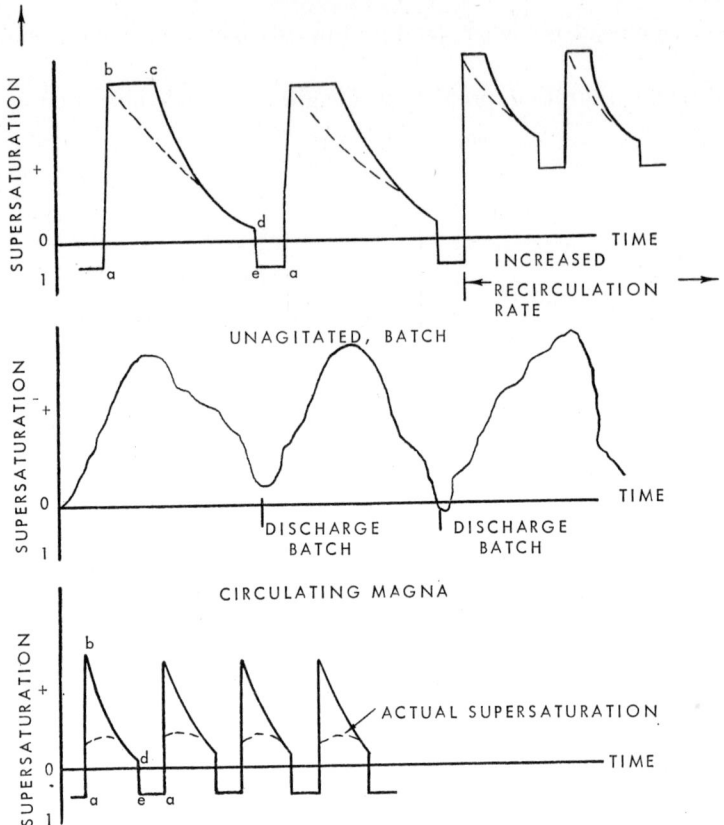

FIG. 33. Supersaturation as a function of time.

Various equipment and process improvements have been introduced in the industrial practice of crystallization from solution (S15, S16). Saeman (S14) has published a comprehensive discussion of crystallizer design principles, extending some of his earlier work cited above (S1, S4). A companion paper by Garrett (G9) considers the application of theory to the selection and operation of commercial equipment. Design and operation of draft tube and baffle type crystallizers is reviewed by Caldwell (C6).

Nomenclature

A Crystal surface area
C Solute concentration in bulk solution
C' Solute concentration in solution at interface
C_s Solute concentration at saturation
d Distance or diameter
d_e Equivalent spherical diameter based on surface area
d_s Equivalent spherical diameter based on mean volume
D Diffusivity of solute in solution
D_m Diffusivity constant (mean value)
e Natural logarithm base
$f_{A(l)}$ Fugacity of pure liquid A
$f_{A(s)}$ Fugacity of pure solid A
$f(s) = dl/d\theta$ Linear growth rate as a

	function of supersolubility
F	Force
F_d	Mass transfer coefficient in solution
F_0	Over-all mass transfer coefficient
F_R	Rate coefficient of surface reaction
ΔG	Free energy change
$\Delta G\ddagger$	Free energy of activation
H_c	Critical humidity
J_d	Chilton-Colburn mass transfer factor
k	Planck's constant
k'	Rate constant for surface process
k_a	Dissolution rate coefficient
K	Growth rate constant
l	Crystal size
l_s	Seed crystal size
L	Characteristic linear dimension of crystal product
m	Mass deposited
M	Molecular weight of nucleating phase
M_m	Mean molecular weight of solution
n_0	Nucleation rate (number/unit volume/unit time)
n_0'	Suspension seed rate (number/unit volume/unit time)
N	Number of crystals
$N(l)$	Crystal size distribution function (number basis)
p	Vapor pressure of droplet
p_∞	Saturation vapor pressure of bulk liquid
P	Crystallizer production rate (weight/unit time)
r	Particle radius
r_c	Critical particle radius
r_g	Over-all growth rate coefficient
r_i	Rate coefficient for interface reaction
r_0	Rate coefficient for solution diffusion
s	Surface area of new phase
S	Submergence
T	Absolute temperature
u	Linear velocity
u_m	Maximum superficial velocity of falling particle
v	Volumetric product rate
V	Crystallizer volume
V_B	Volume of liquid molecule
w_a	Molal rate of transfer of solute
W	Crystal weight
W_p	Weight of crystal product
W_s	Weight of crystal seeds
$W(l)$	Crystal size distribution function (weight basis)
w	Work of forming a new phase
$\mathcal{W}$	Work of nucleogenesis
x	Distance in direction of mass transfer
y	Mole fraction of solute in solution
y_v	Mole fraction of solute in bulk solution
y_i	Mole fraction of solute at interface
y_s	Mole fraction of solute at saturation
z	Fraction of salt deposited

GREEK LETTERS

α	Shape factor
α_1	Shape factor relating length and volume
α_2	Shape factor relating length and surface
β	Fraction of product contributed by grown seeds
γ	Ratio of seed weight to product weight
δ	Thickness of laminar layer
ϵ	Ionic charge
θ	Time
θ_R	Crystallizer residence time
μ	Solution viscosity
ν	Frequency factor in rate equation
ρ	Density
σ	Interfacial tension
ϕ	Interfacial contact angle
ψ	Dielectric constant

References

A1. Amsler, J., *Helv. Phys. Acta* **15**, 699 (1942).
A2. Autenrieth, H., Reike, H., and Dust, H., *Chem.-Ing. Tech.* **29**, 709 (1957).
A3. Amelinckx, S., *J. chim. phys.* **47**, 213 (1950).
B1. Badger, W. L., and Banchero, J. T., "Introduction to Chemical Engineering," pp. 520–552. McGraw-Hill, New York, 1955.
B2. Becker, R., and Doering, W., *Ann. Physik* [5] **24**, 719 (1935).
B3. Bertanza, L., and Martelli, G., *Nuovo cimento* [10] **1**, 324 (1955).
B4. Berthoud, A., *J. chim. phys.* **10**, 624 (1912).
B5. Booth, A. H., and Buckley, H. E., *Can. J. Chem.* **33**, 1162 (1955).
B6. Bransom, S. H., Dunning, W. J., and Millard, B., *Disc. Faraday Soc.* **5**, 83 (1949).
B7. Bruner, L., and Tolloczko, S. T., *Z. physik. Chem.* **35**, 283 (1900).
B8. Buckley, H. E., "Crystal Growth." Wiley, New York, 1951.
B9. Buckley, H. E., *in* "Structure and Properties of Solid Surfaces" (R. Gomer and C. S. Smith, eds.), Chapter 8. Univ. of Chicago Press, Chicago, 1953.
C1. Caldwell, H. B., *Chem. & Met. Eng.* **42**, 213 (1935).
C2. Campbell, A. N. and Smith, N. O., Eds., 9th ed., Alexander Findlay's "The Phase Rule and Its Applications," Dover, New York, 1951.
C3. Chilton, T. H., and Colburn, A. P., *Ind. Eng. Chem.* **26**, 1183 (1934).
C4. Christiansen, J. A., and Nielsen, A. E., *Acta Chem. Scand.* **5**, 673 (1951).
C5. Coulson, J. M., and Richardson, J. F., "Chemical Engineering," Vol. 2, Chapter 22. McGraw-Hill, New York, 1955.
C6. Caldwell, H. B., *Ind. Eng. Chem.* **53**, 115 (1961).
C7. Cartier, R., Pindzola, D., and Bruins, P. F., *Ind. Eng. Chem.* **51**, 1409 (1959).
D1. Davies, C. W., Jones, A. L., and Nancollas, G. H., *Trans. Faraday Soc.* **51**, 812, 1232 (1955).
D2. Davion, M., *Ann. chim. (Paris)* **8**, 259 (1953).
D3. Davion, M., *Compt. rend. acad. sci.* **242**, 1222 (1956).
E1. Ergun, S., *Chem. Eng. Progr.* **48**, 227 (1952).
F1. Frank, F. C., *Phil. Mag. Suppl. (Advances in Phys.)* No. 1, 91–109 (1952).
F2. Frenkel, J., "Kinetic Theory of Liquids," Chapter 7. Oxford Univ. Press, London and New York, 1946.
F3. Freundlich, H., "Kapillarchemie," p. 144. Leipzig, 1909.
F4. Fullman, R. L., *Sci. American* **192**, No. 3, 74 (1955).
G1. Gaffney, B. J., and Drew, T. B., *Ind. Eng. Chem.* **42**, 1120 (1950).
G2. Garrett, D. E., and Rosenbaum, G. P., *Chem. Eng.* **65**, No. 16, 124 (1958).
G3. Gibbs, J. W., "Collected Works," Vol. 1, p. 94. Longmans, New York. 1928.
G4. Glasstone, S., "Textbook of Physical Chemistry," 2nd ed., p. 344. Van Nostrand, Princeton, New Jersey, 1946.
G5. Griffiths, H., *J. Soc. Chem. Ind. (London)* **44**, No. 2, 7T (1925).
G6. Grove, C. S., Jr., Ph.D. Dissertation, Univ. of Minnesota, Minneapolis, 1940.
G7. Grove, C. S., Jr., and Jelinek, R. V., *Intern. Chem. Eng.* **1**, 6 (1961).
G8. Grove, C. S., Jr., and Schoen, H. M., *Ind. Eng. Chem.* **51**, 346 (1959).
G9. Garrett, D. E., *Ind. Eng. Chem.* **53**, 623 (1961).
H1. Hanratty, T. J., *A.I.Ch.E. Journal* **2**, 359 (1956).
H2. Heath, W. S., Ph.D. Dissertation, Syracuse University, Syracuse, New York, 1958. Univ. Microfilms, Ann Arbor, Michigan, No. 58-2302.
H3. Herdan, C., "Small Particle Statistics." Elsevier, Amsterdam, 1953.
H4. Hester, A. S., and Diamond, H. W., *Ind. Eng. Chem.* **47**, 672 (1955).
H5. Hirano, Y., *Kagaku (Tokyo)* **25**, 311 (1955).
H6. Hixson, A. W., and Crowell, J. H., *Ind. Eng. Chem.* **23**, 923, 1002, 1160 (1931).

H7. Hixson, A. W., and Knox, K. L., *Ind. Eng. Chem.* **43**, 2144 (1951).
H8. Hixson, A. W., and Wilkens, G. A., *Ind. Eng. Chem.* **25**, 1196 (1933).
H9. Hodgman, C. D., ed., "Handbook of Chemistry and Physics," 30th ed., p. 1389. Chemical Rubber Publ., Cleveland, Ohio, 1947.
H10. Hooks, I. J., and Kerze, F., *Chem. & Met. Eng.* **53**, No. 7, 140 (1946).
H11. Hougen, O. A., and Watson, K. M., "Chemical Process Principles," Part I, pp. 129–33. Wiley, New York, 1943.
J1. Jang, J. J., Ph.D. Dissertation, University of Minnesota, Minneapolis, 1943.
K1. Kantrowitz, A., *J. Chem. Phys.* **19**, 1097 (1951).
K2. Kossel, W., *Nachr. Ges. Wiss. Göttingen (Math-physik Kl. Fachgruppen)* **135**, 135 (1927).
L1. LaMer, V. K., *Ind. Eng. Chem.* **44**, 1270 (1952).
L2. LaMer, V. K., and Dinegar, R., *J. Am. Chem. Soc.* **72**, 4847 (1950); **73**, 380 (1951).
L3. Linton, W. H., and Sherwood, T. K., *Chem. Eng. Prog.* **46**, 258 (1950).
L4. Lonsdale, K., "Crystals and X-Rays," pp. 50–51. Van Nostrand, Princeton, New Jersey, 1949.
M1. McCabe, W. L., *Ind. Eng. Chem.* **21**, 30, 112 (1929).
M2. McCabe, W. L., *in* "Chemical Engineers' Handbook" (J. H. Perry, ed.), 3rd ed., pp. 1050–1072. McGraw-Hill, New York, 1950.
M3. McCabe, W. L., and Smith, J. C., "Unit Operations of Chemical Engineering," Chapter 14. McGraw-Hill, New York, 1956.
M4. McCabe, W. L., and Stevens, R. P., *Chem. Eng. Progr.* **47**, 168 (1951).
M5. McCune, L. K., and Wilhelm, R. H., *Ind. Eng. Chem.* **41**, 1124 (1949).
M6. Marc, R., *Z. physik. Chem.* **67**, 470 (1909).
M7. Matz, G., *Chem.-Ing.-Tech.* **27**, 18 (1955).
M8. Miers, H. A., *J. Inst. Metals* **37**, 331 (1927).
M9. Miller, P., and Saeman, W. C., *Chem. Eng. Progr.* **43**, 667 (1947).
M10. Montillon, G. H., and Badger, W. L., *Ind. Eng. Chem.* **19**, 809 (1927).
M11. Montmory, R., and Jaffray, J., *Compt. rend. acad. sci.* **246**, 1391 (1958).
M12. Murphree, E. V., *Ind. Eng. Chem.* **15**, 148 (1923).
M13. Mullin, J. W., "Crystallization." Butterworths, London, 1961.
N1. Nernst, H. W., *Z. physik. Chem.* **47**, 52 (1904).
N2. Newkirk, J. B., and Turnbull, D., *J. Appl. Phys.* **26**, 579 (1955).
N3. Newman, H. H., and Bennett, R. C., *Chem. Eng. Progr.* **55**, No. 3, 65 (1959).
N4. Nielsen, A. E., *Acta Chem. Scand.* **11**, 1512 (1957).
N5. Nielsen, A. E., *Acta Chem. Scand.* **12**, 951 (1958).
N6. Nikolskii, B. P., *Vestnik Leningrad Univ.* **1**, 67 (1946).
N7. Noyes, A. A., and Whitney, W. R., *Z. physik. Chem.* **23**, 689 (1897).
O1. Ostwald, W., *Z. physik. Chem.* **22**, 289 (1897).
O2. Ostwald, W., *Z. physik. Chem.* **34**, 493 (1900).
P1. Palermo, J. A., Ph.D. Dissertation, Syracuse University, Syracuse, New York, 1952.
P2. Perry, J. H. (Editor), "Chemical Engineers' Handbook," 3rd ed., pp. 197–198. McGraw-Hill, New York, 1950.
P3. Pound, G. M., *Ind. Eng. Chem.* **44**, 1278 (1952).
P4. Preckshot, G. W., and Brown, G. G., *Ind. Eng. Chem.* **44**, 1314 (1952).
P5. Prelat, C. E., and Gomez, E. M., *An. assoc. quím. Argentina* **35**, 90 (1947).
R1. Ranz, W. E., *Chem. Eng. Progr.* **48**, 274 (1952).
R2. Read, W. T., "Dislocations in Crystals." McGraw-Hill, New York, 1953.
R3. Ricci, J. E., "The Phase Rule and Heterogeneous Equilibrium." Van Nostrand, New York, 1951.

S1. Saeman, W. C., *A.I.Ch.E. Journal*, **2**, 107 (1956).
S2. Saeman, W. C., U.S. Patent 2,737,451 (March, 1956).
S3. Saeman, W. C., U.S. Patent 2,856,270 (October, 1958).
S4. Saeman, W. C., Presented at A.I.Ch.E. Meeting, Salt Lake City, Utah, Sept. 21, 1958.
S5. Saeman, W. C., McCamy, I. W., and Houston, E. C., *Ind. Eng. Chem.* **44**, 1912 (1952).
S6. Schaefer, V. J., *Ind. Eng. Chem.* **44**, 1300 (1952).
S7. Schierholtz, O. J., *Can. J. Chem.* **36**, 1057 (1958).
S8. Schlichtkrull, J., *Acta Chem. Scand.* **11**, 299 (1957).
S9. Schoen, H. M., Ph.D. Dissertation, Syracuse University, Syracuse, New York, 1957. Univ. Microfilms, Ann Arbor, Michigan, No. 20829.
S10. Schoen, H. M., Grove, C. S., Jr., and Palermo, J. A., *J. Chem. Educ.* **33**, 373 (1956).
S11. Seitz, F., "The Modern Theory of Solids," pp. 1–2. McGraw-Hill, New York, 1940.
S12. Stillwell, C. W., "Crystal Chemistry." McGraw-Hill, New York, 1938.
S13. Stranski, I. N., *Z. physik. Chem.* **136**, 259 (1928).
S14. Saeman, W. C., *Ind. Eng. Chem.* **53**, 612 (1961).
S15. Schoen, H. M., *Ind. Eng. Chem.* **52**, 173 (1960).
S16. Schoen, H. M., and Van den Bogaerde, J., *Ind. Eng. Chem.* **53**, 155 (1961).
T1. Tammann, G., "States of Aggregation." Van Nostrand, New York, 1926.
T2. Telkes, M., *Ind. Eng. Chem.* **44**, 1308 (1952).
T3. Thompson, A. R., *Chem. Eng.* **57**, No. 10, 125 (1950).
T4. Thomson, Sir William, (Lord Kelvin), *Phil. Mag.* [4] **42**, 448 (1871).
T5. Ting, H. H., and McCabe, W. L., *Ind. Eng. Chem.* **26**, 2101 (1934).
T6. Turnbull, D., and Vonnegut, B., *Ind. Eng. Chem.* **44**, 1292 (1952).
T7. Tyler, W. S., "The Profitable Use of Testing Sieves." W. S. Tyler Co., Cleveland, Ohio, 1958.
V1. Valeton, J. J. P., *Z. Krist.* **59**, 135, 335 (1923); **60**, 1 (1924).
V2. Van Hook, A., in "Principles of Sugar Technology" (P. Honig, ed.), Vol. 2, Chapters 3 and 4. Elsevier, Amsterdam, 1959.
V3. Van Hook, A., and Frulla, F., *Ind. Eng. Chem.* **44**, 1305 (1952).
V3a. Van Hook, A., "Crystallization: Theory and Practice." Reinhold, New York, 1961.
V4. Van Krevelen, D. W., and Krekels, J. T. C., *Rec. trav. chim.* **67**, 512 (1948).
V5. Verma, A. R., "Crystal Growth and Dislocations," Academic Press, New York, 1953.
V6. Volmer, M., *Z. Electrochem.* **35**, 555 (1929).
V7. Volmer, M., "Kinetik der Phasenbildung." Steinkopf, Dresden, Germany, 1939. (Reprint: Edwards, Ann Arbor, Michigan, 1945.)
V8. Volmer, M., and Flood, H., *Z. physik. Chem.*, **A170**, 273 (1934).
V9. Volmer, M., and Schultze, W., *Z. physik. Chem.* **A156**, 1 (1931).
V10. Von Weimarn, P., *Chem. Revs.* **2**, 217 (1926).
W1. Wagner, C., *J. Phys. & Colloid Chem.* **53**, 1030 (1949).
W2. Wall, F. T., "Chemical Thermodynamics," pp. 320–321. Freeman, San Francisco, 1958.
W3. Wells, A. F., in "Structure and Properties of Solid Surfaces" (R. Gomer and C. S. Smith, eds.), Chapter 7. Univ. of Chicago Press, Chicago, 1953.
W4. Whetstone, J., *Discussions Faraday Soc.* **No. 5**, 261 (1949).
W5. Wilhelm, R. H., Conklin, L. H., and Sauer, T. C., *Ind. Eng. Chem.* **33**, 453 (1941).
Z1. Zdanovskii, A. B., *J. Phys. Chem.* (U.S.S.R.) **20**, 379 (1946); **25**, 170 (1951).

HIGH TEMPERATURE TECHNOLOGY

F. Alan Ferguson and Russell C. Phillips

Stanford Research Institute
Menlo Park, California

I. Introduction	61
II. Temperature Definitions	62
A. Thermodynamic Temperatures	62
B. Statistical Mechanical Temperatures	63
C. Ionization Temperatures	69
D. Empirical Temperatures	69
III. Temperature Measurements	70
A. Thermodynamic Temperatures	70
B. Statistical Mechanical Temperatures	70
C. Ionization Temperatures	79
D. Empirical Temperatures	81
IV. Means for Attaining High Temperatures	83
A. Chemical Reactions	83
B. Electrical Sources of High Temperature	96
C. Mechanical Sources of High Temperature	101
V. Trends in High Temperature Technology	107
A. Uses of Chemically Generated High Temperatures	109
B. Uses of High Temperatures Generated in Shock Tubes	110
C. Uses of Electrically Generated High Temperatures	110

I. Introduction

The substantial contributions which chemical engineers are making to the problems of propulsion systems and materials for high temperature service in missiles and high-speed aircraft are indications of the important role of their profession in high temperature technology. With increased understanding of high temperature phenomena and enhanced capability for producing and controlling high temperatures economically, application of this knowledge to chemical processing will follow inevitably.

Before such application can occur, chemical engineers must become familiar with concepts of high temperature technology developed by physicists, aerodynamists, spectroscopists, ceramists, and scientists of other disciplines. Concepts that appear to be especially significant to the chemical engineer in this field and which are not yet stressed in his formal curriculum are described in this chapter.

The term "high temperature" has connotations varying from 10^3 °K. for the organic chemist to 10^8 °K. for the nuclear physicist. (Some scientists even believe that temperature is an unsatisfactory parameter for describing the energy level or the distribution of energy levels in a system.) However, for purposes of this discussion, temperatures above 3000°K. are considered to be high. In chemical manufacturing this temperature is approached only in the synthesis of carbides (K2). Chemical reactions have been investigated at temperatures up to 6000°K., and means are readily available for continuous processing at temperatures up to 50,000°K. Still higher temperatures are attainable with some electric discharges, and, of course, with nuclear fission or fusion reactions.

II. Temperature Definitions

Maxwell defined "the temperature of a body as its thermal state considered with reference to its ability to communicate heat to other bodies." Perhaps it would be better to call this a concept of temperature, reserving the term definition for the more exact mathematical descriptions.

A. Thermodynamic Temperatures

1. Carnot Cycle and Planck's Law

The classic definition of temperature is based upon thermodynamics. Any suitable relation, based on the laws of thermodynamics, can be used to describe temperature on a thermodynamic scale. The two most commonly used relations are the efficiency of the reversible engine (the Carnot cycle) and the intensity of blackbody radiation (Planck's Law) expressed mathematically by

$$\frac{Q_{H_2} - Q_{H_1}}{Q_{H_2}} = \frac{W_R}{Q_{H_2}} = f(t_1, t_2) \qquad \text{Carnot cycle} \qquad (1)$$

$$J_{\lambda T} = \frac{c_1}{\lambda^5(e^{c_2/T} - 1)} \qquad \text{Planck's Law} \qquad (2)$$

Q_H is the heat given off or absorbed during the engine cycle; W_R is the work done by the engine; $f(t_1, t_2)$ is a function relating the engine efficiency to the temperature of the isothermals; $J_{\lambda T}$ is the radiant energy per unit of wavelength (λ), time, and area at temperature T; c_1 and c_2 are constants for the system.

2. International Scale Temperatures

In adopting the International Temperature Scale in 1927, the Conference of Weights and Measures stated: "The experimental difficulties incident to the practical realization of the thermodynamic scale have made it expedient to adopt for international use a practical scale designated as the International Temperature Scale. This scale conforms with the thermodynamic scale as closely as is possible . . ." The International Temperature Scale assigns numbers to six basic fixed points, the highest being the "equilibrium temperature between liquid and solid gold (gold point) at normal atmospheric pressure," i.e., 1063.0°C. International Scale temperatures above the gold point are measured optically. The optical pyrometer is used to determine the ratio of the radiant flux of a monochromatic visible radiation J_2, of wavelength λ cm. emitted by a blackbody at temperature t_2, to the flux J_1 of the same wavelength emitted by a blackbody at the gold point, by means of the formula

$$\frac{J_2}{J_1} = \frac{\exp\,[c_2/1336.15\lambda] - 1}{\exp\,[c_2/(t + 273.15)\lambda] - 1} \tag{3}$$

which is based on Planck's Law, $c_2 = 1.438$ cm.-deg. While the equation can be applied to measurement of any temperature above the gold point, practically, the problems of obtaining blackbody conditions restrict measurement of International Scale temperatures to below 4000°K.

B. Statistical Mechanical Temperatures

In order to measure temperatures above 4000°K., it is often convenient to consider the properties of the molecules, atoms, and charged particles in the system. Since the laws of thermodynamics apply only to systems that are macroscopic, it is desirable to find another definition of temperature which can be based on measured microscopic properties of these particles. One definition, supplied from statistical mechanics is: temperature is the measure of broadness of a certain kind of distribution (F6).

(One kind of distribution can be expressed by the Maxwell-Boltzmann Law,

$$\frac{N_i}{N_0} = \frac{g_i}{g_0} e^{-(\epsilon_i - \epsilon_0)/kT} \tag{4}$$

where N_i and N_0 are the number of particles in the ith and the ground energy states respectively, ϵ_i is the energy of the ith state, and k is Boltzmann's constant. Another conceivable distribution is the Gaussian, which defines the normal probability curve. Two other distributions occasionally referred to are the Poisson and δ-function.)

According to statistical mechanics, a system composed of independent particles (this limits the discussion that follows to gases; the particles of solids and liquids cannot be considered independent of one another) will have energies distributed in distinct levels ranging from a nearly continuous array of levels to levels that are well separated. If the energy levels are well separated, the relative number of particles (N) occupying levels s and j is given by the equation:

$$\frac{N_s}{N_j} = \frac{g_s}{g_j} e^{-(\epsilon_s - \epsilon_j)/kT} \qquad (5)$$

where g and ϵ are respectively the statistical weight and the energy of the levels indicated by the subscripts (H4). Statistical weight is a term that describes the number of ways (degeneracy) that particles can form a given state (i.e., the j level of a carbon atom may be the ground, or lowest energy, state, and this level can be composed of three configurations, each with equal energy. In this case, the statistical weight of the j level of the carbon atom is 3.) Solving Eq. (5) for temperature gives,

$$T = \frac{\epsilon_s - \epsilon_j}{k} \left(\ln \frac{N_j}{g_j} - \ln \frac{N_s}{g_s} \right) \qquad (6)$$

A few simple calculations would show that, unless both the energy levels are present, the statistical mechanical definition of temperature has no meaning. Also, the following relations must hold, or Eq. (5) will be violated:

$$\frac{N_s}{g_s} \leq \frac{N_j}{g_j}, \quad \text{and} \quad \epsilon_s > \epsilon_j$$

High temperatures, then, imply relatively large differences (or broadness) in the quantum energy levels between particles in equilibrium with one another. (If these quanta are large in number and vary only slightly from one another a continuum is approximated, and a similar development can be made.)

Approaching the problem of measurement of high temperature in a gas from the statistical standpoint, we can calculate theoretically several temperatures (where temperature is defined as a measure of broadness of a distribution) depending upon how the energy is distributed in the gas. The energy modes available in a molecular gas are translational, rotational, vibrational, and electron-excitational, and before attempting a description of temperature measurements, it will be necessary to define, and point out the interrelations between these energies.

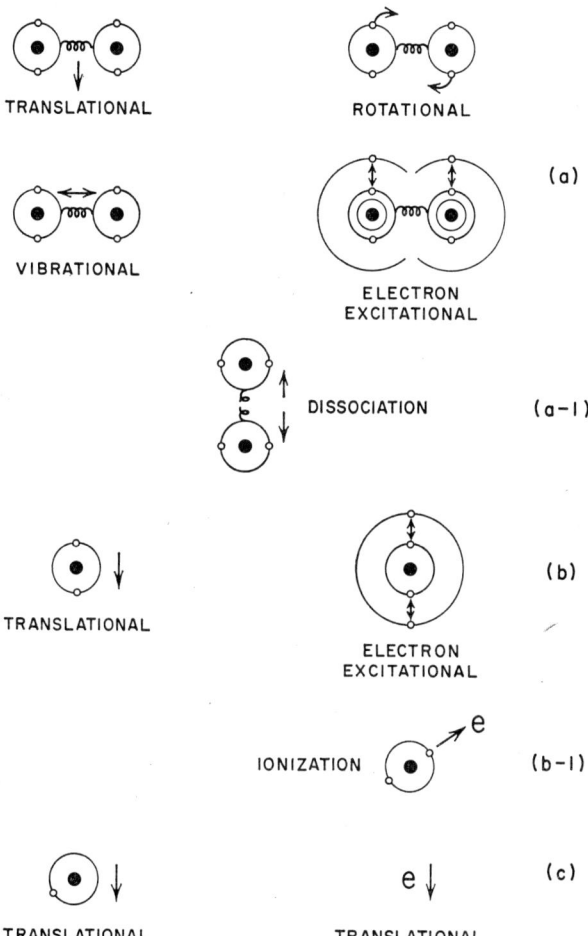

FIG. 1. Graphic representation of energy modes. (The authors are indebted to H. C. Rodean of Chance Vought Aircraft, Inc. for permission to use this graphic presentation.)

1. *Statistical Energies*

Figure 1 is a physical representation of the way in which energy is contained in a diatomic (or polyatomic) molecule. As the energy in the molecule is increased, the energy distribution within each mode (rotational, etc.) becomes broader (more energy levels and more uniform distribution of the particles between levels). However, if sufficient energy is supplied to the molecule, the vibrational- or rotational-mode energy exceeds the strength of the bond, and dissociation occurs (a-1). Similarly, if even more energy is supplied, the electron-excitational energy exceeds the bond energy between the atom and the electrons, and ionization occurs (b-1). Any further increase in energy is accommodated by: (1) broadening of the electronic-excitational energy distribution; (2) increased translational energy of the ions and electrons; and (3) further ionization.

Note that the terms vibrational, rotational, and electronic-excitational energies are in common usage. However, there is apt to be some question of the use of the term translational energy to describe the energy of random motion of the molecule. Occasionally this random motion is called kinetic energy, but we prefer to reserve that term for all the motion energy in the molecule (translational plus vibrational plus rotational, etc). The confusion in terminology occurs when considering rocket propulsion and hydrodynamic problems. In rocketry, directed velocity energy is called translational energy and in hydrodynamics, kinetic energy.

2. *Forms of Statistical Temperatures*

There are at least nine forms of energy distribution in a hot, diatomic, elemental gas, and temperatures can be defined for each form. Table I indicates the interrelationship of energy mode and particle type and the nine definable temperatures. All nine of these temperatures must be the same if thermal equilibrium exists.

TABLE I

Possible Statistical Temperatures in Hot, Diatomic, Elemental Gas

Energy modes	Species of particles			
	Molecular	Atomic	Ionic	Electron
Translational	x	x	x	x
Rotational	x	—	—	—
Vibrational	x	—	—	—
Electron-excitational	x	x	x	—
	+4	+2	+2	+1 = 9

The fact that molecules, atoms, ions, and electrons can exist in equilibrium with one another can be seen by referring to Fig. 2, which shows the concentrations of these species for nitrogen as a function of temperature (F1).

As the temperature of the system goes higher, the number of species and possible statistical temperatures does not increase. For, in most cases, the diatomic species will have disappeared before the doubly ionized species appears in measurable quantities, and, similarly, the monatomic species is gone before the triply ionized species appears.

An obviously more complex system that would yield even more distribution temperatures would be one composed of a mixture of diatomic gases (e.g., 2 elemental gases—18 distribution temperatures).

3. Factors Affecting Temperature Equilibrium

As stated previously, all the statistical temperatures (also the thermodynamic and ionization temperatures; see Sections II,A and II,C) must be equal for thermal equilibrium to exist in the system. Statistical temperatures, however, being defined by microscopic ensembles within the system, can be equilibrium temperatures for their ensemble, but may be entirely different from the equilibrium temperature of any adjoining or

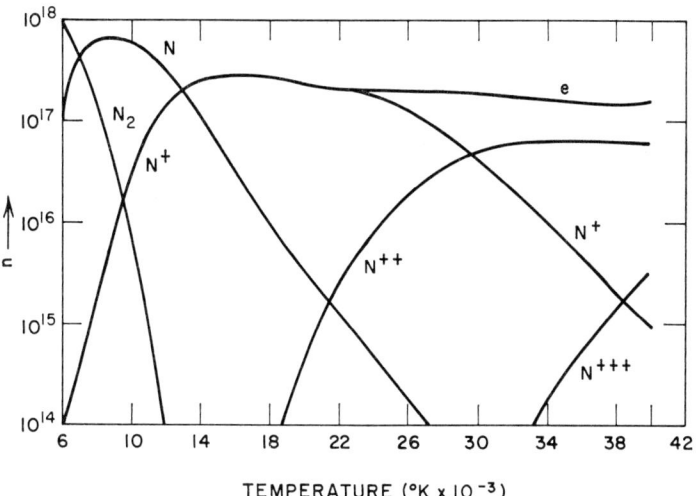

Fig. 2. Particle densities at atmospheric pressure in a nitrogen plasma as a function of temperature (F2).

intermixed ensemble. This microscopic nonequilibrium condition is created when the energy in one microscopic ensemble cannot distribute itself to equilibrium among the other energy modes. Any factors that tend to slow this energy transfer may create different statistical temperatures within the same system.

a. *Pressure Effects.* If the particles are not permitted to collide with one another while energy is being added to one mode, nonequilibrium is likely to result. For example, at low gas pressures in an electric arc (where the energy converted to heat must be transferred initially into electron translational energy), the mean free path between collisions of electrons with other gaseous species is so large that equilibrium between the various translational temperatures is never reached. The electrons can have temperatures of 80,000–130,000°K. while the molecules, atoms, and ions present are only slightly hotter than the container walls (C4, S4, M1). As the pressure increases and the mean free path decreases, the differences between the various temperatures become less until the means of temperature measurement can no longer distinguish between them. Other translational or kinetic energies can act as the modes of transfer, with similar results.

b. *Effect of Excited Mode.* The rate at which thermal equilibrium is established, however, also varies with the excited mode. Qualitatively, it has been shown that vibration modes are the slowest in coming to equi-

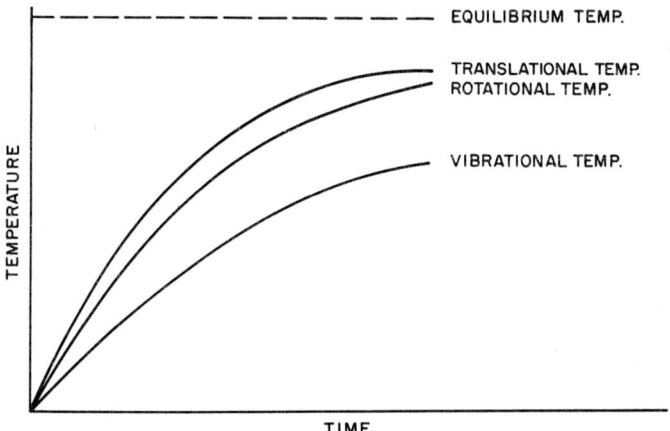

Fig. 3. Relative rates at which energy modes reach equilibrium.

librium, whereas modes such as rotational and translational are relatively fast (G5, L4).

Extremely rapid changes in the temperature of a gas, in the order of 10^6–10^{13}°K. per second give rise to heterogeneous, nonequilibrium systems, with some molecules excited to much higher energy levels than others. For instance, as a gas is heated in a shock tube, the vibrational mode does not respond nearly as rapidly as the rotational mode, which in turn is about one-tenth as fast as the rate of translational heating. The relative responses of these modes to rapid changes in temperature are indicated in Fig. 3. Since energy transfer reactions among gaseous molecules occur only through collisions, one way to express the deviation of a thermally excited species from its equilibrium energy level is in terms of the number of collisions it must experience before it attains equilibrium with the system.

C. Ionization Temperatures

In addition to the statistical mechanical temperatures, a distribution temperature, called the ionization temperature, and defined by the Saha equation, is frequently referred to (D2, S3). In its simplest form—for a monatomic gas—the Saha equation states that the equilibrium established between the species present [positive ions (+), electrons (e), and neutral atoms (0)] is a function of temperature,

$$\frac{(n_+)(n_e)}{(n_0)} = S(T) \tag{7}$$

where n is the concentration of the species present. The total equation is derived to be (A1):

$$K_c = \frac{(n_+)(n_e)}{(n_0)} = \frac{(2\pi m_e kT)^{3/2}}{h^3} \cdot \frac{2Q_+}{Q_0} e^{-E_i/kT} \tag{8}$$

where m_e is the mass of the electron. The factor 2 appears with the partition function (Q_+) because there are two possible electron-spin orientations in the ion ground-state.

The equation can be written in terms of K_p by replacing the n with the corresponding pressure term

$$n_x = p_x g_c / kT$$

$$K_p = \frac{(p_+)(p_e)}{(p_0)} = \frac{g_c(2\pi m_e)^{3/2}(kT)^{5/2}}{h^3} \cdot \frac{2Q_+ e^{-(E_i/kT)}}{Q_0} \tag{9}$$

(The term E_i/kT can also be written in the equivalent form, H_0^0/RT.) E_i is not a constant at high temperatures. Its diminution has been estimated by Unsöld (U1) to be $\Delta E_i = 7.0 \times 10^{-7} n_e^{1/3}$ volt. For very high temperatures, the exponential term in the Saha equation should read

$$e^{-(E_i - \Delta E_i)/kT}$$

D. Empirical Temperatures

All of the temperature definitions discussed so far are founded on mathematical representations of physically proved laws. In chemical engineering, particularly in its application to process chemistry, the prime objective is not to determine a theoretical temperature with known limits of precision, but to obtain some value that can be repeatedly reproduced and that gives reproducible process results. Such temperatures can be as diverse as the means available to measure them. Two or three such temperatures have reached the point of common usage, and are discussed in Section III, D.

III. Temperature Measurements

The engineer has methods available to measure each of the temperatures inherent in high temperature systems. These methods will be outlined in the following section. Also, since statistical mechanical and ionization temperature measurements are not in common use, examples are given of calculations used to determine these temperatures.

A. Thermodynamic Temperatures

In the regions below 4000°K., the disappearing-filament optical pyrometer, the two-color pyrometer, or the photoelectric pyrometer can be used to measure International Scale Temperatures. Since optical pyrometry has been well covered in the literature (F5, R1), further discussion here would serve no purpose.

B. Statistical Mechanical Temperatures

1. *Translational Temperatures*

Of the nine statistical temperatures categorized in Table I, four are defined by the distribution of translational energy of the four types of particles (molecules, atoms, ions, and electrons) present in the system. The electron translational temperature is unique among the four, since translational energy (excluding for all this discussion any possible directed energy of the gas particles) is the only mode available to the electron. Therefore, this temperature is generally called the electron temperature. Electron temperatures should not be confused with the electronic excitational temperatures of the molecules, atoms, and ions defined by the distribution of energy in excited electronic states of these particles. To aid in making this distinction, the electron temperature will be called the free-electron temperature.

It is common practice to group the three remaining translational temperatures and call them gas temperatures (M6). The reason for this will become obvious when it is shown in the following four sections that only one of the methods for determining the translational temperatures of the molecules, atoms, and ions makes possible a distinction between the kinds of particle whose translational energy distributions are being measured.

a. *Free-Electron Temperatures.* The measurement of free-electron temperatures assumes its greatest importance in low-pressure electric arcs. As is explained in Section II, B, 3, relatively few of the larger particles are present in the path of a low-pressure electrical arc discharge. The energy is transferred mainly by electrons, and the energy losses occur at the electrodes, in conversion of the directed translational energy of the electrons to random translational energy, and, to a lesser extent, by collision of the electrons with the heavier particles. This disparity in the particles absorbing the energy creates the large difference in temperatures between the electrons (up to 130,000°K.) and the rest of the system (near room temperature).

In one technique for determination of the electron temperature, an electrode probe (M6), is used. Provided that equilibrium among the electrons exists, the electron current reaching the probe can be represented by the equation:

$$\Pi = \Pi_0 e^{\epsilon \mathcal{V}/kT} \tag{10}$$

Where Π_0 is the current at $\mathcal{V} = 0$, Π is the current at some other voltage, and ϵ is the electron charge. Knowing the current for each of two negative potentials permits determination of the electron temperature according to the following equation:

$$T = \frac{\epsilon}{k} \frac{(\mathcal{V}_2 - \mathcal{V}_1)}{\ln \Pi_1 - \ln \Pi_2} \tag{11}$$

This method has been criticized because the probe, in drawing current, disturbs the normal electron and ion distributions. In one alternative solution that has been suggested, a "floating" double probe is used (J2). Just how important this disturbance is, still seems to be a matter of some discussion. Other methods, such as microwave attenuation, have been tried for measuring electron temperatures, but little has been published about the techniques.

Example: Calculation of a Free Electron Temperature. Calculate the electron temperature in a mercury arc discharge if the probe, at a negative potential of 40 volts, collects 5 ma. of current; and, at 35 volts negative potential, 0.3 ma. Using Eq. (11), where ϵ, the electron charge, equals 1 e.v.:

k, Boltzmann's Constant, equals 8.61×10^{-5} e.v./deg;

$$T = \frac{5040(40 - 35)}{(\log 5 - \log 0.3)} = 20,600°K.$$

b. *Gas Temperatures.* Four methods of measuring gas temperatures have some utility. These are: sound-velocity measurements, pneumatic apparatus techniques, gas density measurements, and measurement of broadening of emission lines (Doppler effect). (In effect, each of the first three techniques measure the gas density.)

i. *Sound velocity.* There are several review articles on the use of sound-velocity measurements to determine temperatures (H2, L1, S7, P6). It will suffice here to outline the principals and the applications. The velocity of a sound wave or any small disturbance propagating through an elastic medium can be represented by the equation

$$a^2 = \frac{dP}{d\rho} \tag{12}$$

If the pressure changes created by the sound waves are adiabatic, this equation becomes

$$a^2 = \frac{RT}{M}(\gamma) \tag{13}$$

where γ is ratio of specific heats C_p/C_v. Determination of the sound velocity, provided that the constants, R, M, and γ, which properly describe the medium, are used, will allow determination of the gas temperature. This temperature measuring procedure has the advantages of being very rapid and of not disturbing the medium, and it measures the translational energy of the system directly. The shortcomings are mainly mechanical, since the average sound velocity from transmitter to receiver is the quantity measured. These instruments must be either extremely rugged so they can be exposed to the high temperatures or they must be located out of the hot zone where some corrections must be made for the cool gas in proximity with them.

Example: Gas Temperature Calculation by Use of Sound Velocities. What is the air temperature if the sound velocity through it is 1.53×10^5 cm./sec.? By means of Eq. (13), one can calculate the translational temperature of a gas from the measured velocity of sound transmitted through it. However, this relationship also contains the temperature-dependent variables γ and M, so it is not possible to calculate the temperature explicitly from the sound velocity. In Fig. 4, the

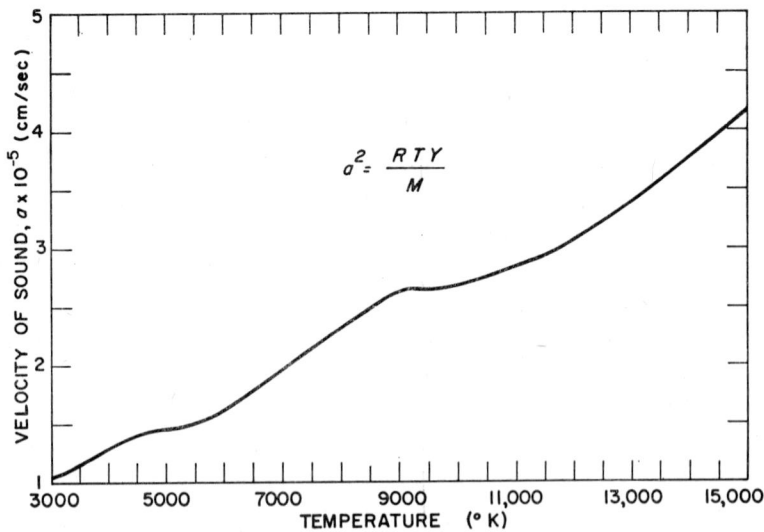

FIG. 4. Velocity of sound through dry air at high temperatures and atmospheric pressure (H1).

velocity of sound through dry air at atmospheric pressure is shown as a function of temperature. Specific heat ratio and apparent molecular weight data were obtained from reference G4. At $5500°K.$, $\gamma = 1.182$ and $M = 23.2$ g./mole, so that

$$a^2 = (8.314 \times 10^7 \times 5500 \times 1.182)/23.2$$
$$a = 1.53 \times 10^5 \text{ cm./sec.}$$

Therefore, a sound velocity of 1.53×10^5 cm./sec. in air establishes a gas temperature of $5500°K$.

ii. Pneumatic apparatus. The pneumatic apparatus method for measuring temperatures is based on the fact that after a critical pressure ratio is reached across an orifice or nozzle, through which a gas is flowing, the local velocity of the gas reaches a limit that cannot be exceeded by any

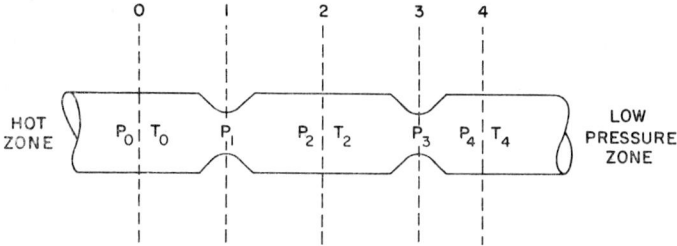

Fig. 5. Schematic of pneumatic apparatus.

further increase in the pressure ratio (W2). The critical pressure ratio is a function of the ratio of the specific heats of the gas:

$$P_c = P_0 \left(\frac{2}{\gamma_0 + 1}\right)^{\gamma_0/\gamma_0 - 1} \tag{14}$$

Lapple (L2) described the mass rate of gas flow w through the nozzles by the equation

$$w = C_1 A_1 \left(\frac{2}{\gamma_0 + 1}\right)^{[(\gamma_0+1)/2(\gamma_0-1)]} P_0 \left(\frac{g_c \gamma_0 M_0}{R}\right)^{1/2} T_0^{-1/2}$$

$$= C_3 A_3 \left(\frac{2}{\gamma_2 + 1}\right)^{[(\gamma_2+1)/2(\gamma_2-1)]} P_2 \left(\frac{g_c \gamma_2 M_2}{R}\right)^{1/2} T^{-1/2} \tag{15}$$

where C and A are the orifice coefficient and area; g_c the acceleration constant; γ the ratio of specific heats C_p/C_v; M the average molecular weight of the gas. The subscripts refer to the zone where the property is being measured (see Fig. 5). P_0 and P_2 are related by the equation

$$\frac{P_0}{P_2} = \frac{C_3 A_3 \left(\frac{2}{\gamma_2 + 1}\right)^{[(\gamma_2+1)/2(\gamma_2-1)]} \left(\frac{g_c \gamma_2 M_2}{R}\right)^{1/2} T_0^{1/2}}{C_1 A_1 \left(\frac{2}{\gamma_0 + 1}\right)^{[(\gamma_0+1)/2(\gamma_0-1)]} \left(\frac{g_c \gamma_0 M_0}{R}\right)^{1/2} T_2^{1/2}} \tag{16}$$

and for most cases this simplified to

$$\frac{P_0}{P_2} = \frac{C_3 A_3 (T_0)^{1/2}}{C_1 A_1 (T_2)^{1/2}} \tag{17}$$

The hot zone temperature can thus be determined from Eq. (17), knowing the hot and intermediate zone pressures, orifice areas and coefficients, and the temperature of the gas in the intermediate zone.

The greatest assets of this technique are its simplicity and sensitivity. The two orifices need only be holes drilled in relatively thick plate, in order to insure a uniform orifice coefficient (L2). DeLaval nozzles, although more difficult to fabricate, are more sensitive. The pressure measurement should easily be accurate to within 0.1 mm., which, provided that the pressures being measured are not too low, should permit temperature accuracies in the order of 0.1%. The technique is hampered by the severity of conditions to which the materials of construction are subjected; the materials are required to be nonreactive, high melting, erosion-resistant, and for these reasons the technique has limited utility above 3000°K.

Example: Calculation of Gas Temperature from Pneumatic Apparatus Measurements. What is the temperature of air in the hot zone (see Fig. 5) if the pressure is 1 atm., the areas of the two restrictions are equal and the coefficients C are 0.61 and 0.90 for the first (orifice) and the second (nozzle) respectively. The pressure of the intermediate zone (3) is 0.332 atm. and the temperature is 88°C. Assuming that γ is the same in the hot zone and in the intermediate zone, Eq. (17) is applicable,

$$\frac{1}{0.332} = \frac{0.90 (T_0)^{1/2}}{0.61 (88 + 273)^{1/2}}$$

and solving for T_0,

$$T_0 = 1500°K.$$

iii. Gas density. The third method that has found some use in measuring gas temperatures is direct determination of the density (S7). The basis for the use of density to measure temperature stems from the equation of state:

$$PV = \tfrac{1}{3} Nm(v^2)_{av} = RT \tag{18}$$

then

$$\tfrac{1}{2} Nm(v^2)_{av} = \tfrac{3}{2} RT = \text{translational energy} \tag{19}$$

Since $\tfrac{1}{2} Nm(v^2)_{av}$ is the translational energy of the system, T is the translational distribution temperature for an ideal gas, and the density (ρ) of the gas is (Nm/V).

$$P\left(\frac{Nm}{\rho}\right) = RT \tag{20}$$

The variation of gases from the ideal state is small at high temperature and moderate pressures. The density of gas can be measured by the absorption of X-rays, alpha rays, or electrons passing through a known path length (l) of the gas. The absorption for any of the above particles is given by the equation:

$$\frac{I}{I_0} = e^{-a\rho l} \tag{21}$$

Since the cross sections a of the atoms are known (C6), the density of the gas can be determined from Eq. (21). By measuring the pressure of the hot gas, its molecular weight can be determined. The temperature

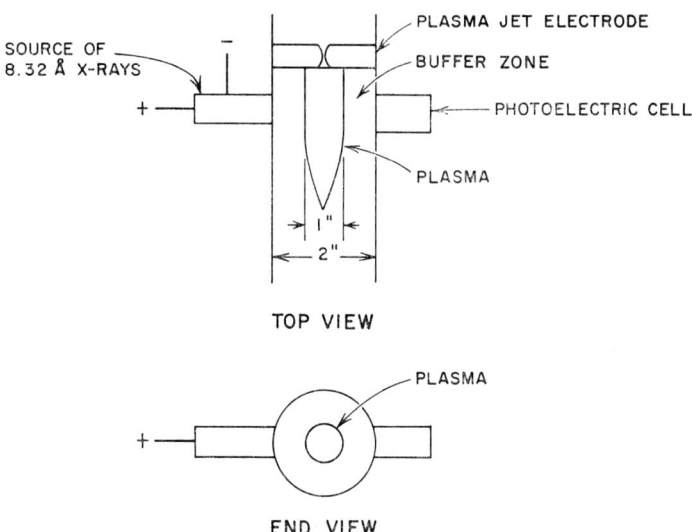

FIG. 6. Gas density measuring instrument.

can then be calculated using Eq. (20). This method, if it could be applied, would be particularly useful in the temperature region where dissociation of molecules takes place because the absorption of radiation in most cases is determined by the number of atom nuclei in the radiation beam-path whether the nuclei are present as molecules, atoms or ions. If the dissociation energies and the equilibrium constants were known for a molecular gas (and they are for most gases), the density ρ could be plotted as a function of temperature, including the dissociation effect, and the temperature could be taken directly from the plot once the density had been experimentally determined. This would eliminate the tedious trial-and-error calculations often required when dissociation is important. The biggest drawback again is the difficulty in obtaining a path length of uniform temperature or eliminating the end effects if it is necessary to keep the emitter and receiver in a lower temperature region.

Example: Calculation of Gas Temperature from Gas Density Measurements. A jet of nitrogen heated, and partially ionized, in an enclosed electric arc issues from a 1 in. diameter orifice and passes into a 2 in. diameter chamber, (Fig. 6). What is the average translational temperature of this nitrogen in the chamber? The jet of hot nitrogen emitted from the orifice has a diameter of 1 in. To determine the temperature of the nitrogen, X-rays are passed across the 2 in. diameter

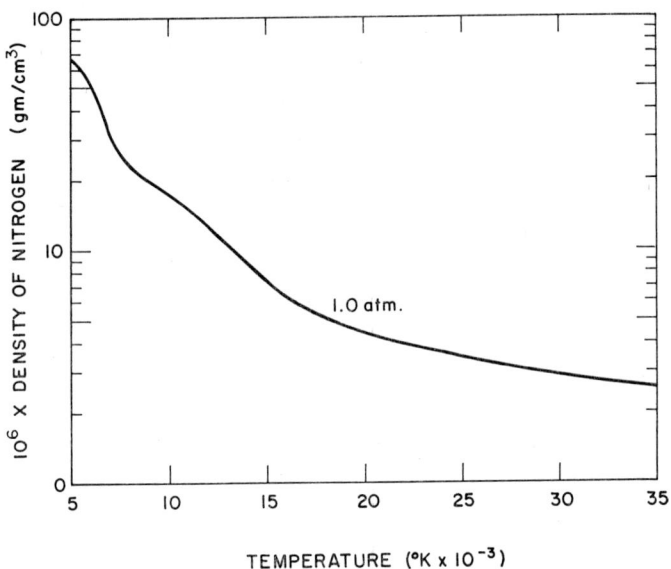

FIG. 7. Density of nitrogen at one atmosphere as a function of temperature (M4).

chamber under three conditions: (1) a very low vacuum, (2) with the chamber filled with nitrogen at the temperature and pressure of the buffer nitrogen during the run, and (3) at the same pressure as (2), but with the plasma jet operating. The average currents (detected by a scintillation crystal coupled with a phototube) for the three conditions are (1) $I_1 = 1000$ μa., (2) $I_2 = 2.7$ μa., and (3) $I_3 = 51.5$ μa. The bulk pressure in the chamber is 1 atm., and the buffer zone around the hot nitrogen is shown to be the same density as the gas under test conditions (2).

1. The density of nitrogen as a function of temperature and pressure has been calculated by Martinek (M4). A portion of his curve is reproduced in Fig. 7.
2. The density of the nitrogen in this example can be determined by the following steps, utilizing Eq. (21):

a. $I_0 = I_1 = 1000$ μa. since the limit of the term $e^{-\alpha \rho_1 l}$ is one at zero pressure.

b. $I_3 = 1000 e^{-(1190)(2.54)(\rho_{31}+\rho_{32})} = 51.5$ where ρ_{31} is the same as ρ_2 and ρ_{32} is the plasma density. $(\rho_{31} + \rho_{32}) = 980 \times 10^{-6}$ g./cm.^{3}

c. $I_2 = 1000 e^{-(1190)(5.08)(\rho_2)} = 2.7$

Solving this equation for ρ_2

$$\rho_2 = 975 \times 10^{-6} \text{ g./cm.}^3$$

d. $\rho_{32} = 5 \times 10^{-6}$ g./cm.^{3}

e. From Fig. 7 the N_2 plasma temperature is 18,000°K.

iv. Doppler effect. The fourth method used to determine the gas temperature measures the effect of translational energy on spectral phenomena. Emission of light energy from a hot gas is caused by transition of the various species involved from higher to lower energy levels. If such an emission manifests itself as a line (i.e., single rotational or electronic-excitational lines), the width of that line will broaden with increased translational energy of the molecule or atom emitting (Doppler broadening). The half-width (b) of the line due to Doppler broadening is given by

$$b = 2\sqrt{\ln 2} \sqrt{\frac{2RT}{Mc}} \lambda \qquad (22)$$

where λ is the wavelength of the emission, M is the molecular weight of the gas, and c is the velocity of light. Only by this last method is it possible to make a distinction between the translational temperatures of the molecules, atoms, and ions. (Measuring the translational temperatures of molecules, would still be relatively difficult because it would mean measuring differences in widths of lines within bands, which would require better instruments than required to measure differences in atom or ion line widths.) The use of this method, like all spectral emission methods, requires the knowledge of an experienced spectroscopist in all but the simplest of cases. For this reason, this and other spectral techniques will be categorized but no attempt at evaluation will be made.

Example: Gas Temperature Calculation from Doppler Broadening Effect. The half-widths of the 4302.19 A. and 4844.02 A. lines for iron were measured on the axes of an iron electrode arc and found to be 0.032 and 0.037 A., respectively. What is the translational temperature of the gas? Equation (22) simplifies to:

$$b = 7.16 \times 10^{-7} \lambda \sqrt{\frac{T}{M}} \qquad (23)$$

where $M = 55.85$. Solving for T gives

$$T_{4302} = 6030°K. \quad \text{and} \quad T_{4844} = 6350°K.$$

The difference of 320°K., as determined by the independent calculations using each line, is well within the limits of accuracy of this technique, and gas temperature equilibrium probably exists.

2. *Other Kinetic Temperatures*

The five kinetic temperatures other than translational—$T(n_2, \text{rot.})$, $T(n_2, \text{vib.})$, $T(n_2, \text{elect.})$, $T(n_0, \text{elect.})$, and $T(n_+, \text{elect.})$—can each be calculated from a knowledge of the Boltzmann distribution of energies in the modes involved, assuming equilibrium within the modes. This distribution is described by Eq. (4) and the radiant energy emitted per cm.^{3}-sec. in the transition between the states (s and j) is described mathematically by the equation:

$$\frac{J}{h\nu} = N_s A_{s,j} \tag{24}$$

J is radiant flux, $A_{s,j}$ is the transition probability, h is Planck's constant, and ν is the frequency. Combining Eqs. (4) and (24) gives:

$$\frac{J}{h\nu} = \frac{g_s}{g_j} N_j A_{s,j} e^{-[(\epsilon_s - \epsilon_j)/kT]} \tag{25}$$

or, if the radiation is given as intensity I,

$$I = C A_{s,j} e^{-[(\epsilon_s - \epsilon_j)/kT]} \tag{26}$$

(See Dieke (D2), Lochte-Holtgreven (L3), Mohler (M6) for several examples of the use of this relationship.) In Eq. (26) $(A_{s,j})$ is defined in such a way as to include the statistical weight and $h\nu$ terms.

The spectrum of radiation from electronically excited states of atoms appears as lines, when the emission from a hot gas is diffracted and photographed, whereas radiation from these excited states of molecules appears as bands because of emission from different vibrational and rotational energy levels in the electronically excited state. Equation (26) shows that the intensity of radiation from a line or band depends upon the temperature and concentration of the excited state and the transition probability (the rate at which the excited state will go to the lower state). Since the temperature term appears in the exponential, as the temperature rises the exponential term approaches unity, as does the ratio of the concentration of the excited (emitting) state to the ground state (as T approaches ∞, $N_g = N_j$). The concentrations of both the ground and excited states, however, reach a maximum, and then decrease due to the formation of other species. The line or band intensity must also reach a maximum and then decrease as a function of temperature. This relationship can be used to determine the temperature of a system.

a. *Electronic Excitation Temperatures.* This method of temperature measurement can be used on any spectral line provided that the transition probability (independent of temperature) and the concentration variation with temperature are known, but it finds its greatest application in the regions above 15,000°K. where only atoms exist and the emission from the electronically excited state appears as a line. Measuring the line intensity as a function of temperature is, then, a way of determining the electronic excitation temperature.

b. *Rotational and Vibrational Temperatures*. Rotational and vibrational temperatures are measured by determining relative intensities (O2, B6). Since the spectral lines within a band are emissions due to the presence of different rotational levels (i.e., J levels) the ratio of the intensities of the lines can be related to temperature, concentration of the emitting mode, and transition probability, so that

$$\ln \frac{I_{j''}}{gA_{j'',j''-1}} = \frac{\epsilon}{kT} \qquad (27)$$

where the primes refer to different J energy levels. If the left-hand term is plotted as a function of a relative rotational energy, a straight line should result, the negative slope of which would be the rotational temperature.

One series of rotational bands for which the relative intensities have already been determined are those for the CN molecule (S3). The calculation of rotational temperatures is simple using the CN data.

Example: Rotational Temperatures from Measurements of the CN Spectrum. The CN band spectrum was taken in a carbon arc from 25740 to 25800 cm.$^{-1}$ (0-0 band). The relative intensities at 25760, 25770, and 25780 cm.$^{-1}$ were found to be 21, 13.5, and 8.5%, respectively. What is the rotational temperature? Figure 8 shows that $T = 4000°K$. Similarly, the vibrational temperatures could be determined by comparing the intensities of whole bands or small portions of the bands.

C. Ionization Temperatures

Determination of ionization temperatures requires solution of the Saha equation [(8) or (9)]. As an example, consider a monatomic gas (A) partially ionized according to the reaction, $A \rightleftharpoons A^+ + e^-$. For a given equilibrium temperature we can write

1. The Saha equation (8),
2. $n_A + n_A{}^+ + n_e = N = f(P)$, and $\qquad (28)$
3. $n_A{}^+ = n_e$ $\qquad (29)$

We thus have three equations with four unknowns, n_A, $n_A{}^+$, n_e, and T; one more equation relating these terms is sufficient to permit calculation of the ionization temperature. The most commonly used phenomena are those which will determine the electron concentration (n_e) by measuring the effect the concentration has on some variables.

Two such variables are the broadness and position of spectral lines (Stark Effect). According to theory each emission line has a characteristic width determined by energy of radiation, and if the emitting particle is disturbed by collisions with other particles, the spectral line is broadened and shifted. The concentration of the electrons can be calculated, knowing the magnitude of the concurrent broadening and the shift (L3). A second assumption regarding broadening (statistical) is that the emitting particle is always in a field of perturbing changes. This also leads to

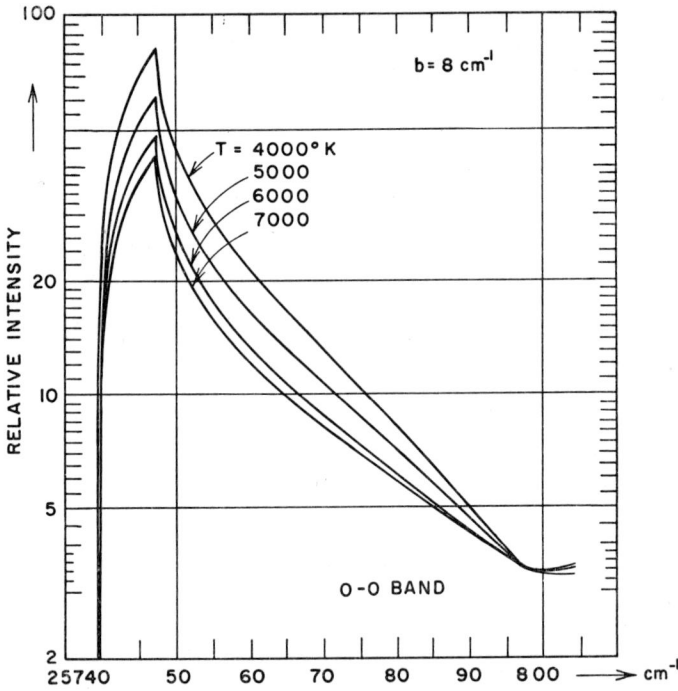

FIG. 8. Logarithmic intensity profiles of the CN rotational bands for four temperatures, O — O Band (S3).

a number for the electron concentration. A third electron concentration effect is that the broadening of the line increases as the energy of the line increases. Simultaneously the lines of higher energy appear closer together until a continuum is reached. The effect of the electrons then is to make the continuum or series limit appear at a lower energy value. This shift in the series limit has been equated to the electron concentration (see the following example). The fourth method also applies the continuous emission from the region beyond the series limit to determine the electron concentration (L4).

Example: Ionization Temperature from Spectral Data on the Series Limit. The ionization temperature of a hydrogen arc operating at 0.5 atm. is desired. An analysis of the spectrum shows that the last line discernible in the Balmer (Hydrogen) series has a quantum number of 5. According to the adjusted Inglis-Teller equation (L3)

$$\log n_e = 23.46 - 7.5 \log n_m \tag{30}$$

where n_m is the number of the last discernible line in the series. Equation (29) gives

$$n_e = \frac{1.66 \times 10^{18}}{\text{cm.}^3} = n_{H^+}$$

and Eq. (28) gives

$$n_{\text{total}} = \frac{(0.5)(6.03 \times 10^{23})}{22.4 \times 10^3} = 1.345 \times 10^{19} \text{ particles/cm.}^3$$

$$n_H = 13.45 \times 10^{18} - 3.32 \times 10^{18} = 10.13 \times 10^{18}$$

The Saha equation (8) becomes

$$\frac{[n_e][n_{H^+}]}{[n_H]} = \frac{(2\pi m_e kT)^{3/2}}{h^3} \frac{2Q_+ e^{-[E_i - \Delta E_i]/kT}}{Q_0}$$

$$\Delta E_i = 7.0 \times 10^{-7}(1.66 \times 10^{18})^{1/3} = 0.826 \text{ e.v.}$$

$$\frac{-(E_i - \Delta E_i)}{kT} = \frac{(13.587 - 0.826)1.6 \times 10^{-12}}{(1.38 \times 10^{-16})T} = \frac{-1.47 \times 10^5}{T}$$

$h = 6.624 \times 10^{-27}$ erg-sec.

$m_e = 9.107 \times 10^{-28}$ g.

$2\pi m_e kT = (6.29)(9.107 \times 10^{-28})1.38 \times 10^{-16} = 78.96 \times 10^{-44}T$

$(7.896 \times 10^{-43}T)^{3/2} = 7.018 \times 10^{-64}T^{3/2}$

$$\frac{[1.66 \times 10^{18}]^2}{10.13 \times 10^{18}} = \frac{7.018 \times 10^{-64}T^{3/2}}{2.91 \times 10^{-79}} \frac{2Q + e^{-(1.47 \times 10^5)/T}}{Q_0}$$

$$\frac{Q_+}{Q_0} = \frac{1}{2}$$

$$\frac{[1.66 \times 10^{18}]^2[2.91 \times 10^{-79}]}{[10.13 \times 10^{18}][7.018 \times 10^{-64}]} = 112.9$$

$$2.3[\log 112.9 - \log T^{3/2}] = \frac{-1.47 \times 10^5}{T}$$

$T[4.70 - 3.45 \log T] = -1.47 \times 10^5$

$T = 15{,}100°K.$

D. Empirical Temperatures

To discuss two of the remaining temperature measurements properly, it will be necessary to make some clarifying statement regarding blackbody or graybody temperature and emissivities. Whenever spectral phenomena are used to measure temperatures, consideration must be given to whether the sensing element is seeing the emitted light or whether some of the emitted light is reabsorbed along the path. One way of avoiding this problem is to be sure that the emitting element and the light path are of uniform temperature throughout, as is the case when looking at the element through a small hole when the element and the walls surrounding it are all at the same (blackbody) temperature. (A blackbody is defined as a body having an emissivity equal to one, while a graybody has an emissivity less than one.) When comparing spectral line intensities with temperature, the degree to which reabsorption occurs is described by considering the emitting gas to be optically thick or optically thin. If the gas is optically thin, emission is complete and, therefore, the emissivity remains equal to one. In the lower temperature ranges (1000–2000°K.), the emitted light is so weak that, if the emission can be photographed, reabsorption is important. However, as the temperature is increased the relative importance of reabsorption is lessened until at about 10,000°K. many light sources are optically thin. If the gas is optically thick, and no cool gases reabsorb the radiation between the source and the sensing element, the radiation will obey Planck's Law

$$I(\nu) = \frac{2(kT)^3}{h^2 c^2} \frac{h\nu^3/kT}{[e^{h\nu/kT} - 1]} \tag{31}$$

and the temperature can be measured directly.

1. *Total Brightness Temperature*

Integration of Eq. (31) gives the Stefan-Boltzmann equation

$$E = \frac{2\pi^5 k^4}{15 c^2 h^3} T^4 \tag{32}$$

The temperature derived from this equation is called the total brightness temperature and is measured by determining the total illumination (**E**) emitted by the hot source. If reabsorption by cool gases does occur then corrections must be made.

2. *Line Reversal Temperature*

Line reversal temperatures can be obtained if the hot gas is optically thick in the wavelength of some particular line (P2, P3). With a light source of known temperature (i.e., a calibrated tungsten lamp) placed behind the gas emitting this line, if the gas is hotter than the source it will appear brighter. At the same temperature both blend.

3. Calorimetric Temperature

One other temperature that has found some utility, in ablation studies using a plasma jet, has been referred to as the calorimetric temperature. The enthalpy of the stream of hot gas is determined with a calorimeter and the known heat content of the gas makes possible a calculation of the original temperature of the gas.

IV. Means for Attaining High Temperatures

High temperature for the discussions in this chapter has been defined as over 3000°K. but less than nuclear reaction temperatures, $\sim 10^8$ °K. These temperatures may be attained by sustained and explosive chemical reactions, mechanical shock tubes, electrical resistance heating, low- and high-current arc-discharges, plasma arcs, and exploding wires. Selection of the most appropriate means for creating a specific high temperature environment should be based upon the characteristics and limitations of the various methods available. Therefore, each of the useful means for attaining high temperatures is described in some detail below.

A. Chemical Reactions

There are relatively few chemical reactions capable of heating matter to temperatures greater than 3000°K. Table II contains a list of some of these reactions and the theoretical flame temperatures attainable. These reactions have two characteristics in common: (1) high exothermic heats of reaction; and (2) stable molecular products with low heat capacities, since dissociation consumes energy and results in additional products which must be heated to the flame temperature.

TABLE II
Theoretical Flame Temperatures of Some Highly Exothermic Reactions (A2)

Reaction	Temperature, °K.
1. $H_2 + \frac{1}{2} O_2 = H_2O$	3120
2. $C_{(s)} + O_2 = CO_2$	3150
3. $C_{(s)} + \frac{1}{2} O_2 = CO$	3500
4. $C_2H_2 + \frac{3}{2} O_2 = 2\ CO + H_2O$	3500
5. $H_2 + F_2 = 2\ HF$	3950
6. $C_2N_2 + O_2 = 2\ CO + N_2$	4850
7. $C_4N_2 + 2\ O_2 = 4\ CO + N_2$	5260
8. $C_4N_2 + \frac{4}{3} O_3 = 4\ CO + N_2$	5515

The importance of dissociation and the heat capacity of the product gases can be illustrated by comparing the reaction resulting in the highest reported flame temperature (G9)

$$C_4N_2 + \tfrac{4}{3} O_3 \to 4\, CO + N_2 \qquad H^0_{298} = 299.9 \text{ kcal.} \tag{33}$$

with what was once considered a very high temperature reaction,

$$H_2 + \tfrac{1}{2} O_2 \to H_2O_{(g)} \qquad H^0_{298} = 57.8 \text{ kcal.} \tag{34}$$

The reaction energy available per mole of product gas is approximately the same,

$$E_{33} = \frac{299.9}{5} = 60.0 \text{ kcal./mole product gas}$$

$$E_{34} = \frac{57.8}{1} = 57.8 \text{ kcal./mole product gas}$$

$$\frac{E_{33}}{E_{34}} = 1.04$$

Assuming ideal gas behavior and no dissociation, the theoretical temperatures, T', attainable from these two reactions should be proportional to the available heat of reaction divided by the average heat capacity per mole of product (C_p for CO and $N_2 = 8$ cal./mole-°K., C_p for $H_2O = 11$ cal./mole-°K.) (N1). Therefore,

$$T_{33}' = \frac{299.900}{5 \times 8} = 7500°K.$$

$$T_{34}' = \frac{57,800}{1 \times 11} = 5250°K.$$

and

$$\frac{T_{33}'}{T_{34}'} = 1.43$$

The actual flame temperatures are $T_{33} = 5515°K.$, $T_{34} = 3120°K.$ (2); $T_{33}/T_{34} = 1.76$.

The increase in the temperature ratio comes about because nitrogen and carbon monoxide are only slightly dissociated, even at the 5515°K. temperature, while the water molecule would be more than 19% dissociated at 3120°K., and the hydrogen molecules would also be more than 20% dissociated. The ideal chemical reaction for generation of high temperature, then, is between unstable species to form very stable products (i.e., undissociated) that have a small heat capacity.

Table III contains the bond energies of some gaseous diatomic molecules. The heat capacities (C_p) for such molecules all range from near 7 cal./mole-°K. at room temperature to 9 cal./mole-°K. at 3700°K.

Some of these high temperature reactions are utilized industrially for cutting, welding, and refining metals and for spray-coating metals and ceramics. The most sophisticated applications of exothermic chemical reactions are in rocket propulsion. Therefore, it is appropriate to consider how the performance of a rocket fuel can be evaluated.

TABLE III
Bond Energies for Selected Diatomic Molecules (A2)

Bond	Bond energy, kcal./mole
HF	135
AlO	138
LiF	139
PO	143
NO	150
BO	161
CN	175
SiO	184
NN	225
CO	256

1. *Determination of Combustion Temperature and Equilibrium Composition of Combustion Products*

Since the temperature and molecular weight of the exhaust gas are the controlling parameters in the performance of a rocket engine, the technology of high temperature combustion reactions has been developed

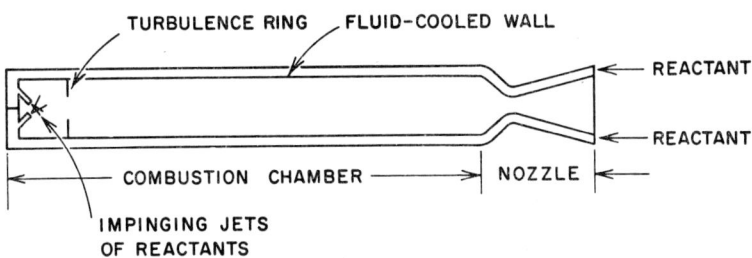

Fig. 9. Schematic diagram of rocket engine.

largely around rocket applications. However, the computation methods for analysis of combustion of rocket propellants can be applied to any high temperature flow reactions, both endothermic and exothermic.

A simplified diagram of a liquid rocket engine is shown in Fig. 9. The present state of the art requires a number of simplifying assumptions regarding the combustion process to permit calculation of rocket engine performance, even with high-speed computers. The validity of these assumptions is indicated by the small deviation ($\pm 10\%$) usually experienced between measured and calculated performance. The idealized conditions normally assumed in combustion calculations are listed below.

(1) Flow is steady and one-dimensional, along the axis of the combustion chamber.
(2) The combustion products are homogeneous and constant in composition throughout the chamber.
(3) The combustion products obey the perfect gas laws. (At the 2500° to 4000°K. combustion temperatures, the perfect gas laws apply quite rigorously.)
(4) Flow is adiabatic and isentropic (typically only 2 or 3% of the energy of the gas is transferred as heat through the rocket chamber or dissipated as friction losses.)

The calculation of temperatures and equilibrium compositions of gas mixtures involves simultaneous solution of linear (material balance) and nonlinear (equilibrium) algebraic equations. Therefore, it is necessary to resort to various approximate procedures classified by Carter and Altman (C1) as: (1) trial and error methods; (2) iterative methods; (3) graphical methods and use of published tables; and (4) punched-card or machine methods. Numerical solutions involve a four-step sequence described by Penner (P4).

(1) Assume an adiabatic flame temperature T_c and calculate the corresponding equilibrium composition with the proper set of equilibrium and material balance equations.
(2) Use the calculated composition to determine the available heat of the reaction Q_R from the difference between the heats of formation of the products and of the reactants.
(3) Calculate from enthalpy tables or charts the heat absorbed, ΔH, when the indicated equilibrium gases are heated from the reference temperature, on which the enthalpy data are based, to the assumed combustion temperature T_c.
(4) If $Q_R > \Delta H$, then the actual value of T_c is higher than assumed. Conversely, if $Q_R < \Delta H$, the assumed T_c is too high. Steps 1 through 4 are repeated until $Q_R = \Delta H$ for the correct value of T_c.

Consider a reaction between the gases A and B, yielding the ideal gases C and D.

$$n_A(A) + n_B(B) \rightleftharpoons n_C(C) + n_D(D) + \text{energy}$$

where n_A, n_B, n_C, and n_D indicate the number of moles of the respective components.

The heat of reaction Q_R is related to the heats of formation of the components:

$$Q_R = N_C(\Delta H_f^0)_C + N_D(\Delta H_f^0)_D - N_A(\Delta H_f^0)_A - N_B(\Delta H_f^0)_B \qquad (35)$$

The equilibrium constant K_p for the above reaction among perfect gases is related to the partial pressures of the components according to the following relation,

$$K_p = \frac{(p_C)^{n_C}(p_D)^{n_D}}{(p_A)^{n_A}(p_B)^{n_B}} \qquad (36)$$

Since K_p is dimensionless, the partial pressures may be expressed in any consistent units. Also, since the partial pressures are directly proportional to the volumetric concentrations, the partial pressure of a component is the product of the molar or volumetric concentration of that component and the total pressure P, e.g.,

$$p_A = x_A P \qquad (37)$$

where x_A is the mole fraction of component A.

Equilibrium constants may also be expressed in terms of the mole fractions,

$$K_n = \frac{(x_C)^{n_C}(x_D)^{n_D}}{(x_A)^{n_A}(x_B)^{n_B}} \qquad (38)$$

and, for any general reaction, the conversion between these two equilibrium constants is,

$$K_p = K_n P^{n'} \qquad (39)$$

where n' is the number of moles of products diminished by the number of moles of reactants ($n' = C + D \ldots - A - B \ldots$).

The heat absorbed when the product gases are heated to the equilibrium temperature may be expressed as

$$\Delta H = \Sigma[n_C(H_{T_c} - H_{T_r})_C + n_D(H_{T_c} - H_{T_r})_D \cdots] \qquad (40)$$

where H_{T_c} is the enthalpy at the combustion temperature and H_{T_r} is the enthalpy at the reference temperature.

Note: If the initial temperature of the reactants is different from the reference temperature, then ΔH must be corrected by adding the heat absorbed, or subtracting the heat evolved, in bringing the reactants from the initial temperature to the reference temperature.

2. Combustion of Ethylene with Oxygen

The calculation of combustion temperature for even the most simple fuel system is tedious, but an example is outlined below to clarify the procedure. A reaction of general interest is the combustion, at atmospheric pressure, of gaseous ethylene with a stoichiometric ratio of gaseous oxygen, both initially at 298°K. (25°C.),

$$C_2H_4 + 3\,O_2 \rightarrow n_{CO_2}CO_2 + n_{CO}CO + n_{H_2O}H_2O + n_{OH}OH + n_O O + n_H H$$

This example was also used by Gaydon (G1). The dissociation equilibria with which we are concerned are listed below:

$$CO_2 \rightleftharpoons CO + \tfrac{1}{2} O_2, \qquad K_1 = \frac{(p_{CO})\sqrt{p_{O_2}}}{p_{CO_2}}$$

$$H_2O \rightleftharpoons H_2 + \tfrac{1}{2} O_2, \qquad K_2 = \frac{(p_{H_2})\sqrt{p_{O_2}}}{p_{H_2O}}$$

$$H_2O \rightleftharpoons \tfrac{1}{2} H_2 + OH, \qquad K_3 = \frac{(p_{OH})\sqrt{p_{H_2}}}{p_{H_2O}}$$

$$\tfrac{1}{2} H_2 \rightleftharpoons H, \qquad K_4 = \frac{p_H}{\sqrt{p_{H_2}}}$$

$$\tfrac{1}{2} O_2 \rightleftharpoons O, \qquad K_5 = \frac{p_O}{\sqrt{p_{O_2}}}$$

The conservation equations may be expressed in terms of the number of atoms of each element present in the system. In the process under consideration, the hydrogen is present in four species at combustion temperatures—water vapor, molecular and atomic hydrogen, and hydroxyl radicals. The number of hydrogen atoms n_H is proportional to the sum of the partial pressures of each of these four components, adjusted for elemental hydrogen content,

$$n_H = \frac{V}{RT}(2p_{H_2O} + 2p_{H_2} + p_H + p_{OH})$$

Stein introduced the concept of fictitious partial pressures of the elements to eliminate the V/RT term (S4). The fictitious pressure $\bar{p}$ is the partial pressure each element would exert if it were present as a monatomic gas; thus, we have the following set of three conservation equations.

$$\bar{p}_C = p_{CO_2} + p_{CO}$$
$$\bar{p}_H = 2p_{H_2O} + 2p_{H_2} + p_H + p_{OH}$$
$$\bar{p}_O = 2p_{CO_2} + p_{CO} + p_{H_2O} + 2p_{O_2} + p_{OH} + p_O$$

The ratios of the fictitious partial pressures of the elements are constants for the above stoichiometric reaction:

$$\frac{\bar{p}_C}{\bar{p}_H} = 0.5, \qquad \frac{\bar{p}_O}{\bar{p}_H} = 1.5$$

The total pressure is:

$$P = p_{CO_2} + p_{CO} + p_{H_2O} + p_{O_2} + p_{H_2} + p_{OH} + p_H + p_O$$

We can now use the five equilibrium equations, the two ratios of fictitious partial pressures, and the total pressure equation to solve for the equilibrium among the eight species at any given temperature. The initial flame temperature assumed is 3000°K. At this temperature the values of the equilibrium constants (N2) are:

$$K_1 = 0.3395 \quad K_4 = 0.1576$$
$$K_2 = 0.0476 \quad K_5 = 0.1201$$
$$K_3 = 0.0499$$

If we also assume that $p_{CO_2}/p_{CO} = 1.000$ and $p_{H_2O} = 0.300$, it follows from the sequence of calculations indicated in Fig. 10 (G1) that

$$\sqrt{p_{O_2}} = K_1 \frac{p_{CO_2}}{p_{CO}} = 0.3395 \times 1.000 = 0.3395$$

$$p_{O_2} = 0.1153$$

$$p_O = K_5\sqrt{p_{O_2}} = 0.1201 \times 0.3395 = 0.0408$$

$$p_{H_2} = K_2 \frac{p_{H_2O}}{\sqrt{p_{O_2}}} = 0.0476 \times \frac{0.300}{0.3395} = 0.0421$$

$$p_{OH} = K_3 \frac{p_{H_2O}}{\sqrt{p_{H_2}}} = 0.04990 \times \frac{0.300}{\sqrt{0.0421}} = 0.0730$$

$$p_H = K_4\sqrt{p_{H_2}} = 0.1576 \times \sqrt{0.0421} = 0.0323$$

$$\bar{p}_H = 2p_{H_2O} + 2p_{H_2} + p_H + p_{OH}$$
$$= 2(0.300) + 2(0.042) + 0.032 + 0.073$$
$$= 0.789$$

$$\bar{p}_C = 0.5\bar{p}_H = 0.5 \times 0.789 = 0.395$$

$$p_{CO_2} = \bar{p}_C - p_{CO} = p_{CO}\left(\frac{p_{CO_2}}{p_{CO}}\right)$$

$$p_{CO} = \frac{\bar{p}_C}{1 + (p_{CO_2}/p_{CO})}$$

$$p_{CO} = \frac{0.395}{1 + 1.000} = 0.198$$

$$p_{CO_2} = 0.395 - 0.198 = 0.197$$

$$\bar{p}_O = 2p_{CO_2} + p_{CO} + p_{H_2O} + 2p_{O_2} + p_{OH} + p_O$$
$$= 2(0.197) + 0.198 + 0.300 + 2(0.115) + 0.073 + 0.041$$
$$= 1.236$$

$$\frac{\bar{p}_O}{\bar{p}_H} = \frac{1.236}{0.789} = 1.566$$

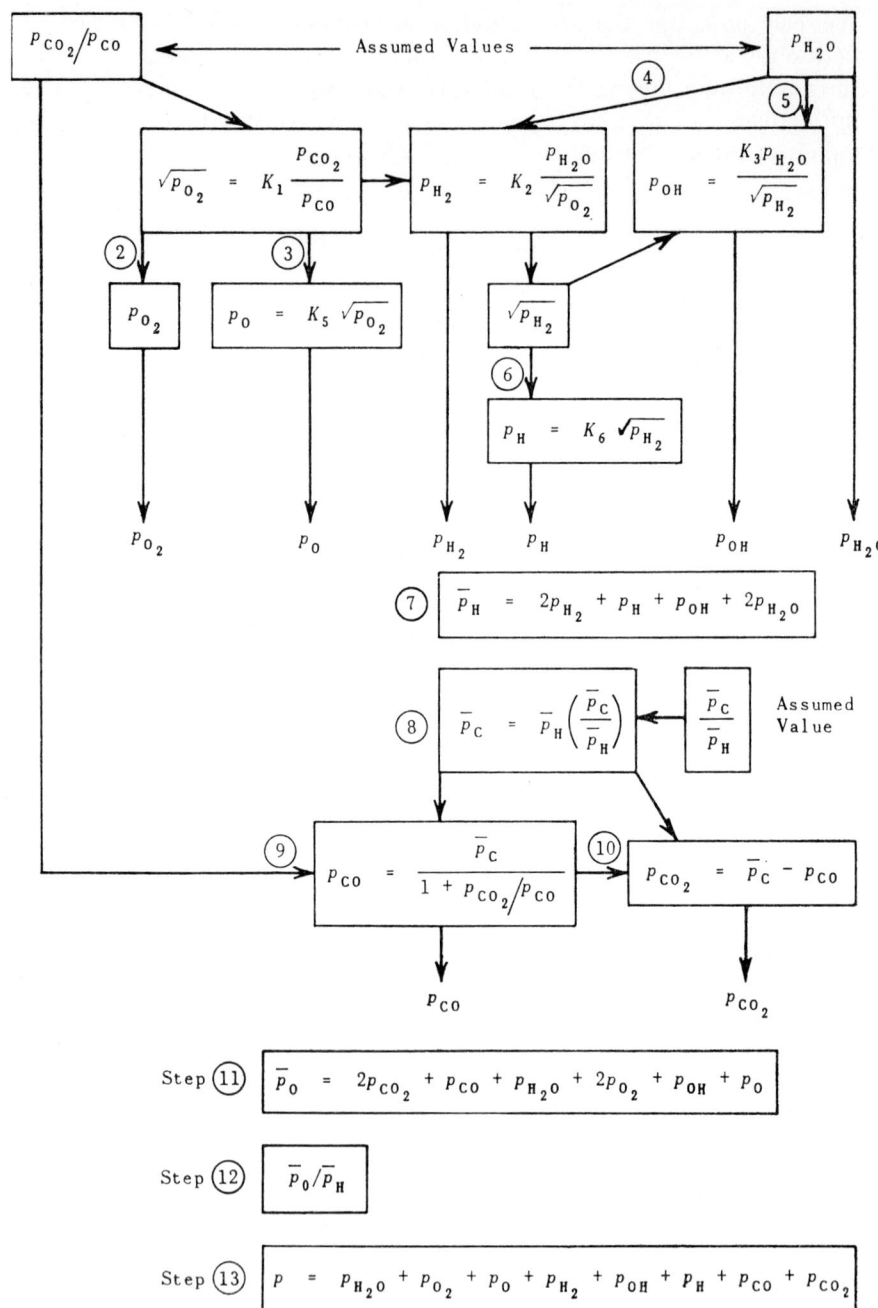

Fig. 10. Sequence of calculations for equilibrium gas composition (G1).

The calculated values are entered as Trial No. 1 in Table IV since a change in the ratio p_{CO_2}/p_{CO} mainly affects the ratio $\bar{p}_O/\bar{p}_H$ and a change in the assumed p_{H_2O} affects primarily the total pressure, the assumed $p_{H_2O} = 0.300$ must be reasonably close to the equilibrium value, and the next trial should be an attempt to approach a $\bar{p}_O/\bar{p}_H$ of 1.5 by assuming a lower p_{CO_2}/p_{CO} ratio (=0.990). It is apparent from Trial 2 that the actual p_{CO_2}/p_{CO} value must be substantially lower and also that p_{H_2O} must be greater than 0.300 to obtain $p = 1$. Therefore, for Trial 3, the assumed values were $p_{CO_2}/p_{CO} = 0.930$ and $p_{H_2O} = 0.305$. The last trial may not be justified because of the uncertainties in the equilibrium constants and the idealized basis for the calculations.

TABLE IV

Calculation of Equilibrium Composition of $C_2H_4 + 3\,O_2$ Flame

	3000°K.				3250°K.
	Trial 1	Trial 2	Trial 3	Final	Final
p_{CO_2}/p_{CO}	1.000	0.990	0.930	0.932	0.463
p_{H_2O}	0.300	0.300	0.305	0.306	0.244
p_{O_2}	0.005	0.113	0.100	0.100	0.130
p_O	0.041	0.040	0.038	0.038	0.095
p_{H_2}	0.042	0.042	0.046	0.046	0.071
p_{OH}	0.073	0.073	0.071	0.071	0.011
p_H	0.032	0.032	0.034	0.034	0.085
$\bar{p}_H$	0.789	0.789	0.807	0.809	0.726
$\bar{p}_C$	0.395	0.395	0.404	0.405	0.363
p_{CO}	0.198	0.199	0.210	0.209	0.248
p_{CO_2}	0.197	0.197	0.194	0.195	0.115
$\bar{p}_O$	1.236	1.224	1.212	1.214	1.088
$\bar{p}_O/\bar{p}_H$	1.566	1.551	1.502	1.501	1.500
P	0.999	0.997	0.998	0.999	0.999

Due to dissociation, the four moles of reactants yield more than the four moles of CO_2 and H_2O which result at low temperatures. Since the partial pressures are numerically equivalent to mole fractions of the various species, the true number of moles of products is obtained by dividing the mole number of any atom introduced as a reactant by its fictitious partial pressure. In theory, therefore, $n_C/\bar{p}_C = n_H/\bar{p}_H = n_O/\bar{p}_O$

$$\frac{n_C}{\bar{p}_C} = \frac{2}{0.405} = 4.94 \text{ moles products}$$

It follows that,

$$C_2H_4 + 3\,O_2 \rightarrow 4.94(0.306\,H_2O + 0.199\,O_2 + 0.038\,O + 0.046\,H_2 \\ + 0.071\,OH + 0.034\,H + 0.209\,CO + 0.195\,CO_2)$$

The standard heats of formation at 298°K. of the products (N1) from the respective elements are then used to obtain the heat of reaction at 3000°K.

$$Q_R = 4.94[(0.306)(57.80) + (0.038)(-59.16) + (0.071)(-10.06)$$
$$+ (0.034)(-52.09) + (0.209)(26.42) + (0.195)(94.05)]$$
$$= 181.9 \text{ kcal.}$$

In comparison the heat absorbed by the products in raising their temperature from 298°K. to 3000°K. is derived from enthalpy data (N2).

$$\Delta H = 4.94[(0.306)(32.16 - 2.37) + (0.100)(25.52 - 2.07)$$
$$+ (0.038)(15.13 - 1.61) + (0.046)(23.19 - 2.02)$$
$$+ (0.071)(23.56 - 2.11) + (0.034)(14.90 - 1.48)$$
$$+ (0.209)(24.43 - 2.07) + (0.195)(38.94 - 2.24)]$$
$$= 129.2 \text{ kcal.}$$

Since $Q_R > \Delta H$, the actual value of the flame temperature is higher than the 3000°K. assumed initially. It is logical to choose $T_c = 3250°$K. as the next trial value, because the equilibrium constants are tabulated by NBS with this temperature interval. At this temperature, the values of the equilibrium constants are:

$$K_1 = 0.7780 \qquad K_4 = 0.3205$$
$$K_2 = 0.1044 \qquad K_5 = 0.2638$$
$$K_3 = 0.0123$$

A series of trials with different assumed values of p_{CO_2}/p_{CO} and p_{H_2O} were made until a balance was obtained, the final trial showing the equilibrium composition at 3250°K. is recorded in Table IV. In this case, the 4 moles of reactants yielded 5.51 moles of products, and $Q_R = \Delta H = 117.6$ kcal. Therefore, the adiabatic flame temperature of the $C_2H_4 + 3\ O_2$ reaction is 3250°K. (2977 ± 10°C.).

3. Alternative Calculation Procedures

The method described above for calculating high temperature equilibria is straightforward and was selected to demonstrate the basic principles. There are, however, several techniques which reduce the number of numerical operations. Such procedures are of particular value when nitrogen is present (as N_2 and NO) or if the fuel mixture is so rich that elemental carbon is deposited. The well-known Hottel charts (H5) contain the equilibrium compositions at many temperatures and pressures for the $H + O + C + N$ system. An excellent approach to the slide-rule calculation of high temperature equilibria was developed by

Weinberg (W1) as an extension of a method developed by Winternitz (W3). When applied to the C + H + O system, Weinberg's involves the following steps:

(1) Estimate rough values for the partial pressures of three of the eight components (CO, CO_2, O, O_2, H, H_2, H_2O, OH)
(2) Calculate the corresponding pressures of the other five components, using equations expressed in terms of the three assumed values and the equilibrium constants
(3) Use a series of 12 linear equations involving the assumed and calculated partial pressures to evaluate, by determinants, the errors in the three partial pressures assumed initially
(4) Add each error to the corresponding assumed partial pressure to obtain a good approximation of the equilibrium partial pressure. The other five partial pressures are calculated with the equations used in step 2.

The procedure can be repeated, using the new values for deriving a more accurate equilibrium composition. This technique gives rapid convergence. If reasonable discretion is used in making initial assumptions, the first approximation may be as accurate as the values for the high temperature equilibrium constants. Weinberg derived a similar series of explicit linear equations for the C + H + O + N system. Certainly, high-speed computers are the preferred means for repetitive calculations of high temperature equilibria and thermodynamic properties of high temperature gas systems. For some systems (C + H + O and C + H + O + N) punched cards containing all the equilibrium constants usually needed are available for machine computation (B4, B5). Myers (M7) and his associates at the Bureau of Mines developed a technique for punched-card machine calculation of equilibrium composition and enthalpy of gaseous products of combustion of solid, liquid, and gaseous fuels with air in industrial furnaces. These data can be used for determination of adiabatic flame temperatures between 1300° and 2200°C.

4. Expansion of Gas Through De Laval Nozzle

The high temperature gas generated in the combustion chamber of a rocket engine is expanded through a De Laval nozzle to convert a major portion of the enthalpy to mechanical thrust. The thrust F of a rocket engine is defined by the following conservation of momentum equation:

$$F = (\tfrac{1}{2} + \tfrac{1}{2}\cos \boldsymbol{a})(wv_e/g_c) + (p_e - p_0)A_e \qquad (41)$$

where $\boldsymbol{a}$ is one-half the divergence angle of the nozzle, w is weight-rate of flow, v_e is linear flow velocity of gas from nozzle, g_c is gravitational constant, p_e is pressure at nozzle exit, p_0 is external atmospheric pressure, and A_e is the cross sectional area of the nozzle exit.

If the nozzle is designed so that α is relatively small and $p_e = p_0$ (the condition of maximum thrust), then Eq. (41) becomes

$$F = \frac{wv_e}{g_c} \qquad (42)$$

The common term, specific impulse, is the thrust per unit weight-rate of flow designated by I_{sp}. Therefore, from equation (42)

$$I_{sp} = \frac{F}{w} = \frac{v_e}{g_c} \qquad (43)$$

For adiabatic, nonviscous, one-dimensional flow through a rocket nozzle,

$$\tfrac{1}{2}v_e^2 = (\Delta h)_{c \to e} \qquad (44)$$

where $(\Delta h)_{c \to e}$ is the enthalpy change per gram of gas as it expands from the combustion chamber temperature T_c to the nozzle exit temperature T_e. Moreover, the change in enthalpy can be expressed in terms of γ^*, the average ratio of specified heats; W^*, the average "molecular weight" of the exhaust products; the flame temperature; the molar gas constant; and the ratio of pressures p_e/p_c,

$$v_e^2 = \frac{2\gamma^*}{\gamma^* - 1} \cdot \frac{RT_c}{W^*} \left[1 - \left(\frac{p_e}{p_c}\right)^{(\gamma^* - 1)/\gamma^*} \right] \qquad (45)$$

The value of γ^* varies only slightly among propellants. Consequently, for a fixed ratio p_e/p_c, the velocity is proportional to $(T_c/W^*)^{\frac{1}{2}}$. Since $I_{sp} = v_e/g_c$, the maximum impulse is obtained for propellants which have high flame temperatures and low molecular weight products.

It is common practice to assume values for γ^* and W^* to calculate v_e. On the other hand it is also possible to calculate more precisely the change in state of the expanding gas through the nozzle by iterative methods involving very small increments of distance along the nozzle or of time. Such a detailed procedure is beyond the scope of this discussion. As first approximations, Penner (P4) has considered two limiting conditions for analysis of chemical reactions during adiabatic expansion through a De Laval nozzle. In one case, he assumes that the reaction kinetics are so rapid that the composition approximates the equilibrium composition at all positions in the nozzle (near-equilibrium flow). At the other extreme is the treatment based upon the assumption that reactions occur so slowly that negligible deviations in the composition of the combustion chamber gas occur during flow through the nozzle (near-frozen flow). The actual conditions are always intermediate between equilibrium and frozen flow.

For the near-frozen (or constant composition) case, the calculation procedure involves determination of the exhaust temperature T_e, by trial and error, from the following relationship:

$$x_A(S^0_{A,T_c} - S^0_{A,T_e}) + x_B(S^0_{B,T_c} - S^0_{B,T_e})$$
$$+ x_C(S^0_{C,T_c} - S^0_{C,T_e}) + \cdots = R \ln \frac{p_c}{p_e} \quad (46)$$

where S^0_{A,T_c} = absolute molar entropy of component A at the flame temperature, S^0_{A,T_e} = absolute molar entropy of component A at the exhaust temperature, and B, C, etc. are the other species present in the gas.

Knowing T_e, one may then calculate the molar enthalpy change $(\Delta H)_{c \to e}$ between nozzle entrance and exit positions from the changes in molar heat capacities c_p of the components,

$$(\Delta H)_{c \to e} = x_A \int_{T_e}^{T_c} C_{p,A} \, dT + x_B \int_{T_e}^{T_c} C_{p,B} \, dT + x_C \int_{T_e}^{T_c} C_{p,C} \, dT + \cdots \quad (47)$$

Note: The molar entropy values and molar heat capacities may be obtained from reference (N3).

The exit velocity, and hence the specific impulse, may then be calculated from Eq. (44) since

$$(\Delta h)_{c \to e} = \frac{(\Delta H)_{c \to e}}{W} \quad (48)$$

For the near-equilibrium flow case, it is also necessary to calculate T_e by trial and error. Each assumed value of T_e will have a unique gas composition. The correct value will satisfy the criteria for isentropic flow,

$$\overline{S}_{T_e,p_e} = \frac{\overline{W}_c}{\overline{W}_e} \overline{S}_{T_c,p_c} \quad (49)$$

where $\overline{S}_{T,p}$ is the molar entropy of an ideal gas mixture at temperature T and pressure p. The next step is determination of the enthalpy change for a given weight of the gas, since the molecular weight is variable. The value of v_e is then calculated from Eq. (44).

Near-equilibrium flow conditions generally yield the maximum thrust for rocket propulsion, because partial recombination of the dissociated atoms, as the temperature falls, releases additional kinetic energy. On the other hand, when the rocket engine is considered for high temperature chemical processing, it is invariably desirable to freeze the composition attained in the combustion chamber. From both theoretical and practical standpoints, it is not always possible to predetermine the flow conditions in the De Laval nozzle; as the foregoing discussion indicates, the flow conditions are functions of initial conditions and composition of the gas, its chemical and the thermodynamic characteristics, and the geometry of the nozzle.

B. ELECTRICAL SOURCES OF HIGH TEMPERATURE

1. *Atom-Atom Recombination*

Since a major proportion of the heat required to raise the temperature of matter in conventional chemical reactions comes from the formation of stable molecular species, the temperatures at which these species begin to dissociate is the upper temperature limit attainable with these reactions. Conversely, very high temperatures can be obtained if gaseous atoms formed in one zone are then allowed to recombine in a separate zone. Langmuir's atomic hydrogen torch and Cobine and Wilbur's electronic torch (C4) operate on this principle by producing the atoms in an electrical field and then passing them into a neutral zone where they recombine to evolve the quantities of energy required to heat the recombined gases to high temperatures. The final temperature is directly related to this bond energy. The theoretical maximum temperature attainable with the electronic torch would result from recombination of atomic, gaseous carbon and oxygen atoms into carbon monoxide (see Table III; $C + O \rightarrow CO$, $-\Delta H = 256$ kcal./mole). If the atoms are at room temperature when they reach the recombination zone, 7600°K. is the theoretical temperature of the recombination zone gases (A2). So far, the use of recombination energies has not approached the efficiency required to produce this temperature, because of the problem of isolating the atoms in their environment yet containing a high enough density of atoms to allow rapid and complete recombination. Cobine and Wilbur (C4, C5) report a gas flame temperature of 3000°K. by recombination of nitrogen atoms. Nitrogen theoretical recombination temperature is 7100°K. (A2).

2. *Ion-Electron Recombination*

A logical extension of this type of energy transfer to obtain even higher temperatures would be electrical neutralization of ions by electrons. A theoretical gas-stream temperature attainable from the reaction $e^- + N^+ \rightarrow N$ can be calculated in the following manner, assuming that the electron and ion are initially at 298°K. The heat of ionization at 298°K., $\Delta H_{298}^0 = -337$ kcal./mole, is the sum of the ΔH^0's for the two reactions

$$\tfrac{1}{2}N_2 \rightleftharpoons N, \qquad \Delta H_{298}^0 = +85 \text{ kcal./mole}$$
$$e^- + N^+ \rightleftharpoons \tfrac{1}{2}N_2, \qquad \Delta H_{298}^0 = -422 \text{ kcal./mole}$$

[See NBS Circular 500, Table 18 in Series I (N1) and also (S6).]

The equilibrium constant as a function of temperature is determined from the free energy function (fef) values tabulated by Woolley (W4, W5) and from a calculated free energy function for the electron.

$$\frac{(F_T^0 - E_0^0)_N}{RT} - \frac{(F_T^0 - E_0^0)_{N^+}}{RT} - \frac{(F_T^0 - H_0^0)_{e^-}}{RT} = \frac{\Delta(F - E_0^0)}{RT} \quad (50)$$

where

$$\frac{(F_T - H_0^0)_{e^-}}{RT} = -\left(\frac{H_T^0}{RT}\right)_{e^-} + \left(\frac{S_T}{R}\right)_{e^-} \quad (51)$$

Note: Woolley uses E_0^0 instead of H_0^0 but the terms are synonymous.

The entropy S of a particle can be represented in statistical mechanical terms:

$$S = k \ln \frac{Q^N}{N!} + RT \left(\frac{\partial \ln Q}{\partial T}\right)_v \quad (52)$$

where N is the number of particles in the system and Q is the partition function of the species of particle. If the particles are atoms of a gas, the above equation simplifies to

$$\frac{S}{R} = k \ln \left[Q_e \frac{(2\pi MRT)^{3/2}}{NL^3} \cdot \frac{RT}{p} \right] + \frac{5}{2}$$

(the Sackur-Tetrode equation) (53)

where Q_e is the electronic partition function. The simplification occurs because the atom has no vibrational or rotational energy, and so these terms drop out of Eq. (52). If the particle is an electron the term is further simplified, since only translational energy is involved. It is also assumed that the standard state of electron is gas at $0°K.$, and the equation becomes:

$$\frac{S}{R} = \frac{3}{2} \ln M_e + \frac{5}{2} \ln T - \frac{2.314}{1.987} + \ln 2 \quad (54)$$

where $M_e =$ atomic weight of the electron and

$$\frac{H_T^0}{RT} = \frac{(5/2)RT}{RT} \quad (55)$$

(assuming perfect gas relationships for the electron). Taking the atomic weight of an electron (M_e) as 0.00055 g., the fef for the electron can be calculated as a function of temperature. See Margrave (M3) for more examples of calculations using the fef.

Having obtained a value for the fef of the reaction, the equilibrium constant can be calculated,

$$\ln K_p = -\frac{\Delta F}{RT} = \Delta \text{ fef} + \frac{\Delta E_0^0}{RT} \tag{56}$$

Once the equilibrium constant is determined, the relative proportions of the three gases can be established by assuming for calculation one atmosphere total pressure;

$$P = p_N + p_{N^+} + p_{e^-}; \qquad p_{N^+} = p_{e^-}$$

$$p_N = P - 2p_{N^+} \quad \text{and} \quad K_p = \frac{[p_N]}{[p_{N^+}]} = \frac{[P - 2p_{N^+}]}{[p_N]^2}$$

The theoretical temperature, 15,200°K., will be reached when the heat of reaction (heat of ionization times the fraction of product formed) equals the increase in the heat contents of the constituent gases. The heat contents are also given in the tables, usually as a function similar to the fef. Obviously, correspondingly higher gas-stream temperatures can be obtained by neutralization of ions in multiply-ionized states. Such reactions can occur, using ions produced in radio-frequency discharges.

3. *Resistance and Induction Heating*

The more frequently used forms of electrical heating, such as resistance and induction, have limited use above 3000°K. (M1, D1, C2, F1, F4, B3). Tungsten, carbon, and some of the carbides are the only solid materials capable of use as resistors or susceptors above that temperature. Conducting liquids, although potentially capable of reaching higher temperatures than these solids as electrical resistance elements, suffer from problems of containment and have not been used above 3000°K. except in levitation melting (O1, W6).

4. *Conventional and High Intensity Arcs*

Arcs can be considered as gaseous resistance heaters and offer temperatures up to 50,000°K. The sustained temperatures realizable from electric arcs can be divided into three general regions according to the current density of the conducting path. The lower temperatures (up to ~4000°K. and a current density of 60 amp./cm.^{2}) make the anode material incandescent, but as the current density is increased beyond a critical level the voltage drop shifts suddenly from a uniform drop between cathode and anode to a drop concentrated at the anode surface (for a dc arc—at both electrodes for an ac arc). The transition from the conventional to a high intensity arc is marked by pronounced increases in brilliance and temperature; the arc path becomes distorted, and a jet of plasma, called the tail flame, issues from the rapidly vaporizing anode (see Fig. 11). The high intensity arc was first described analytically by Finkelnburg (F1) (F2) (F3).

As the current intensities increase further, the potential drop, and hence energy transfer, shift more and more toward the anode crater. As much as 70% of the energy transfer has already been made to occur in this anode region, creating temperatures up to 20,000°K. at the core of arc path. The third region (>20,000°K.) is reached by a further increase in the current density. Gerdian and Lotz (G2) found that this could be done by constraining the arc within a narrow tube of flowing liquid. The gas inside the tube near the fluid was cooled sufficiently to

FIG. 11. Tail flame from high-intensity electric arc (F1).

concentrate the current-carrying gas (now ionized) in a narrow region at the core. It was found that, in addition to the thermal pinch just described, an electromagnetic pinch created by the current flow through the gas served to restrict the cross-sectional area of the arc stream further. Continuous temperatures up to 50,000°K. have been achieved with this technique (B8, B9, L5, P7).

5. *Plasma Jet*

A useful modification of the Gerdian arc first described by Peters (P5) and called a plasma arc is shown schematically in Fig. 12 (G3, J1, M5, B2). (Plasma is defined as an electrically neutral gas composed of ions, with one or more positive charges, and electrons. Since the gas is heterogeneous, the definition applies to a finite volume.)

The electrical energy being transported between the electrodes serves to ionize the gases. The zone of ionized gases is constricted by a continuous flow of cold gas or liquid being fed in such a way as to form a nonconducting wall around the column of plasma. An orifice in one electrode serves as an outlet for the plasma and, again, a protective shield of cool gas surrounds the plasma as it is ejected through the orifice.

The value of the plasma jet for applied research on thermally resistant materials and for fundamental research on the properties of matter at

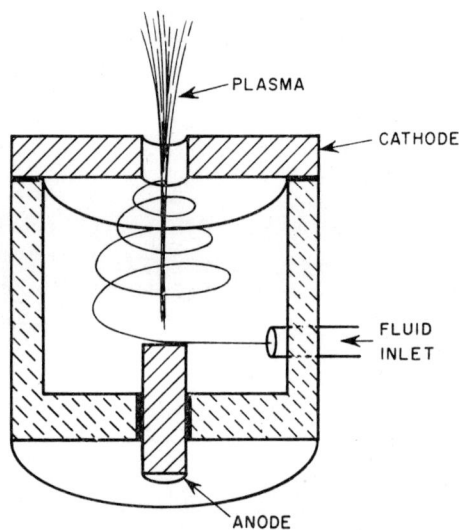

Fig. 12. Schematic diagram of plasma jet.

extremely high temperatures was recognized promptly, and plasma jets proliferate throughout research laboratories today. Plasma jets are presently manufactured for sale by Avco, Plasmadyne Corporation, and Thermal Dynamics Corporation.

6. *Exploding Wires*

Exploding wires by passing large surges of electrical energy through them is another means of creating extremely high temperatures. Energy from the batteries (or transformer) is stored in a large condenser bank. The condensers are discharged through the fine wire, causing its disintegration (A3).

The purpose of this technique was to create high temperatures for brief periods by dumping large amounts of energy into a relatively small space. The hoped-for temperatures ($>10^5$ °K.) that could be reached if all the stored energy was consumed by the material in the wire were not realized. In actuality the energy is dissipated in two stages. The first stage energy goes into the melting and evaporation of the wire. At this point the current ceases momentarily due to the decreased conductivity of the path, with a resultant surge in overvoltage at the wire terminals. The second stage begins with the initiation of an arc that permits the remainder of the energy on the condenser to discharge (C7, L3).

C. Mechanical Sources of High Temperature (Shock Waves)

The chemical engineer's interest in the shock phenomena stems from its capability for heating gas to temperatures as high as 10^6 °K., then quenching them rapidly. The shock tube has demonstrated its utility for

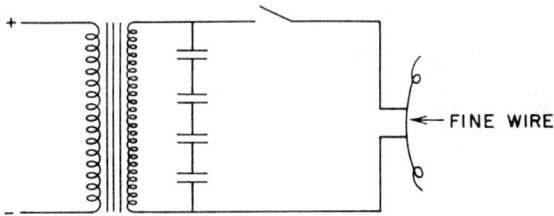

Fig. 13. Exploding wire apparatus.

investigation of reaction kinetics at high temperatures; in addition, it is a conceivable process device.

If two gases at largely different pressures ($P_0/P_1 \gtrsim 4$) are suddenly allowed to mix with one another, a steep-fronted (shock) wave develops as the high pressure (driver) gas moves into the zone occupied by the low pressure (experimental) gas. The pressure difference can be created by an explosion, an electrical discharge, or simply by varying the differential pressures across a removable diaphragm (D3). The latter case, in which the diaphragm separates two sections of pipe, reduces to the simplest case from a mechanical and mathematical standpoint and is the only one considered here. Ideally, when the diaphragm is broken in such a system, the high pressure gas acts like a piston traveling into the low pressure zone with a contact surface between it and a portion of the experimental gas that is now at the same pressure as the driver gas. The shock front occurs entirely in the experimental gas and separates the low from the high pressure region. Simultaneously a rarefaction wave begins at the diaphragm, traveling back into the driver gas zone.

Also considerable work has been done using a metal (ballistic) piston in place of the driver gas (L6).

The shock front created in the experimental gas is a physical and mathematical discontinuity that requires irreversible thermodynamics for description. For convenience the shock wave system is divided into three parts: the parts before and after the shock front are considered to obey the laws for reversible processes ($dS = O$, etc.) so only what occurs at the discontinuity is described as an irreversible process ($dS > O$). However, even at the front the laws of conservation (mass, momentum, and energy) still hold for the nonuniform, unidimensional flow of the shock wave when confined in a tube:

Conservation of mass,
$$\rho_1 U_1 = \rho_4 U_4 \tag{57}$$

Conservation of momentum,
$$P_4 - P_1 = \rho_1 U_1 (U_1 - U_4) \tag{58}$$

Conservation of energy,
$$E_4 + P_4 v_4 + \tfrac{1}{2} U_4^2 = E_1 + P_1 v_1 + \tfrac{1}{2} U_1^2 \tag{59}$$

where U is the velocity of the fluid, P is the pressure, v is the specific volume, and ρ is the density. The utility of these equations for describing what happens in the shock front is covered by Courant and Friedricks (C8) along with derivations of equations of state for polytropic gases.

The sequence of events after the diaphragm is broken, and their ramifications, are explained by Glass (G5) from an aerodynamic standpoint. Our interest in using the shock tube, however, stems from its ability to heat gases rapidly and then quench them over large temperature ranges, up to 10^6 °K. The shock tube is now used as a tool to study reaction kinetics at high temperatures and conceivably could be used as a process tool. In order to justify this last statement, it will be necessary to describe accurately the temperature-time relation and pressure profile of the experimental gas during the period when the shock is traveling through it. This description is frequently made in graphical form similar to Figs. 14 and 15 for the ideal gas case. It is necessary to know the temperature of the experimental gas between the normal shock front and the contact surface, the time the experimental gas is subjected to this temperature, the rate at which the gas is heated and cooled, and any end effects caused by finite boundaries.

The change in temperature across the normal shock front at any position in time and space is represented by the equation

$$\frac{T_2}{T_1} = \frac{P_2}{P_1} \left[\frac{1 + \alpha(P_2/P_1)}{P_2/P_1 + \alpha} \right] \tag{60}$$

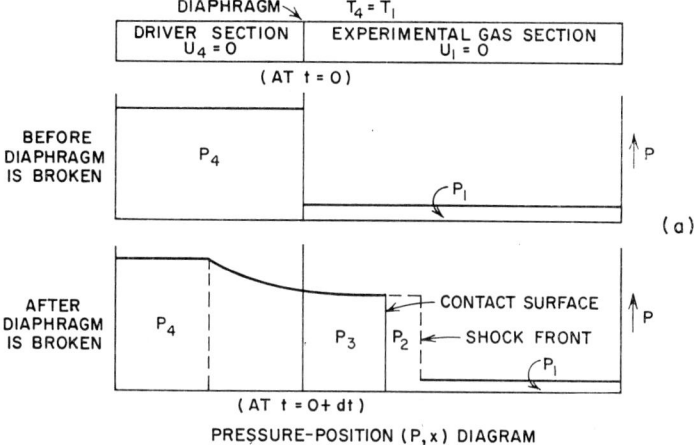

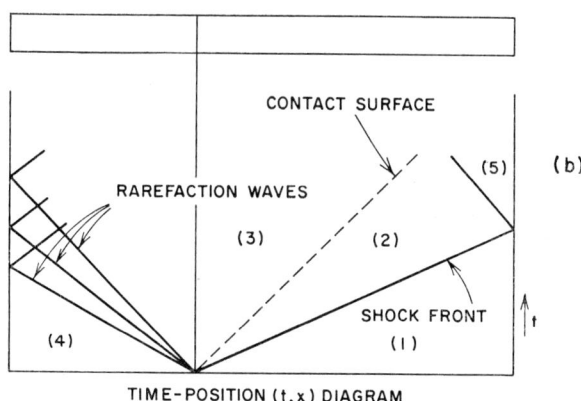

Fig. 14. P, x (a) and t, x (b) diagrams for shock tube.

where the subscripts 1 and 2 indicate the conditions before and after the shock front, respectively, and

$$\alpha = (\gamma_1 - 1)/(\gamma_1 + 1)$$

As shown in Fig. 14a, P_2 is only momentarily the initial pressure of the driver gas, P_4. The value of P_2/P_1 in turn is determined by the equation

$$\frac{P_2}{P_1} = [1 + \alpha]\mathsf{M}^2 - \alpha \tag{61}$$

where

$$\mathsf{M} = \frac{V}{a_1} = \frac{\text{shock front velocity}}{\text{velocity of sound in experimental gas at pressure, } P_1}$$

The value of M, however, is dependent upon the pressure ratio and the position of the normal shock front in the tube (G6). If $(P_2/P_1) < 3$, the variation of M as a function of x (the distance along the tube) is described in Fig. 16a, but if $(P_2/P_1) > 3$, M is described by Fig. 16b. For the case $(P_2/P_1) > 3$ the equation

$$\log \frac{P_4}{P_1} = 1.33 x^{0.03} (M - 1) + 0.08 \tag{62}$$

has been shown to hold for air.

If the shock wave is very strong, the temperature behind the shock front becomes high enough to alter the specific heat of the gas, and

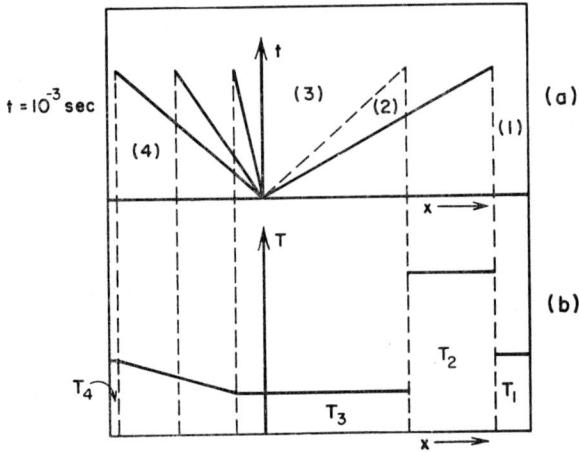

FIG. 15. Typical t, x and T, x diagrams for shock tube.

this must also be taken in consideration in using Eqs. (60) and (61).

The rate at which the gases are heated depends upon the velocity of the particles (U_1) relative to the shock front and the rate at which the energy is converted. For the fast processes (translational and rotational energy changes), the energy conversion occurs almost instantaneously with the change in state of the gas. (It requires approximately 3 collisions between particles at the shock wave pressures to establish equilibrium for such fast processes.) The rate $dT/d\theta$ of temperature change for such a process can be represented by the equation

$$\frac{dT}{d\theta} = \frac{V_1^2 - V_2^2}{2} \left(\frac{\gamma - 1}{R\gamma} \right) \lambda v \tag{63}$$

where λ is the mean free path required to establish equilibrium, V is the gas velocity and **v** is the specific volume of gas. The temperature rise is given by

$$T_2^* - T_1 = \frac{V_1^2 - V_2^2}{2}\left(\frac{\gamma - 1}{R\gamma}\right) \tag{64}$$

where T^* is the maximum temperature of the active energy. T_2^* is

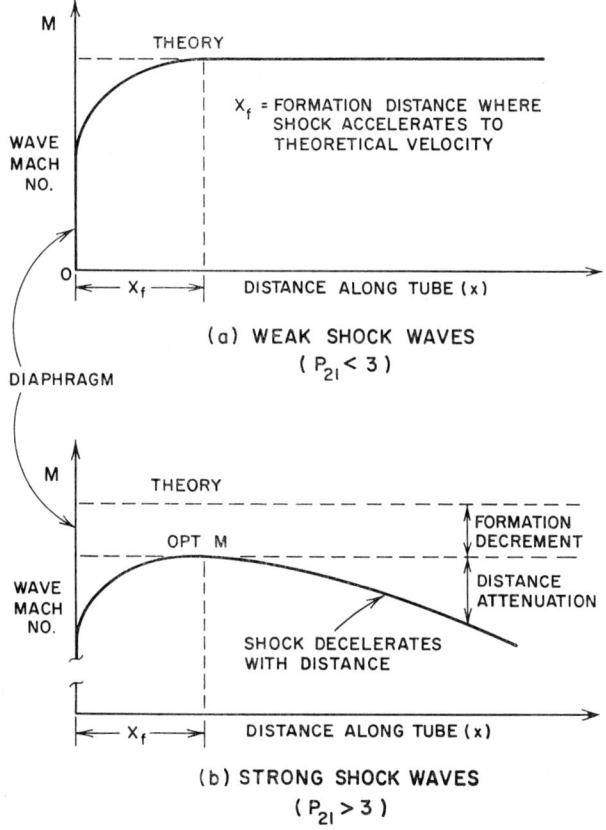

Fig. 16. Shock front velocity as a function of distance from diaphragm (G6).

higher than the equilibrium temperature behind the shock front in contrast with the less active species (vibrational, electron-excitational) which are at a lower temperature than at equilibrium. A typical temperature profile across the shock front is shown in Fig. 17.

The time that the experimental gas is at the elevated shock wave temperature and pressure depends upon the geometry of the tube and the velocities of the shock front, the contact surface, and the rarefaction waves. Another important effect in practical systems is what happens at the ends of the tubes. As shown in Fig. 14, both the shock and rarefaction waves will be reflected from such surfaces. The pressure on the gases in the zone (5) behind the reflected shock will increase from 2 to 8 times depending upon the Mach number **M** of the normal wave (see Fig. 36 in G5) and the gas temperature in the same zone will be a function of the pressure,

$$\frac{T_5}{T_2} = \frac{P_5}{P_2}\left(\frac{1 + (P_5/P_2)\alpha}{\alpha + (P_5/P_2)}\right) \tag{65}$$

The theoretical maximum increase in temperature for most diatomic

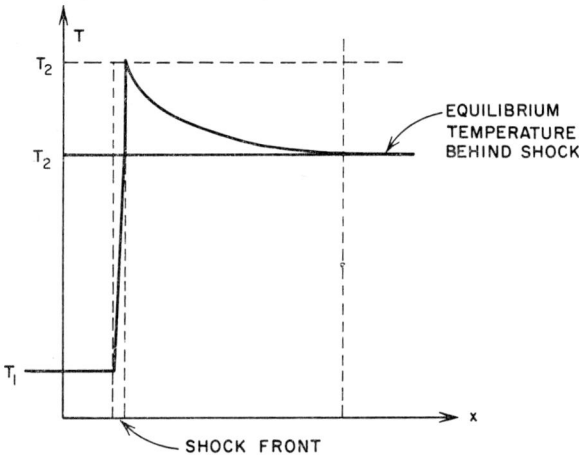

FIG. 17. Temperature profiles across shock front (G5).

gases $[\alpha = (1.4 + 1)/(1.4 - 1)]$ due to reflection would be a factor of 2.28 times the temperature behind the shock before reflection.

Cooling the shock gases is the most difficult process to describe, since the number of possible interactions is so large. Depending on length of driver and experimental sections of the tubes, initial pressure ratios, and types of gases used, the gases between the contact surface and the shock front could be cooled:

1. by a rarefaction wave before the shock front reaches a reflecting surface;
2. by a rarefaction wave after the shock has been reflected, but before the reflected shock returns to the contact surface;
3. by a rarefaction wave any time after the shock front has reflected and met the contact surface, but before the interaction (either a

shock or rarefaction wave resulting from the interaction of the shock and contact surface) returns to the reflecting surface. In this case, the hot experimental gases would be between the two discontinuities;
4. by interaction between the shock and rarefaction waves and contact surfaces until equilibrium is established.

In each case, the cooling rate should be as great or greater than the heating rate occurring across the shock front. If the shock is not allowed to reflect (case 1) before cooling, the extra heating obtainable by such reflection will not be realized. However, if the last sequence is allowed to occur, the gases will be subjected to a number of heatings and coolings, each of lesser magnitude, until the shocks and rarefactions have died out, and the rapid-cooling-rate effect will be lost.

Example: Calculation of Pressure and Temperature of Shock Zone Gas. What is the temperature and pressure behind a shock front in air if $P_1 = 10$ mm. Hg, $P_4 = 10,000$ mm. Hg, and $T_1 = 300°K$. By Eq. (62) the shock front velocity can be calculated for an ideal case (G6),

$$\log \frac{P_4}{P_1} = 1.34(M - 1) + 0.176$$

$$\frac{3 - 0.176}{(1.34)} + 1 = M = 3.11$$

$$\gamma_{air} \cong 1.39 \qquad \alpha = \frac{\gamma - 1}{\gamma + 1} = \frac{0.39}{2.39} = 0.163$$

From Eq. (61)

$$\frac{P_2}{P_1} = [1 + 0.163] \quad 9.66 - 0.163 = 11.07$$

$$P_2 = 110.7 \text{ mm. Hg.}$$

From Eq. (60)

$$\frac{T_2}{300} = 11.07 \left[\frac{1 + 0.163(11.08)}{(11.08) + 0.163} \right], \qquad T_2 = 830°K.$$

V. Trends in High Temperature Technology

There are no reported commercial processes now operating in which the temperature exceeds 3000°K. The highest process temperature, about 2800°K., is used in production of boron carbide from boron oxide and carbon (K1), and to raise this 2800°K. ceiling will require a good deal of ingenuity. It is reported (B7) that more than 3000°K. can be attained in the process used for production of synthetic diamond, although the process temperature is not given. The fact that extremely high pressures

(>50,000 kg./sq. cm.) are also required in the process makes achievement of these temperatures even more impressive.) Problems of high temperature processing, and possible solutions that have already been considered, are included in part C of this section, which describes uses of electrically generated high temperatures. Higher temperatures, how-

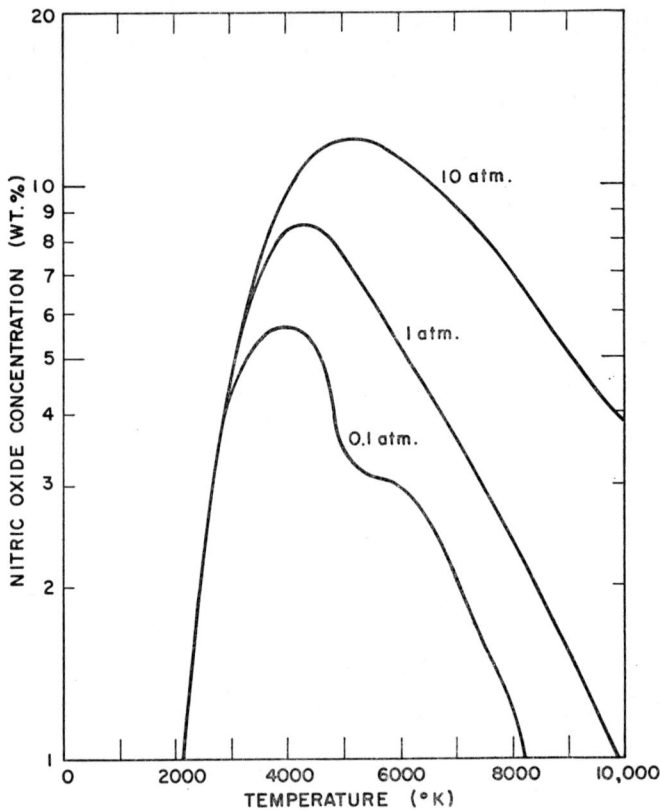

Fig. 18. Nitric oxide concentration of air as a function of temperature at several total pressures (G4).

ever, are used by industry and science in areas other than processing, such as metal coating and rocketry, and this technology will be discussed along with the testing and experimental research being conducted above 3000°K.

Probably the one high temperature chemical process which has received the most attention is the fixation of atmospheric nitrogen as nitric oxide. The advantage of high temperatures is apparent from Fig. 18, which relates the equilibrium nitric oxide concentration with temperature. It was not possible to produce nitric acid competitively in the 40 tons per day Wisconsin process plant at Sunflower, Kansas, primarily because the concentration of nitric oxide obtained from pebble-bed furnaces operating at 2400°K. was too low (never over 2%) for economic recovery. At 3200°K. the concentration of nitric oxide would be three times as high as at 2400°K. (P1). However, the Wisconsin process, as such, could not be operated at 3200°K. for two reasons: (1) combustion of natural gas, used as the fuel, will not yield that high a flame temperature, and (2) the magnesia refractories used would not tolerate this temperature. Studies of nitrogen fixation, using the shock tube, have shown that conversion to nitric oxide can be good (G7). Also, the walls of the shock tube remain cool, while the gases in the shock front easily reach temperatures in excess of 3200°K. However, the mechanical preparation required for each run, such as removing the old and inserting the new diaphragm, and the fact that only a small amount of gas is reacted each run, still disqualifies the shock tube as a process reactor. The electric arc readily provides operating temperatures of 3200°K. and higher, and involves no serious containment problems, because the flow of air can keep the walls below 1500°K. Yet the last commercial arc furnace for production of nitric oxide was shut down nearly 30 years ago, because it could not compete with the oxidation of ammonia from the technically advanced Haber process (K1). It may be advisable to review electrical processes in the light of recent advances in this area of high temperature technology.

A. Uses of Chemically Generated High Temperatures

The highest temperatures created chemically for commercial applications are used in welding and cutting. This use dates from the advent of the oxyacetylene torch. Now, not only metals are cut, but refractories as well, with torches supplying greater quantities of energy and higher temperatures (A5). There are concrete cutting torches in which metal-powder oxygen reaction systems are used (G8). Penner (P4) has proposed that hydrazine might be produced by partial combustion of ammonia. It is also conceivable that higher temperature reactions can be used for rocket propulsion. The difficulties in handling many of the more energetic fuels, such as ozone and liquid fluorine, is one deterrent to their use. Another problem is finding materials (for parts like the rocket nozzle) that can withstand the higher temperatures. The most exotic of the possible chemical propulsion systems envisions trapping free ions in space, then concentrating them until the mean free path is short enough to force recombination. The energy from this recombination would supply the heat and expand the gases to drive the ship.

B. Uses of High Temperatures Generated in Shock Tubes

A refractory coating device which has been developed by Linde Company is a unique application of the shock tube. A mixture of acetylene, oxygen, and the powder to be metallized are fed into an open-ended tube where the acetylene-oxygen mixture is detonated. The detonation creates a shock wave that both heats the powder and propels it out of the tube and against the piece to be coated. Both the 4500°C. temperature of the gas and the 4000-ft./sec. exit velocity aid in attaining a thin, uniform, adherent coating. Shock tubes are also being used in several laboratories (A6, C3) for studying chemical reaction kinetics and thermodynamics at high temperatures. The shock tube is particularly applicable for this use because, with proper design such as that proposed by Hertzberg *et al.* (H3) (see Fig. 19), the gases under consideration can be heated and then cooled very rapidly. This is accomplished by first

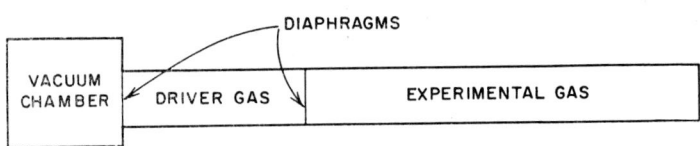

Fig. 19. Chemical shock tube (H3).

breaking the diaphragm between the driver and the experimental gas, then breaking the second diaphragm between the large vacuum chamber and the driver chamber. Breaking the second diaphragm removes the pressure gradient, damps the shock wave, and cools the experimental gas, as soon as the rarefaction wave catches up with the shock front.

C. Uses of Electrically Generated High Temperatures

Among the means considered here for heating material to high temperatures, electrical energy is the most suitable, in that relatively large amounts of material can be heated to extremely high temperatures (50,000°K.) in a carefully controlled way. These attributes make electrical energy the most likely to be used in chemical processing above 3000°K. For instance, a plasma jet operating at 70 atm. pressure with air as the coolant could produce a gas stream containing over 10% NO by weight (see Fig. 17). However, as in most high temperature chemical syntheses, the endothermic reactions are rapidly reversible upon cooling. Very high cooling rates, in the order of 10^7 °C./sec., are required to "freeze" chemical compositions obtained at high temperatures. One possible solution to the problem of maintaining the high temperature compositions is by expansion of gases through the De Laval nozzle, which can provide the rapid cooling rates required. (In fact, the De Laval nozzle can be designed for a wide range of cooling rates.) It should be recognized, however, that the nozzle basically converts thermal energy only to directed translational energy, a reversible process which can reheat the effluent gas from the nozzle if this energy is not used immediately. Therefore, it is necessary to couple the nozzle with a device, such as an impulse turbine, which extracts work from the gases and contributes to the thermal efficiency of the process. Although no commercial synthesis process uses the nozzle-turbine combination to freeze the equilibrium of a synthesis gas and to recover useful work concurrently, the possibilities have stimulated several investigations. The principal deterrent to industrialization has been the low working temperature limits imposed by conventional turbine blades. A secondary but nevertheless significant reason for the tardy adoption of this technique is the dearth of high temperature kinetics data needed for design.

New materials of construction and methods of cooling turbine blades resulting from research on jet engines and secondary power plants for missiles point the way toward much higher working temperatures. Small turbines used for secondary power generation operate at 3000°C. for short periods of time. Ceramic blades are under development, and transpiration cooling of the blades offers still another way of accommodating the higher temperatures which would be most useful in the production of chemicals.

Applications of the high intensity arc to pyrometallurgical and chemical processing have been investigated most intensively by Sheer and Korman (S1, A7) at the Vitro laboratories. The Sheer-Korman high intensity arc employs consumable anodes composed of the nonconductive material being processed and a small proportion (15–25%) of carbon. The process has been used on a pilot plant scale for the recovery of manganese from domestic rhodonite. This arc contains up to 35% manganese as a complex silicate which is intractable in conventional roasting-leaching processes. In the anode tail flame, the refractory arc vaporizes and decomposes to the elements. At the end of the tail flame, the elements recombine into simple oxides (MnO_2 and SiO_2). These are condensed to a fume of submicron particles by rapid cooling, and manganese is recovered from the dust by conventional chemical leaching. Power consumption is in the range of 3 to 7 kw.-hr./lb. of ore processed. The Sheer-Korman arc process has also been evaluated for the carbothermic reduction and the vapor phase chlorination of metallic oxides (S2). Vaporization of refractory materials in the arc and subsequent condensation is a means for generation of submicron spheroidal particles.

The primary industrial applications of the plasma jet are in the fabrication of metals (cutting, welding, flame hardening, and application of refractory coatings). Heat transfer rates attainable with the plasma jet are about eight times those of an oxyacetylene torch. Therefore, cutting rates are much faster, and stainless steels, which are cut with the oxyacetylene flame only by the addition of iron powder, yield to the plasma jet. The operating cost of the plasma jet is half that of the oxyacetylene torch. However, it has a higher first cost and, at its present state of development, is somewhat less convenient to use.

In flame spraying of refractory coatings with the plasma jet, the refractory metal or ceramic is normally fed as a powder (occasionally as a rod or wire) into the plasma flame as it issues from the orifice. The fused and/or vaporized coating impinges at a high velocity on the cooled surface of the specimen being processed. High density, uniform coatings are obtained. Two or more refractory compounds can be co-deposited. When Stanford Research Institute scientists injected hafnium carbide and hafnium boride concurrently into a plasma flame, a solid solution of the two compounds was deposited (E1).

The plasma jet has an established position in the evaluation of materials at high temperatures. In one of the more dramatic applications, it is being used to study the problem of re-entry of ballistic missiles. The nose cone of a missile traveling at supersonic speeds through the atmosphere acts very much like a ballistic piston in that it creates a shock front ahead of it. Between the shock front and the nose, the gases in the atmosphere are compressed and heated to the high temperatures attainable with such phenomena. This is called the stagnation region (B2). (For instance, at Mach 5.8 the stagnation temperature is 2000°K.). Currently, there are three ways of preventing complete destruction of the missile, and application of all of them requires techniques similar to those used in other fields by chemical engineers:

(1) The nose cone can be made of refractory material, in which case the material must be capable of withstanding the final temperature attained by absorbing the heat from the stagnation region.
(2) The nose cone can be cooled by feeding a liquid through its porous skin to dissipate the heat by vaporizing.
(3) Nose cones fabricated of reinforced plastics can be cooled by ablation. The plastics are relatively good insulators, so most of the heat is concentrated in the skin. This skin is decomposed, but this

evolves gases that dilute the hot shock gases and help carry away some of the heat. Since the time that the cone is subjected to the high heats during re-entry is relatively short (~30 sec.) only a small fraction of the plastic is consumed.

The plasma jet does not exactly simulate re-entry conditions, but it is capable of supplying approximately the same gas temperatures and heating rates to a specimen nose cone.

Since temperatures above 3000°K. are seldom encountered in chemical processing, it is not surprising that there are as yet no industrial applications of the plasma jet for chemical synthesis. Still, the potential of the plasma jet for this purpose has been demonstrated. At the Research Institute of Temple University, Dr. Grosse and his associates have synthesized refractory nitrides by introducing powdered metals into a nitrogen plasma, and they have shown that reaction between graphite powder and a nitrogen plasma at 12,000°K. yields substantial amounts of cyanogen (G10). The Linde Company has produced acetylene from natural gas with a plasma (A6). Other potential applications for petroleum or petrochemical processing include oil gasification, recovery of oil from shale, carbon black manufacturing, and cracking of hydrocarbons. Dr. Baddour of Massachusetts Institute of Technology has investigated high temperature reactions between hydrogen and carbon (B1).

Another potential application of the plasma jet in the chemical industry was demonstrated at the National Bureau of Standards recently; Margoshes and Scribner (M2) developed a plasma jet spectrograph. Water solutions of metal ions where atomized in a helium plasma, and the emission spectra were recorded. The primary advantage of the plasma jet as a spectroscopic source is its ability to excite elements which are stable at the highest attainable temperatures of chemical flames.

Nomenclature

A Area, cm.^{2}
A, B, C Components A, B, and C
A_c Cross-sectional area of exhaust nozzle, sq. in.
$A_{s,j}$ Transition probability
a Sound velocity, cm./sec.
b Half-width of spectral line due to Doppler broadening, A.
C Orifice or nozzle coefficient
C_p Specific heat at constant pressure, cal./mole-°K.
C_v Specific heat at constant volume, cal./mole-°K.
c Velocity of light, cm./sec.
c_1 $= 0.885 \times 10^{-12}$ (cal.)(sq. cm.)/sec.
c_2 $= 1.438$ cm.-°K.
E Illumination, erg/cm.^{2}-sec.
E Internal energy, kcal./mole
E_i Energy of ionization, ev.
E_0 Standard energy content at 0°K., kcal./mole

F	Thrust, lb.	p_e	Pressure at rocket nozzle exit, p.s.i., g./cm.^{2}
F^0	Free energy of formation, kcal./mole	p_0	Atmospheric pressure, p.s.i.
g	Statistical weight factor, dimensionless	Q	Partition function for particle, dimensionless
g_c	Gravitational constant, 32.18 ft./sec.^{2} or 980 cm./sec.^{2}	Q_e	Electron partition function, dimensionless
H	Molar enthalpy, kcal./mole	Q_H	Heat absorbed or given off in Carnot cycle, cal.
ΔH	Change in molar enthalpy, kcal./mole	Q_0	Atom partition function, dimensionless
Δh	Change in enthalpy, kcal./g.		
ΔH^0	Standard heat of formation, kcal./mole	Q_R	Heat of reaction, kcal./mole
H_0^0	Standard heat of formation at 0°K., kcal./mole	Q_+	Ion partition function, dimensionless
h	Planck's constant, 6.624×10^{-27} erg-sec.	R	Gas constant, 8.205×10^{-2} liter atm./mole °K. 1.987 cal./mole °K. 8.314×10^7 erg/mole °K.
I	Intensity of radiation: μa. or photons/sec.	S	Entropy, e.u./mole
I_{sp}	Specific impulse, thrust per unit weight—rate of flow, normally lb.-sec./lb.	$\bar{S}$	Molar entropy of ideal gas e.u./mole
		T_c	Combustion temperature, °K. or °R.
J	Rotational quantum number; radiant flux, joules/sec.	t	Temperature, °C. or °F.
$J_{\lambda T}$	Monochromatic radiant energy source at temperature T, ergs/cm.-sec.-ster.	U	Velocity of gas relative to a shock front, ft./sec.
		$\mathcal{U}$	Voltage, volts
k	Boltzmann's constant, 1.38×10^{-16} erg/°K.	V	Velocity, cm./sec.
		V	Volume, liters
K_c	Equilibrium Constant based on particle concentration	v	Particulate velocity, cm./sec.
		v	Specific volume $(1/\rho)$, cm.^{3}/g.
K_n	Equilibrium constant based upon mole fractions	v_e	Velocity of gas from rocket nozzle, normally ft./sec.
K_p	Equilibrium Constant based upon partial pressures	W_R	Work of reversible engine, ft.-lb.
		$\bar{W}$	"Molecular weight" of a gas mixture, g./mole
l	Length of X-ray beam, cm.		
M	Molecular or atomic weight, g./mole	W^*	Weighted average molecular weight of a gas mixture during change of state, g./mole
M	Ratio of shock to sound velocity in a gas		
		w	Weight rate of flow, g./sec.
m	Mass of a molecule, g.	x	Mole fraction, dimensionless
m_e	Mass of an electron, g.	α	Ratio $(\gamma - 1)/(\gamma + 1)$;
N	Number of particles, Avogadro's number	α	One-half the divergence angle of rocket nozzle, degrees
n	Concentration of particles, number/cm.^{3}	a	Atom cross section, cm.$^{-2}$
		γ	Ratio of specific heats, C_p/C_v
P	Pressure, g./cm.^{2}	γ^*	Weighted average ratio of specific heats during change of state
p	Partial pressure, atm. p.s.i., or g./cm.^{2}	$d\theta$	Differential time
$\bar{p}$	Fictitious partial pressure of an element, p.s.i. or g./cm.^{2}	λ	Wavelength of radiation, A; mean free path of particle, cm.

ν	Frequency of radiation line; wave number, cm.$^{-1}$	t_c	Combustion zone
		t_r	Reference temperature
ϵ	Energy of a quantum state or electron charge, e.v.	e	Rocket nozzle exit
		s	Energy level
Π	Current, amp.	j	Energy level
ρ	Density, g./cm.^{3}	i	ith state
		p	Pressure
Subscripts		v	Volume
		0	Ground state, initial conditions
A	Component A	+	Positive ions

REFERENCES

A1. Aller, L. H., "Astrophysics: The Atmospheres of the Sun and Stars." Ronald Press, New York, 1953.
A2. Altman, D., *in* "High Temperature—A Tool for the Future," pp. 47–52. Stanford Research Institute, Menlo Park, California, 1956.
A3. Anderson, J. A., and Smith, S., *Astrophys. J.* **64**, 295–314 (1926).
A4. Anonymous, *Missiles and Rockets* **4**, p. 27 (1959).
A5. Anonymous, *J. Metals* **11**, 40–42 (1959).
A6. Anonymous, *Chem. Eng.* **66**, 78–80 (1959).
A7. Anonymous, *Chem. Week*, p. 28 (March 9, 1957).
B1. Baddour, R. F., Private communication. (1958).
B2. Bond, J. W., Jr., *Jet Propulsion*, pp. 228–235 (1958).
B3. Brewer, L., "The Use of Laboratory, High Frequency Induction Furnaces." UCRL—653 (April, 1950).
B4. Brinkley, S. R., Jr., *Ind. Eng. Chem.* **43**, 2471 (1951).
B5. Brinkley, S. R., Jr., and Lewis., B., *Chem. Eng. News* **27**, 2540 (1949).
B6. Broida, H. P., *in* "Temperature: Its Measurement and Control in Science and Industry," Vol. II, pp. 265–286. Reinhold, New York, 1955.
B7. Bundy, F. P., Hall, H. T., Strong, H. M., and Wentorff, R. H., *Nature* **176**, 51–55 (1955).
B8. Burhorn, Fr., *Z. Physik* **140**, 440–448 (1955).
B9. Burhorn, Fr., Maeker, H., and Peters, T., *Z. Physik* **131**, 28 (1951).
C1. Carter, J. M., and Altman, D., "Combustion Processes," p. 7. Princeton Univ. Press, Princeton, New Jersey, 1956.
C2. Chesnut, F. T., *in* "High Temperature—A Tool for the Future," pp. 35–38. Stanford Research Institute, Menlo Park, California, 1956.
C3. Clouston, J. G., *J. Imp. Coll. Chem. Eng. Soc.* **2**, 123–30 (1957).
C4. Cobine, J. D., and Wilbur, D. A., *Electronics* **24**, 92–93 (1951).
C5. Cobine, J. D., and Wilbur, D. A., *J. Appl. Phys.* **22**, 835–841 (1951).
C6. Compton, A. H., and Allison, S. K., "X-rays in Theory and Experiment." Van Nostrand, Princeton, New Jersey, 1935.
C7. Conn, W. M., "Conference on Extremely High Temperatures." Wiley, New York, 1958.
C8. Courant, R., and Friedrichs, K. O., "Supersonic Flow and Shock Waves." Interscience, New York, 1948.
D1. Davis, T. P., *in* "High Temperature—A Tool for the Future," pp. 10–15. Stanford Research Institute, Menlo Park, California, 1956.
D2. Dieke, G. H., *in* "Temperature: Its Measurement and Control in Science and Industry," Vol. II, pp. 19–30. Reinhold, New York, 1955.

D3. Duvall, G. E., and Kells, M. C., *in* "High Temperature—A Tool for the Future," pp. 53–58. Stanford Research Institute, Menlo Park, California, 1956.

E1. Engelke, J. L., *in* "High Temperature Technology," p. 343. Stanford Research Institute, Menlo Park, California, 1960.

F1. Finkelnburg, W., *in* "High Temperature—A Tool for the Future," pp. 34–44, Stanford Research Institute, Menlo Park, California, 1956.

F2. Finkelnburg, W., *Z. Physik.* 112, 305–325 (1939).

F3. Finkelnburg, W., *J. Appl. Physics* 20, No. 5 (1949).

F4. Finlay, G., *in* "High Temperature—A Tool for the Future," pp. 45–46. Stanford Research Institute, Menlo Park, California, 1956.

F5. Forsythe, W. E., *in* "Temperature: Its Measurement and Control in Science and Industry," Vol. I, pp. 1115–1131. Reinhold, New York, 1941.

F6. Fowler, R. N., "Statistical Mechanics." Cambridge Univ. Press, London and New York, 1929.

G1. Gaydon, A. G., and Wolfhard, H. G., "Flames." Chapman and Hall, London, 1953.

G2. Gerdien, H., and Lotz, A., *Wissenschaftliche Veröffentlichungen Siemens-Konz.* 2, 489, 1922.

G3. Giannini, G. M., *in* "Proceedings: High Intensity Arc Symposium," S. D. Marks, Jr., ed., pp. 29–48. Carborundum Co., Niagara Falls, New York, 1958.

G4. Gilmore, F. R., "Equilibrium Composition and Thermodynamic Properties of Air at 24000°K." RAND Corp., Santa Monica, California, Report RM-1543 (August, 1955).

G5. Glass, I. I., Martin, W. A., and Patterson, G. N., "A Theoretical and Experimental Study of the Shock Tube." UTIA Report No. 2 (November 1953).

G6. Glass, I. I., and Martin, W. A., *J. Appl. Phys.* 26, 113–20 (1955).

G7. Glick, H. S., Squire, W., and Hertzberg, A., *in* "Fifth Symposium on Combustion," p. 393. Reinhold, New York, 1955.

G8. Grosse, A. V., *in* "High Temperature—A Tool for the Future," pp. 59–68. Stanford Research Institute, Menlo Park, California, 1956.

G9. Grosse, A. V., and Kirschenbaum, A. D., "Study of Ultra High Temperatures: The Combustion of Carbon-Subnitride, C_4N_2, and a Chemical Method for the Production of Continuous Temperatures in the Range of 5000–6000° Kelvin or 9000–11,000° Rankine." A.D. 84316 (December, 1955).

G10. Grosse, A. V., *in* "High Temperature Technology," p. 342. Stanford Research Institute, Menlo Park, California, 1960.

H1. Hansen, C. F., "Approximations for Thermodynamic and Transport Properties of High Temperature Air." NACA TN-4150 (March, 1958).

H2. Hedrich, A. L., and Pardue, D. Q., *in* "Temperature: Its Measurement and Control in Science and Industry," Vol. II, pp. 383–392. Reinhold, New York, 1955.

H3. Hertzberg, A., Glick, H. S., and Squire, Wm., U. S. Patent 2,832,665 (April, 1958).

H4. Herzfeld, K. F., *in* "Temperature: Its Measurement and Control in Science and Industry," Vol. II, pp. 233–248. Reinhold, New York, 1955.

H5. Hottel, H. C., Williams, G. C., and Satterfield, C. N., "Thermodynamic Charts for Combustion Processes." Wiley, New York, 1949.

J1. Jacobs, K. H., Bonin, J. H., and Dickerman, P. J., *in* "Proceedings: High Intensity Arc Symposium," S. D. Marks, Jr., ed., pp. 116–37. Carborundum Co., Niagara Falls, New York, 1958.

J2. Johnson, E. O., and Malter, L., *Phys. Rev.*, **80**, 58 (1950).
K1. Kirk, R. E., and Othmer, D. F., eds., "Encyclopedia of Chemical Technology," Vol. 9, p. 413. Interscience, New York, 1952.
K2. Kobe, K. A., "Inorganic Process Industries." Macmillan, New York, 1948.
L1. Landenburg, R. W., "Physical Measurement in Gas Dynamics and Combustion," pp. 198–200. Princeton Univ. Press, Princeton, New Jersey, 1954.
L2. Lapple, C. E., et al. "Fluid and Particle Mechanics." Univ. of Delaware, Newark, Delaware (March, 1951).
L3. Lochte-Holtgreven. W., *in* "Reports on Progress in Physics." Vol. XXI, A. C. Stickland, ed. The Physical Society, London, 1958.
L4. Lochte-Holtgreven, W., *in* "Temperature: Its Measurement and Control in Science and Industry," Vol. II, 1955, pp. 413–27. Reinhold, New York.
L5. Lochte-Holtgreven, W., *VDI Zeitschrift* **97**, 785–788 (1955); Royal Aircraft Establishment transl. No. 630 (1957).
L6. Longwell, P. A., Reamer, H. H., Wilburn, N. P., and Sage, B. H., *Ind. Eng. Chem.* **50**, 603–610 (1958).
M1. MacPherson, H. G., *in* "High Temperature—A Tool for the Future," pp. 31–34. Stanford Research Institute, Menlo Park, California, 1956.
M2. Margoshes, M., and Scribner, B. F., *Spectrochim. Acta*, p. 138 (1959).
M3. Margrave, J. L., *in* "High Temperature—A Tool for the Future," pp. 87–106. Stanford Research Institute, Menlo Park, California, 1956.
M4. Martinek, F., *in* "Thermodynamic and Transport Properties of Gases, Liquids and Solids" (Am. Soc. Mech. Engrs.), pp. 130–156, McGraw-Hill, New York, 1959.
M5. McGinn, J. H. and Wachman, H., *in* "Proceedings: High Intensity Arc Symposium." S. D. Marks, Jr., ed., pp. 98–115. Carborundum Co., Niagara Falls, New York, 1958.
M6. Mohler, F. L., *in* "Temperature: Its Measurement and Control in Science and Industry"; Vol. I, pp. 734–744, Reinhold, New York, 1941.
M7. Myers, J. W., Goldberg, S. A., and Smith, R. W., *Trans. Am. Soc. Mech. Engrs.* **80**, 202 (1958).
N1. National Bureau of Standards, "Selected Values of Chemical Thermodynamic Properties." *Natl. Bur. Standards Circ.* **No. 500**, Ser. I and II (February, 1952).
N2. National Bureau of Standards, "Selected Values of Chemical Thermodynamic Properties." *Natl. Bur. Standards* Ser. III, Vol. I (March, 1947–March, 1956).
N3. National Bureau of Standards, "Tables of Thermal Properties of Gases." *Natl. Bur. Standards Circ.* **No. 564** (November, 1955).
O1. Okress, E. K., and Wroughton, D. M., *Iron Age* **170**, 83–86 (1952).
O2. Ornstein, L. S., *Physik. Z.* **32**, 517 (1931).
P1. Peck, A. C., *in* "Nitrogen Symposium, A Review of the Wisconsin Process for Nitrogen Fixation." Food Machinery Corp., San Jose, California. (Held at Rye, New York, November, 1955), p. 2.
P2. Penner, S. S., *Amer. J. Phys.*, **17**, 422–429 (1949).
P3. Penner, S. S., *Amer. J. Phys.* **17**, 491–500 (1949).
P4. Penner, S. S., "Chemistry Problems in Jet Propulsion." Pergamon, New York, 1957.
P5. Peters, T., *Naturwissenschaften* **41**, 571–572, 1954.
P6. Poritsky, H., and Suits, C. G., *Physics* (now *J. Applied Phys.*) **6**, 196 (1935).

P7. Preining, O., "Generation of High Temperatures (up to 55,000°) in the Laboratory." Österr. Chemiker-Ztg. **55**, 67–72 (1954); Royal Aircraft Establishment transl. No. 629 by M. Squires (1957).

R1. Russell, H. W., Lucks, C. F., and Turnbull, L. G., "Temperature: Its Measurement and Control in Science and Industry," Vol. I, pp. 1158–1163. Reinhold, New York, 1941.

S1. Sheer, C., and S. Korman, "Arcs in Inert Atmospheres and Vacuum." Wiley, New York, 1956.

S2. Sheer, C., A. W. Diniak, J. A. Dyer, and S. Korman, "The Production of Boron Trichloride by Means of the High Intensity Arc Process." WADC Tech. Rept. 56–487 [ASTIA Document No. AD 118305] (March, 1957).

S3. Smit-Miessen, M. M., and Spier, J. L., *Physica* **9**, 193 (1942).

S4. Steffans, C., *in* "High Temperature—A Tool for the Future," pp. 190–194. Stanford Research Institute, Menlo Park, California, 1956.

S5. Stein, M. R. von, *Forsch. Ing. Wes.* **B14**, 113 (1943).

S6. Stull, D. R., and Sinke, G. C., Thermodynamic Properties of the Elements. *Advances in Chem. Ser.* **No. 18** (1956).

S7. Suits, C. G., *in* "Temperature: Its Measurement and Control in Science and Industry," Vol. I, pp. 720–733. Reinhold, New York, 1941.

U1. Unsöld, A., *Z. Astrophys.* **24**, 355 (1948).

W1. Weinberg, F. J., *Proc. Roy. Soc.* **A241**, 132 (1957).

W2. Wildhack, W. A., *Rev. Sci. Instr.* **21**, 25 (1950).

W3. Winternitz, P. F., *in* "Third Symposium on Combustion and Flame and Explosive Phenomena," p. 623. Williams & Wilkins, Baltimore, Maryland, 1949.

W4. Woolley, H. W., *J. Natl. Bur. Standards* **61**, 469–90 (1958).

W5. Woolley, H. W., "Thermodynamic Functions for Atomic Ions." AFSWC-TR-S6-34 [ASTIA Document No. AD 96302] (April, 1957).

W6. Wroughton, D. M., Okress, E. K., Brace, P. H., Comenetz, G., and Kelley, J. C. R. *J. Electrochem. Soc.* **99**, 205–211 (1952).

MIXING AND AGITATION

Daniel Hyman

Stamford Research Laboratories
American Cyanamid Company
Stamford, Connecticut

I. Introduction	120
II. General Characteristics of Mixing Processes and Agitated Vessels	121
A. Fundamentals of the Mixing Process	121
B. Descriptive Studies	125
C. Quantitative Experimental Studies	128
III. One-Liquid-Phase Systems	133
A. Power Requirements	134
B. Performance Studies: "Mixing Time"	146
IV. Gas-Liquid Systems	157
A. Basic Studies	157
B. Power Requirements	161
C. Performance Studies	162
D. Discussion	166
V. Liquid-Liquid Systems	167
A. Basic Studies	168
B. Power Requirements	171
C. Performance Studies	173
VI. Solid-Liquid Systems	176
A. Basic Studies	177
B. Power Requirements	177
C. Performance Studies	178
VII. Heat Transfer	183
VIII. Scale-Up of Heterogeneous Systems	187
A. Introduction	187
B. General Considerations	187
C. Scale-Up Techniques	188
IX. Experimentation with Agitated Systems	190
A. General	190
B. Performance Parameters	191
C. Similarity	192
D. Controlling Variables	193
E. Discussion	196
Acknowledgment	196
Nomenclature	196
References	198

I. Introduction

Mixing and agitation are among the oldest of engineering operations and, together with materials transport and heat transfer, are probably the most universally employed operations in modern chemical technology. It is a rare industrial process whose performance is not dependent in some way on a mixing operation, where several materials must be brought into physical proximity in order to accomplish a specific purpose. The device which ordinarily comes to mind in connection with "mixing" is the conventional vessel-and-rotating-impeller combination; while this review will be mainly concerned with these "agitated vessels," it should be observed that the principles governing the operation of mixing are reflections of the fundamental processes of fluid dynamics which control the performance of so many other chemical engineering operations.

Serious gaps still exist in our knowledge of the performance of agitated vessels, in their varied applications. It is probable that the rapidly improving state of our knowledge about fluid dynamics will dictate a new path in the approach of researchers to mixing and agitation problems, and it seems reasonable to assume that the relatively near future will see considerable increase in our knowledge in these areas. One objective of this chapter is to help provide a background for the engineer who must assess and use these new contributions.

There is an intimate and often very complex interrelation between the performance of an agitated vessel and the system geometry, the properties of the processed materials, and the operating conditions. In some cases the true complexity of a mixing situation may not be appreciated. Particularly in "scale-up" problems, the physical nature of the mixing in a process may be as important a factor as the proper reaction temperature or reactant ratios. Thus, another objective of this review is to demonstrate that agitation is a process variable of major significance.

The main purpose of this work is to review the extensive published literature on mixing and agitation. While a great deal remains to be learned, much of this literature presents useful results and techniques of approach which can be used to achieve understanding of many mixing situations. It is hoped that this chapter will provide a unified approach to a complicated subject, and a source of access to prior knowledge in this important technological area.

The literature published before 1959 will be discussed. Operations carried out in a liquid continuum will be considered, including one-liquid-phase systems, gas-liquid, and solid-liquid mixtures, and systems of immiscible liquids. Several related subjects will not be covered: mixing of heavy pastes, elastomers, and non-Newtonian fluids in general; mixing in packed beds, fluidized systems, and similar devices; the relations between performance and mixing patterns in chemical reactors. The mixing of solids with solids has been reviewed by Weidenbaum (W2). This review will not deal extensively with the mechanical aspects of mixing devices or with mixing in highly specialized pieces of machinery. Standard handbooks provide much valuable information in this regard; Graybeal and Bechtel (G1) provide a useful list of sources of information about the mechanical features of agitation systems.

Agitated systems, regardless of the nature of the phases involved, have a number of characteristics in common. Accordingly, the next section of this chapter will review work applicable to agitated liquid systems in general, and will also discuss certain fundamental aspects of mixing processes. It is hoped that the reader will consult this general Section II, as well as the section covering the specific class of system with which he is concerned. Investigations dealing specifically with heat transfer will be reviewed separately in Section VII. The extremely important problem of scale-up will be discussed in two places. Scale-up for one-liquid-phase systems will be covered in Section III, while the scale-up of heterogeneous systems will be considered in Section VIII. Some comments on experimental or development programs involving agitated systems will be presented in Section IX.

II. General Characteristics of Mixing Processes and Agitated Vessels

Information has been published on the general nature of the mixing process, or the general behavior of material in agitated vessels. Also, a number of investigators have reported on various aspects of the flow patterns which exist in agitated vessels. These studies will be discussed in subsections B and C following. It will be seen that the basic quantitative relations governing the behavior of liquids in agitated vessels have largely been neglected from an experimental standpoint.

A. Fundamentals of the Mixing Process

Consider a vessel which contains two pure liquids, distinguishable by their compositions, which are ultimately miscible in each other. Assume that the fluids are segregated from each other when the observations begin, and that there are no mass-flow currents involving large numbers of molecules (i.e., the fluid is "at rest.") Now it is known from the kinetic theory that the individual molecules are in continuous motion. As time proceeds, this motion will result in an intermingling of the different molecules. This process is the classic one of molecular diffusion, and is regarded as one of "mixing" since the vessel contents eventually become "uniform" or "homogeneous." If samples are taken big enough to contain a large number of molecules but also small in volume relative to the vessel, it will be seen that the concentrations of each component in successive samples from any region in the vessel will approach the original over-all vessel average as the diffusion process continues.

The lack of precision in the foregoing statements results from one of the first problems to arise in a discussion of mixing operations. It is extremely difficult to define precisely the basic concept of "degree of mixedness," since the definitions become interrelated with the particular application for which the mixing operation is being carried out. If the vessel considered above were only used for transporting the different components from one place to another, the "degree of mixing" inside the vessel might be of little importance, so long as it contained the proper amount of each component. On the other hand, if it were necessary to mix the components so that a large number of smaller containers could be filled uniformly, a satisfactory "degree of mixing" would exist only if one of these small containers, used as a sampler anywhere in the original vessel, would collect a volume of material containing the desired proportions of each component. Finally, if the mixing operation required that every molecule of one component would at some time be within a few mean-free-path lengths of a molecule of the other component, an extremely small volume of sample would have to be taken in order to see whether this criterion of "mixedness" had been satisfied. It is seen, then, how necessary it is to specify the size of the region of interest in any discussion of mixing processes.

This problem was considered quantitatively by Danckwerts (D1), who applied statistical concepts previously used in turbulent-flow studies. He considered mixture systems in which the smallest portion of interest would be very large relative to molecular size, with the exclusion of any apparatus effects which might give widely different characteristics in different regions ("dead space" in a mixing vessel, for example). He then defined a "scale of segregation" as a measure of the distance (or volume) in which the mixing has not proceeded very far. In a mixture of A and B, if the "scale" is large, the size of the relatively unmixed "blobs" containing mostly A or mostly B is large. If the "scale" is small, these "blobs" are small, i.e., the unmixed portions are smaller. He further defined an "intensity of segregation," which measures the average extent to which the local concentrations deviate from the over-all average. These definitions utilize statistical parameters derived from the values of local concentrations. For chemical engineering practice, this work provides a method of great potential value for coming to grips quantitatively with many mixing problems. Much remains to be done from the point of view of developing convenient techniques for measuring "scale" and "intensity." More important, the relations of these parameters to the specific performance of mixing vessels as devices for carrying out particular process steps must be established. Weidenbaum (W2) gives a generally accessible detailed description of Danckwerts' treatment and so it will not be presented here.

Consider now another mixing situation, somewhat different from our first example. Assume that the volume of material originally containing two segregated components is actually in a turbulent flow field, and that molecular diffusion does not occur. The effect of the turbulence is to draw out and extend the original "blobs" of each component so that the surface area between them increases rapidly. At the same time, the random turbulent motion twists the stretched-out "blobs" so that if the original volume contained a filament of dye in a colorless solution, the action of turbulence would in time cause the colored component to look something like a tangled jumble of ribbon. If the scale of interest is large, this action of turbulence, alone, may be satisfactory for mixing, since it will have produced a considerable intermingling of the originally separated components. However, as the scale of interest gets smaller, and samples of ever decreasing volume are taken, a point may be reached when the sample is small enough to contain only one or the other component, so that the small-scale mixing is not complete. Due to the nature of turbulent fields, there exists a certain scale below which the random motions are not effective for mixing; no matter how much time elapses, the turbulent stretching and twisting will not bring the average distance between the components' interface below a certain "cutoff" value. If this "cutoff" length is larger than the scale of interest for the mixing, turbulence alone will not be a satisfactory way to mix the components. The "cutoff" exists because viscous forces damp out the turbulent motion of very small eddies. The minimum scale thus increases with kinematic viscosity, although it also depends on the local rate of energy dissipation. This is discussed by Hughes (H14), who shows that in turbulence fields normally encountered this "cutoff" length is on the order of 0.01 to 0.3 mm. The scale of interest in many practical cases, especially in those involving reaction between molecules, is actually smaller than this minimum turbulent-eddy scale, and it must be concluded that turbulence alone will not do a satisfactory mixing job under these conditions.

Fortunately, molecular diffusion, assumed to be absent in the preceding discussion, proceeds simultaneously with the action of turbulence in the true situation. Thus, as the surface of contact between the components becomes stretched and twisted by the turbulence, diffusive molecular interchange causes it to become "fuzzy" and produces an intermingling of the components down to the molecular scale. Molecular diffusion provides the only significant amount of mixing that occurs on the small scale of interest involved in chemical reaction, for example. Particularly in liquids, where molecular diffusion is often slow relative to turbulent mixing, the time-controlling step in many processes may be the molecular-diffusive mixing time. Hughes (H14) suggests that this time can be estimated by using the minimum turbulence scale in the Einstein equation for molecular diffusion. The estimated time is quite small for gaseous systems (order of 0.001 sec.); it can be much larger in liquid systems. The estimated molecular diffusion time for ethanol in an ethanol-water system, for example, is about 2 to 3 sec. for typical turbulent flows in pipelines, where the local energy dissipation rate may be on the order of 1 ft.-lb./sec. per cubic foot of liquid. Even if the local power dissipation were fiftyfold greater, as might be expected in the region close to a rotating impeller, the estimated time would still be about 0.3 sec.

There has been only limited study of the problem of simultaneous turbulent and molecular-diffusive mixing, concerned mostly with the effect of isotropic turbulence on temperature fluctuations. Corrsin (C9) extends such analysis to a mixture in which a chemical reaction is taking place, and derives statistical functions of the concentration of reactant. In another paper (C8), he estimates the rate of decay of concentration fluctuations in an idealized turbulent mixer. Here, he concludes that the mean square concentration fluctuation decreases with time in an exponential manner, the rate of decrease being related to the turbulence integral scale and to the portion of the system power input which goes into the turbulent energy. While the turbulent fields considered are simple compared with those in conventional agitated vessels, this work is a valuable first step for understanding the underlying fundamentals of mixing processes. Application of these fundamental approaches to heterogeneous systems should also be useful, since the factors determining equilibrium drop size, rate of solution of a crystal as a function of its size, etc., are intimately related to the scale of turbulence and other fluid-dynamic parameters. This is a very promising area for additional research.

The preceding discussion has been concerned with relatively small-scale phenomena. If we consider the entire body of liquid in a conventional mixing vessel, we see that intermingling of the unmixed components is also being carried out by the relatively gross transport of material in the large-scale flow patterns set up in the vessel by the action of the impeller. This largely neglected problem of the conventional mixing vessel as a whole can be treated in several ways. One possibility is to consider the large-scale flow patterns as an extension in size of the small-scale turbulent eddies, with the entire vessel one turbulent flow field containing a range of eddy size up to the order of the vessel size itself. Another approach is to consider that a mixing vessel is in effect made up of various zones, each with its own internal fluid regime, with some degree of transfer of material between these zones caused by convective or diffusive types of transport. A simple example of such a model might consist of a zone near the vessel wall in which there is a gross upward flow and a central zone in which the gross flow is downward, each flow with its own degree of turbulence, and with appropriate assumptions about the transfer of material from one zone to the other. Obviously, a large number of different models can be postulated. While this approach could eventually serve to describe various aspects of the behavior of conventionally-agitated mixing vessels, it will be seen that in many vessels the situation is complicated considerably by several factors: the presence of sources of different scales of turbulence (baffles and impeller, for example); possible differences in the properties of the turbulence generated by different impeller types; differences in mean flow patterns caused by differences in vessel geometry; the lack of knowledge about the small-scale turbulence characteristics in the liquid either near the impeller or elsewhere in the vessel.

B. Descriptive Studies

1. *Flow Patterns*

Although workers in the field of mixing and agitation have obviously been concerned with the nature of flow patterns since the earliest studies (W10), two publications in 1938 give the first schematic representations of flow patterns. MacLean and Lyons (M5) show a sketch of currents in a tank with a turbine mixer and Bissell (B4) gives a photograph of striae due to flow currents in a small propeller-agitated vessel. In later papers, Bissell *et al.* (B6), Lyons (L3), and Rushton and Oldshue (R12) give sketches of flow patterns for a wide variety of rotating agitator types in different geometric situations. Sachs and Rushton (S2) present some light-streak photographs made in the course of their study of impeller discharge capacity, and Taylor (T1) also has some very interesting light-streak photographs. These and other published descriptions are in general agreement on the nature of the mean flow patterns in a liquid being mixed by a rotating impeller, and are summarized in the following discussion.

2. *Impellers*

Almost every rotating impeller in common use can be considered to fall into one of three classes, paddle, propeller, or turbine, with certain mean flow patterns typical for each of these groups. The paddle-type stirrers, slow-moving and generally large in diameter relative to the mixing vessel, effectively act to push the liquid in front of them with relatively little flow radially along the blades and thus produce a predominating circular pattern of flow around the paddle axis. Propeller stirrers are designed to produce primarily an axial flow of fluid giving a discharge stream of a generally cylindrical shape, with swirl in it due to the propeller rotation. The turbine stirrer with nonpitched blades (i.e., with the blade faces parallel to the impeller shaft) produces a swirling "pancake" discharge with mean components only in the radial and tangential directions. Pitched-blade turbines are also in use, the pitch of the blades resulting in some degree of axial flow and thus giving a mean flow pattern intermediate between propellers and flat-blade turbines.

Special-purpose impellers (B10, L3, P3) are generally not difficult to classify on the basis above. The essential differences in performance of these special rotors are related to the nature of the turbulence they create (generally in the region near the impeller) rather than to the mean flow patterns.

3. *Effect of Baffles*

Consider now the mean flow patterns in a liquid, open to the air, being agitated by a rotating impeller mounted axially in a cylindrical vessel. In the absence of any form of baffle there will be a strong tendency for the liquid to rotate as a mass with a vortex formed around the tank axis. The vortex will become deeper as the impeller is rotated faster and may reach the impeller itself in certain cases. If the impeller is a propeller or pitched-blade turbine there will be some vertical currents caused by axial flow through the impeller, the extent of these in relation to the rotating current being dependent on the system geometry and operating conditions. The consensus appears to be that the rotating flow about the vessel axis will be the dominating one. This description is considered applicable to liquids with viscosities less than 20,000 cp. For more viscous materials, the mean flow patterns approach the situation described below for baffled flows (L3, R12).

Installation of vertical baffles at or near the walls of a cylindrical tank is a common practice when an axially mounted impeller is used.

Such baffles convert the rotational motion of the fluid into mean flow currents in the vertical and radial directions which result in a "turnover" of the vessel contents. A similar result occurs when a nonpitched blade turbine has a stator ring mounted around it for the purpose of directing the discharge flow so that it is radial with no rotational component. Reavell (R1) describes a cruciform baffle, installed on the tank bottom under an axial propeller, which is also designed to induce vertical circulation instead of rotational flow.

4. Off-Center Impeller Location

A flow pattern characterized by the presence of vertical mean currents, with little or no vortex and no obvious rotational flow, also results in an unbaffled cylindrical vessel if a top-entering propeller is mounted off center and inclined to the vertical, or if a side-entering propeller is mounted horizontally near the tank bottom with its axis at a small angle to the radius. The exact positions required are critical if rotational flow is to be avoided, and Rushton (R12) and Oldshue (O6) give details on these.

Sometimes structural modifications inside the mixing vessel are used to establish desired mean flow patterns. Thus draft tubes (B8, P3) may be used to emphasize an up-and-down flow pattern, and special vessel shapes (B10, L3, P3) may be used for specific applications.

5. Production of Turbulence

The mean flow currents in mixing vessels are supplemented at most points in a mixing vessel by a fluctuating flow whose magnitude and direction will vary with time about the mean value. This effect is most apparent in baffled tanks, where the fluctuating character of the flow (i.e., turbulence) may contribute as much or more to accomplishing the ultimate objective of the mixing operation as the mean flow currents. The influence of these fluctuations was recognized in early work (W10) by Wood et al., in 1922. This subject arose again in articles in 1938 by Bissell (B4) and Gunness and Baker (G3), and in 1944 by Miller and Rushton (M9). In 1951, Newitt et al. (N4) discussed the subject in general terms, and in addition included two sketches of flow around tank baffles showing how they can break up rotational flow streamlines by forming a trail of vortexes behind them. Rushton and Oldshue (R12), in 1953, gave the first extended discussion of turbulence specifically directed to describing the behavior in an agitated vessel; this was based in part on experimental results (S2) to be described below.

6. Jet Mixing

Mixing of liquids in tanks has been accomplished by the use of free jets discharging into the vessel, with no rotating impellers present in the tank (F4, F7, F9). A description of the mean circulation patterns in such systems has been presented by Fossett and Prosser (F7), who find that a regular but nonrotational mean flow pattern can be caused by the jets. They give simplified sketches for a variety of single and multiple-jet arrangements. It should be observed that the subjects of turbulence inside a jet and the induction and mixing of a surrounding fluid by a jet discharging into it are classic problems of fluid dynamics. An immense amount of theoretical and experimental work has been done on them, which the interested reader will find referred to in works by Pai (P1) and Alexander et al. (A5). Krzywoblocki (K10) has prepared an extensive bibliography on jets. All this work is concerned primarily with the characteristics of jets themselves, rather than with turbulent flow in a vessel whose contents are being mixed by the action of free jets. However, Folsom and Ferguson (F4) have studied this last question, and their results will be discussed in Section III.

C. Quantitative Experimental Studies

1. Velocity

Relatively little work has been done toward experimentally defining the basic properties of mixing systems. A major difficulty has been a comparative lack of measurement techniques which do not interfere with the flow being studied.

Rushton et al., in 1946 (R16), measured the flow from marine-type propellers mounted axially in a 36-in. diameter baffled tank. A 21-in. diameter baffled tank with a hole in the bottom was mounted concentrically inside the large tank, around the propeller (Fig. 1). With the propeller rotating, water flowed out of the inner tank, and the rate of water supply required to maintain the same level in each tank was measured. For a given propeller and speed of rotation, the rate of flow was found to depend on the diameter of the hole and its distance below the propeller. The maximum rate of flow, found by variation of these dimensions, was called the rate of discharge. These measurements were strongly affected by the geometry of the measuring device (the inner tank) and hence there is little assurance that the maximum flow observed is the flow which would exist in the absence of the inner tank. The authors of this work, who were fully aware of this limitation, nevertheless found the results useful in the design of large-scale installations.

For eight three-bladed propellers tested, with diameters from 3.6 to 12.4 in. and pitch-to-diameter ratios in the range 0.92–1.54, these authors found that the discharge flow was directly proportional to rotational speed and about proportional to the square of the diameter. Thus:

$$Q = 0.26ND^2 \qquad (1)$$

where Q = cubic feet of flow per minute, N = revolutions per minute, and D = propeller diameter, feet. As an indication of the flows involved, this equation predicts that a 6-in. diameter propeller at 800 r.p.m. will have a discharge on the order of 400 gal./min. If this flow is considered

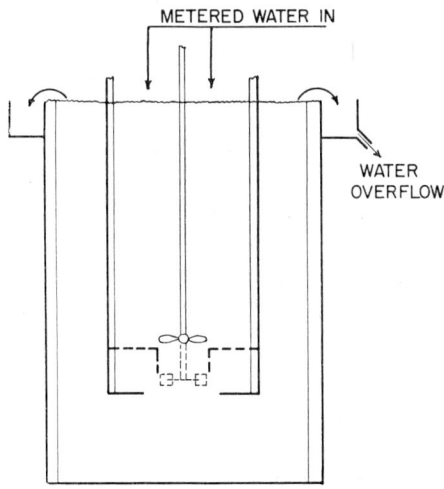

FIG. 1. Experimental scheme of Rushton *et al.* (R16) for measurement of propeller discharge rates. (Dotted lines show arrangement for tests with turbines.)

as flowing uniformly through a cylinder with the same diameter as the propeller, its velocity is about 4.8 ft./sec. The linear velocity of discharge defined in this way depends only on propeller speed, not on propeller diameter.

In a later publication (R12), Rushton states that the flow of water from dimensionally similar three-bladed marine-type propellers, manufactured by the Mixing Equipment Company, is given by the relation:

$$Q = 0.40ND^3 \qquad (2)$$

No further experimental data are reported. The flows predicted by these two equations, for a given propeller in the range of diameters from 4 to 16 in., differ by less than a factor of 2. Theory indicates that a precise relation would probably have a more complicated form (H16); Eq. (2) represents the latest opinion of the only workers who have published on this subject.

Rushton *et al.* (R16) also made a few measurements on turbines, using the same concentric tank arrangement but somewhat modified as shown by the dotted lines in Fig. 1. For turbines with flat blades mounted on the edge of a disk, the discharge was about 0.063 ft.^{3}/revolution for a 6-in. diameter turbine with three blades, and about 0.074 ft.^{3}/revolution for one with four blades. In a later publication Rushton (R11) has stated (without data) that the flow of water from a similar six-blade turbine can be computed by the relation:

$$Q = 0.50ND^3 \tag{3}$$

Oyama and Aiba (O10) observed the mean flow patterns in a mixing tank by use of a small rotating pointer, similar to a weather vane, suspended at the end of a pipe at different locations in the vessel. Their studies on two-blade flat paddles in an unbaffled cylindrical vessel confirm the expected presence of a vortex, deepening with speed of rotation, and of swirling flow about the tank axis with a slight amount of flow up the tank walls and down in the center.

In a recent work, Aiba (A2) studied the flow currents in water, in a mixing vessel 14 in. in diameter, using an axially-mounted two-bladed flat paddle 4.7 in. in diameter. Measurements were made both without baffles and with four baffles $\frac{1}{12}$ tank diameter wide. A sphere about 6 mm. in diameter was suspended by a flexible wire, and its displacement from the equilibrium (no-flow) position was measured. To get the horizontal displacement, cobalt-60 was embedded in the sphere, and a Geiger-Mueller counter approximately 10 mm. in diameter was immersed in the tank 2–5 cm. from the sphere. The vertical movement of the sphere was measured with a cathetometer, and its angular position observed by eye. From the known components of displacement and the assumed drag coefficient of the sphere, values of the radial, tangential, and vertical components of the flow around the sphere were calculated.

In these runs, the presence of the Geiger-Mueller counter may affect the flow pattern significantly. Further, examination of the results presented shows that the radial velocities do not satisfy material balance requirements. Nevertheless, several of Aiba's conclusions are of interest. With baffles, the vertical circulation is large compared with the unbaffled case, but considerable tangential flow remains. In neither the baffled nor unbaffled case do the mean flow velocities approach the tip velocity of the impeller; the ratio of fluid velocity to impeller-tip velocity is essentially independent of impeller rotation speed, varies somewhat with position, and has a mean value of 0.4–0.6 in the unbaffled tank and 0.1–0.3 in the baffled tank.

An experimental technique which has been used in a few investigations involves the observation of the movement of small spherical particles in a liquid in a mixing vessel. The particles may be solids or liquid droplets. Their paths are observed by illuminating them and photographing the resultant reflections as light streaks. If illumination is confined to a relatively narrow plane, it is possible to photograph the action in various regions of a tank. Short exposures of known duration are used to give photographs showing short light streaks whose lengths are related to the particle velocities. Somewhat longer exposures give long streaks which can be used to show effectively the general nature of the flow pattern in the illuminated area. Details of the experimental procedure as applied to mixing vessels are given by Sachs and Rushton (S2). Papers on flow visualization techniques by Roberts (R5) and by Winter and Deterding (W9) present good background discussions on this subject.

Sachs (S1) used the light-streak technique to measure some mean velocities in water agitated in a 11.5-in. diameter tank 16 in. high with four 1-in. baffles at the wall. The turbine used in this work was a 4-in. diameter four-flat-blade type manufactured by Mixing Equipment Company. The turbine disk was located 4 in. off the tank bottom, and measurements were made at speeds from 100 to 200 r.p.m. Analysis was made of the streaks in horizontal slices about $\frac{3}{16}$-in. thick, disposed vertically from just below the center of the turbine blade to just above its upper edge. The velocities were stated to be symmetrical vertically about the midplane of the turbine but no specific measurements in this connection are given. A vertical distribution of radial velocities was found, with a maximum opposite the middle of the turbine blades. This distribution was most peaked at points near the turbine and became less pronounced at radial positions farther from the tank center. The ratio of the maximum radial velocity to the average of the measurements ranged from about 1.7 to about 1.2, at various distances from the tank axis.

The average radial velocity, at each radius studied, increased directly with rotational speed of the turbine. These velocities were in the range of 0.4–0.8 ft./sec.; for comparison, values of impeller tip speed varied from 1.7–3.5 ft./sec. Also, for a fixed impeller speed, the average radial velocity increased somewhat with radial distance from the impeller to a maximum about 1.5 in. from the tank wall. This implies a vertical inflow to the region examined; such an effect was confirmed by the light-streak measurements, which showed vertical components of roughly the correct size to account for this behavior. (This inflow was estimated to account for about half the total radial flow.) Vertical velocities were in the range 0.0–0.2 ft./sec., and radial volumetric flows, also roughly proportional to turbine speed, were in the range of 2–6 ft.^{3}/min. A considerable tangential velocity component was also noted, ranging from about 50% to 200% of the radial velocity component. A considerable variation in velocities was found with angular location with respect to the turbine blades; the conclusions given are based on averages of values from six angular positions between the blades.

Tennant (T2) also studied velocity distributions, using a six-blade turbine and two viscous corn syrup solutions as well as water. For impeller speeds in the range of 100–200 r.p.m., his results generally confirm those of Sachs. A 300-fold increase in viscosity reduces the fluid velocity by about 30%. Comparison with Sachs' data indicates that increasing the number of turbine blades from four to six increases the radial velocities by roughly 10–50%, depending on impeller speed and on radial position in a manner as yet undefined.

Another study using the light-streak technique was made by Taylor (T1), who worked with standard six-flat-blade Mixing Equipment Company turbines of 2-, 4-, and 6-in. diameter, in an 11.3-in. diameter tank, both with and without baffles. The liquids used were corn syrup (viscosity about 70 cp.) and several solutions of carboxymethylcellulose in water, the latter giving non-Newtonian behavior. Photographs of the corn syrup system showed that in a baffled tank at low agitation speeds (60 r.p.m.) only laminar tangential flow occurs, with the tangential velocity approaching zero at the wall. As the impeller speed is increased, radial and vertical flows set in, but the tangential velocity is still important. Even at 600 r.p.m., it is roughly equal to the radial velocity. In this flow region, eddies could be observed near the blade edges. The flow is more strongly tangential if the baffles are removed. In a narrow slice through the midplane of the turbine, Taylor found that the total velocity in the horizontal plane varied directly with speed near the impeller tip and was somewhat more strongly dependent on speed at large radii. The velocity ranged from 10% to 80% of the impeller-tip velocity, depended somewhat on angular position with respect to the turbine blades and the baffles, and decreased sharply with radius at any given speed.

2. *Turbulence Parameters*

Sachs (S1) attempted to establish the extent of the turbulent fluctuations present. He reports root-mean-square fluctuating velocities ranging up to values on the order of 40% of the mean velocities. Although their quantitative significance may be doubtful, these data do represent the first published attempt actually to measure the turbulence intensity in a mixing vessel.

Yamamoto and Kawahigasi (Y1) studied the rate of dispersion of material in an unbaffled tank, agitated by a two-blade flat paddle, in which a swirling and vortexing flow was present. They measured the average spread of an injected dye stream by a photographic technique. In a 33-in. diameter tank, with an 11.3-in. diameter impeller at 109 r.p.m., they found an eddy diffusivity of about 0.0095 ft.^{2}/sec. and a turbulence intensity in the range of 19–21% of the mean velocity. The root-mean-square velocity fluctuation decreased from 0.8 to 0.2 ft./sec. with rise in radial distance. In a smaller unbaffled tank (11.7-in. diameter), with a similar paddle 3.9 in. in diameter and with impeller speeds from 80 to 400 r.p.m., eddy diffusivity ranged from 0.001 to 0.003 ft.^{2}/sec., turbulence intensity from 17% to 12%, and the root-mean-square velocity fluctuation from 0.06 to 0.2 ft./sec. Estimates of energy dissipation and eddy viscosities were also made. Although it seems unlikely that the particular experimental technique could be applied to the more complex flow structure which occurs in a baffled vessel, this type of quantitative description of mixing in terms of known turbulent-flow parameters provides an interesting and significant approach.

3. Discussion

In summary, experimental work shows that even in the baffled tank considerable tangential flow occurs. The fluid velocities observed are usually well below the turbine or paddle tip speeds. With respect to basic experimental investigations, only a small beginning has been made. A prime requirement is instrumentation that will give reliable information about the mean flow patterns, over a wide range of equipment types and operating conditions. The turbulence in mixing vessels must be characterized if completely rational analysis and solution of mixing problems are to be achieved. The application of recently developed techniques of instrumentation and experimentation concerned with the detection and analysis of rapidly-fluctuating phenomena should result in significant progress in the near future.

III. One-Liquid-Phase Systems

The results of studies on flow patterns and other basic flow characteristics in mixing devices operating on one-liquid-phase systems have been discussed in the preceding section. As has been observed, the performance of these operations depends ultimately on phenomena such as turbulent flow and molecular diffusion, but the basic studies of these phenomena in liquid systems have not yet been applied successfully in interpreting or predicting the performance of conventional agitated-vessel mixing systems.

A. Power Requirements

The power requirements of one-liquid-phase systems have been studied extensively. Before we examine the available data, it will be useful to review the methods used for measuring the power input to a rotating impeller in a mixing vessel. Since the power consumed in the relatively small experimental vessels used in most experimental work is not very large, the measurements must have a high degree of accuracy and sensitivity to changes of small magnitude. Workers in this area (W10) recognized early that it was not sufficient to measure electric-power input to the agitator drive motor, even with use of "empty-tank" correction factors, because variations in performance of the entire driving system under varying loads make the corrections uncertain. Consequently, numerous investigators have contributed to the development of dynamometers based on a simultaneous measurement of agitator-shaft speed and torque. In the area of torque measurement, especially, experimenters have applied their ingenuity to produce a wide variety of devices, most of which fall into the following two classes:

In the first type, the agitator shaft is connected to its drive shaft through a device which both resists and measures the rotation of the agitator shaft relative to the drive shaft. Separate static measurements yield the torque required to cause this amount of relative rotation of the driving and driven shafts, and the power being supplied to the agitator shaft is calculated as the product of the speed and the torque. A description of one such device is given by Stoops and Lovell (S8).

In the second type, the agitated vessel is mounted on a table which is free to rotate, or the motor and drive are mounted so that they are free to rotate. The rotation of the vessel or of the motor and drive is prevented by applying a force (measured by weights or a scale) at a known distance from the agitator axis. From this balancing torque and the agitator speed, the power is computed. Hixson and Luedecke (H9) describe a "torque-table" in which the vessel is prevented from rotating; Rushton *et al.* (R13) show a small dynamometer with which the counter-torque required to keep the motor from rotating is measured.

Other devices include a "differential-type automotive dynamometer" (R4, R13) using the rear-axle assembly from an automobile. Power was sent through the differential to the agitator shaft at one end of the axle, while the torque required to prevent rotation of the other end was measured with a system of levers and a scale. More recently (O3) torque measurements have been made by use of a strain gauge mounted on the agitator shaft. Small power inputs have been measured by using a falling weight to turn the impeller (V1).

In most cases, the measuring devices are not described in detail, and little information is given on the calibration techniques which were used. Most workers have recognized that the effects of friction, of static vs. dynamic calibrations, etc., can be very important, and have attempted to minimize errors from these sources. Further quantitative work on the properties of these dynamometer systems would be useful. Nevertheless, from the point of view of providing data suitable for engineering application, these different types of power measurements give results which are reasonably consistent.

1. Data and Correlations

The distinguishing characteristic of each reported study is, usually, the physical design of the apparatus involved. In Table I, a brief outline of the published information on power requirements in one-liquid-phase systems is presented. This table gives the significant mechanical characteristics of the systems studied; the range of liquid viscosities; and the range of values for the impeller Reynolds number, which will be discussed below. In most of these studies the general objective was to relate the power consumption to tank diameter, impeller type and diameter, rotational speed, and liquid properties. Other variables studied are also indicated in the table. The major features of this work will now be reviewed.

White et al. (W5) used paddle stirrers in unbaffled vessels, and presented their results in two different types of correlations. The first had the following form:

$$(HP.) = 0.000129 D^{2.72} N^{2.86} \mu^{0.14} \rho^{0.86} T^{1.1} W^{0.3} H^{0.6} \qquad (4)$$

where

$(HP.)$ = horsepower
D = impeller diameter
N = revolutions per second of impeller
μ = viscosity of liquid
ρ = density of liquid
T = tank diameter
W = impeller blade width
H = liquid height in tank

with units in feet, pounds mass, and seconds. The degree of fit to the

TABLE I
Published Data on Agitator Power Requirements in One-Liquid-Phase Systems

Year	Investigators	Tank diameter (in.)	Impeller type	Baffles	Liquid viscosity (cp.)	Impeller Reynolds number	Other variables
1934	White et al. (W6)	5–51	Paddle	No	1–170	60–5×10^6	Blade width Liquid depth Impeller height[a]
1937	Hixson and Luedecke (H9)	8–18	Turbines, 4 blades, pitched and non-pitched types	No	1–60	10^4–10^7	Liquid depth[a] Impeller height[a]
1942	Hixson and Baum (H7)	6–18	Same as above	No and yes[a]	1–580	3–10^5	Direction of rotation Liquid depth Impeller height
1943	Stoops and Lovell (S8)	38	Propeller, 3-blade	No	1–950	100–5×10^5	Liquid depth Impeller height
1947	Olney and Carlson (O7)	12 10	"Arrowhead" turbine Spiral turbine with stator	Yes No	0.53–1.0 1	9×10^4–3×10^5 6×10^4–1.5×10^5	—
1948	Mack and Kroll (M2)	10–16	Paddle	Yes No	1	4×10^4–2×10^5	Baffle effects
1950	Rushton et al. (R13)	9–96	All types	Yes and No	1–40,000	1–10^6	Liquid depth[a]
1950	O'Connell and Mack (O1)	10	Turbine, flat-blade	Yes	1–500	1–3×10^5	
1954	Oldshue and Gretton (O3)	48	Turbine, 6-flat-blade	Yes	1–400	10^4–10^6	Heating coil
1955	van de Vusse (V1)	6–8	Paddle Turbine, 3 pitched blades Propeller, 3-blade	Yes[a] and No	—	70–8×10^4	Vortexing[a]
1955	Uhl (U1)	24	Turbine, 6 pitched blades Paddle Anchor	No Yes and No No	2,500–30,000 150–20,000 50–6000	2–170 5–2500 50–4000	—
1956	Nagata and Yokoyama (N1)	23–46	Paddle Propeller	No No	1–10,000	10–2×10^5	—

[a] Relatively little of the work was concerned with this variable.

data is estimated at about ±30%. For another form of the same correlation, dimensionless groups can be used:

$$\frac{Pg_c}{\rho N^3 D^5} = 1.77 \left(\frac{D^2 N \rho}{\mu}\right)^{-0.1} \left(\frac{W}{D}\right)^{0.3} \left(\frac{H}{D}\right)^{0.6} \left(\frac{T}{D}\right)^{1.1} \tag{5}$$

The influence of tank diameter may be more complicated than is indicated by these equations. In Eq. (5), the group $D^2 N \rho / \mu$ is a form of Reynolds number, analogous to that used in pipe-flow and other hydraulic problems, with a velocity term proportional to the impeller tip speed (ND), and a length term equal to the impeller diameter (D). This form of the Reynolds number has come into general use for characterizing mixing operations that employ rotating agitators. Likewise, the group $Pg_c/N^3 D^5 \rho$ or N_P, where P is power in foot-pounds per second and g_c is the gravitational conversion factor, is often used and is known as the "power number."

Hixson and co-workers (H7, H10) carried out an extensive investigation of the power requirements for three types of turbine agitator, mostly in unbaffled tanks, with a variety of operating conditions. The bulk of their work was with geometrically similar systems. In such a system, for vessels of different capacities, all corresponding linear dimensions of impeller and tank remain in the same ratio. They also presented their results by means of dimensionless groups, plotting power number as ordinate versus the Reynolds number as abscissa (both logarithmically). The data for each geometrically similar system could be correlated reasonably well (±20%) by such plots, with exceptions to be noted below. These correlation curves showed similar characteristics, i.e., a slope of about −1.0 at low values of the impeller Reynolds number, and a gradual flattening out with increasing Reynolds number to a slope around −0.1. There was a slight "bump" in the curve at Reynolds numbers in the range of about 100–1000. Because of the similarity of these correlation curves to the "friction-factor plot" for pipe flow (P3), the authors concluded that the region where the slope was −1.0 represented a "laminar-flow" region followed by a "transition" region and a "critical" region (around the bump on the curve), followed in turn by a "turbulent flow" region at high Reynolds numbers.

Two situations were encountered which did not give the "typical" correlation curve. First, in unbaffled vessels with a liquid height 50% or 75% of the vessel diameter, a speed is reached at which the vortex formed in the liquid becomes deep enough to reach the impeller, and the power consumption drops very quickly with any further increase in speed. The onset of this deep vortexing was not correlated with the other system variables; in limited tests, even one baffle was found sufficient to destroy the vortex. Second, the correlation curves for vessels with two or four baffles differed from the others in three respects: there was no "critical" region; at high Reynolds numbers the slope of the curve was zero; and tank diameter had an unexplained effect at high Reynolds numbers.

In an effort to generalize their correlations so as to include systems which were not geometrically similar, Hixson and Baum (H7) proposed the concept of a "standard" system, selecting as a standard the system on which they had done the most work. The power requirement at any impeller Reynolds number for a nonstandard system was to be obtained

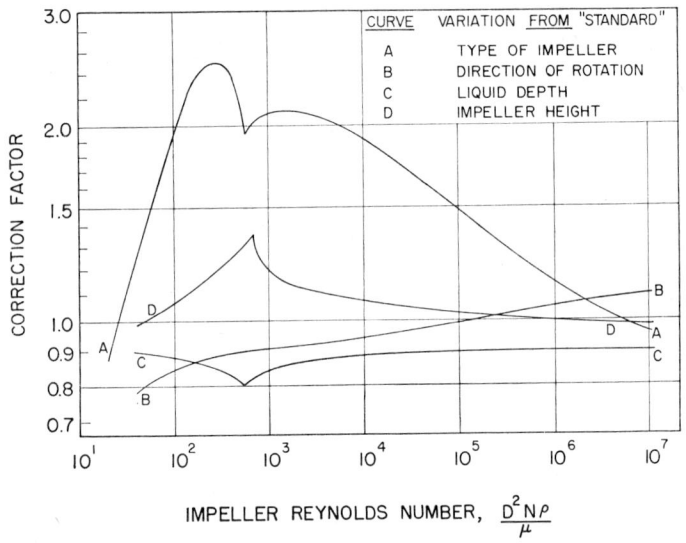

FIG. 2. "Correction Factors" of Hixson and Baum (H7) for variations from "standard" design. (Courtesy of Industrial and Engineering Chemistry.)

by multiplying the "standard" system power at that Reynolds number by a separate "correction factor" for each deviation from the "standard" design. These "correction factors," as functions of Reynolds number, were computed from the experimental data and the variations of some are shown in Fig. 2. These curves lead to two significant conclusions: (a) the change in power requirements with geometric design can be quite large (up to fivefold, in this study); and (b) the erratic variations of these "correction factors" with Reynolds number indicate the irregular differences in the shapes of the correlations of power with Reynolds number for the individual systems. That is, one relatively simple correlation based on the dimensionless groups described above is not sufficient to describe the power requirements of systems which are geometrically different.

Power requirement data for three-bladed marine propellers were published by Stoops and Lovell (S8), who worked with axially-mounted propellers in an unbaffled tank. They found no effect of variations in liquid depth or impeller height, and did not report the extent of vortexing, which must have been present. They correlated their data in a dimensionless form similar to that discussed above, i.e.,

$$\frac{Pg_c}{\rho N^3 D^5} = 18 \left(\frac{D^2 N \rho}{\mu}\right)^{-0.19} \left(\frac{T}{D}\right)^{0.93} \tag{6}$$

although only the speed (N) and viscosity (μ) were varied over an extended region. The effect of the modified Reynolds number in this equation is of roughly the same order as found by previous workers for different impeller types.

An extensive compilation was published by Rushton et al. (R13), giving the results of a program of measurements carried out by the Mixing Equipment Company. The data cover a wide range of impeller types, impeller and tank sizes, liquid properties, and operating conditions. They also presented their results in the form of a function of dimensionless groups:

$$\frac{Pg_c}{\rho N^3 D^5} = K \left(\frac{D^2 N \rho}{\mu}\right)^m \left(\frac{DN^2}{g}\right)^n \tag{7}$$

By assuming that in some cases gravitational forces would be significant, they found it necessary to include the dimensionless group DN^2/g, analogous to the Froude number in hydraulics, in which g is the acceleration of gravity. They further assumed that a relation of the form of Eq. (7) would correlate data from nongeometrically similar systems, if expanded to include dimensionless geometric-ratio terms like $(T/D)^t$ for the effect of tank diameter, $(W/D)^w$ for the effect of blade width, etc. The values of the terms K, m, n, and the other exponents are not constant, but instead depend on the mixing system geometry and on operating conditions.

For values of Reynolds numbers less than 300, it was possible to correlate the results for any given geometrically-similar system without consideration of the Froude group, i.e., exponent n in Eq. (7) was zero. For unbaffled vessels, in which a deep vortex could form, it was found that at higher Reynolds numbers the power number would decrease rapidly as the agitator speed was increased. In Fig. 3, the solid lines in the right-hand part show, for liquids of three different viscosities, how the power number goes down as the impeller speed goes up. Such

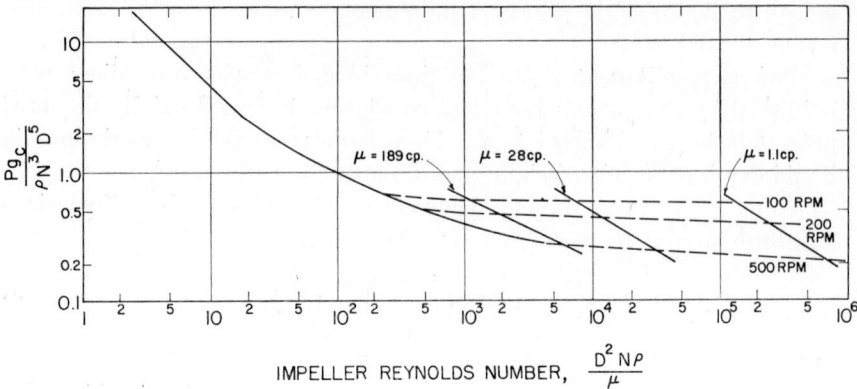

FIG. 3. Power number—Reynolds number correlation for a 12-in. propeller in a tank without baffles. Tank diameter = 54 in. Liquid depth = 54 in. Data of Rushton et al. (R13).

data for an unbaffled system were brought together by using the Froude group as in Eq. (7), provided that the exponent n had the complicated form

$$n = \frac{a - \log(\text{Reynolds Number})}{b} \qquad (8)$$

in which the values of a and b depend on the geometry of the system as indicated in Table II. The effect of this correlation is shown in Fig. 4, where a single curve is given for each system.

For systems in which a deep vortex is prevented by the presence of baffles or the proper nonaxial positioning of the impeller, at all values

TABLE II
VALUES OF "a" AND "b" FOR EQ. (8)[a]

Impeller				
Type	Diameter (in.)	Tank diameter (in.)	a	b
Propeller	4	8.5	2.6	18
	4	13	1.7	18
	12	54	0	18
	18	54	2.1	18
	20	54	2.3	18
Turbine, 6 flat blades	4	13	1.0	40
	6	18	1.0	40

[a] From Rushton et al. (R13).

MIXING AND AGITATION 141

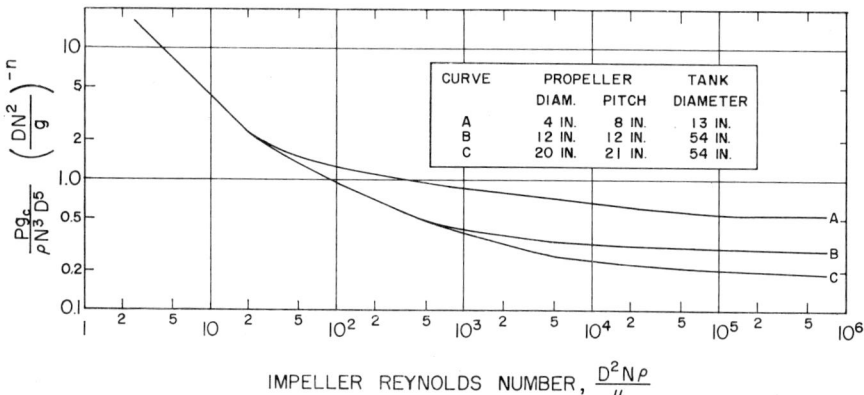

FIG. 4. Use of Froude group in correlation of power data in unbaffled tanks. From Rushton et al. (R13).

of the Reynolds number, correlation is obtained without including the Froude group. Figure 5 shows the power-group variation with Reynolds number, for one type of impeller, with various baffle conditions; it is seen that a single curve is obtained for each system.

Curves of power group as a function of Reynolds number (and Froude number in the case of unbaffled systems) are given in Fig. 6 for several different agitation systems which were studied. These systems and the ranges over which they were studied are described in Table III. Certain generalizations may be made from these results:

(a) Below a Reynolds number of about 10, the slope of each curve is -1.0.

(b) As Reynolds number increases from about 10 to about 10^4, there

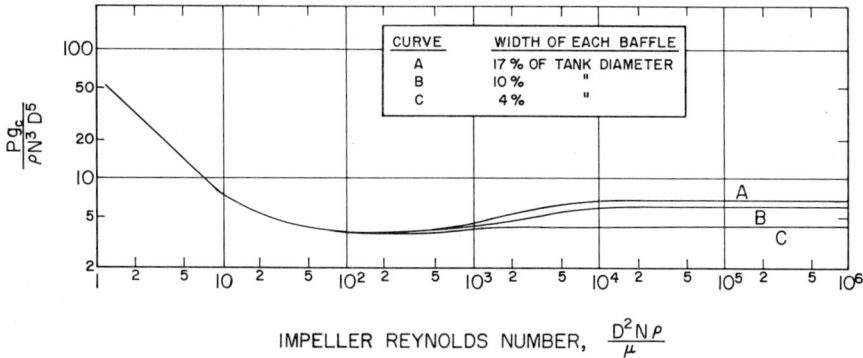

FIG. 5. Power number—Reynolds number correlation for a 6-in. diameter flat-blade turbine in an 18-in. diameter baffled tank. From Rushton et al. (R13).

TABLE III
Descriptions of Systems for Curves in Fig. 6

Curve	Impeller type	Baffles Number	Baffles Width, % of tank diameter	Impeller diameter, (in.)	Tank diameter, (in.)	Ratio, tank diameter to impeller diameter	Ratio, liquid depth to impeller diameter	Ratio, impeller height to impeller diameter	Number blades on impeller
A	Square-pitch propeller	0	—	4–18	13–54	3.3–4.5	3.9–4.5	1.0	3
B	Square-pitch propeller	4	10	4–20	13–96	3.0–4.8	3.0–4.8	1.0	3
C	Flat-blade disk turbine[a]	0	—	3–24	8.5–54	1.7–6.0	2.2–5.4	0.7–1.7	6
D	Flat-blade disk turbine[a]	4	10	3–36	8.5–96	1.7–6.0	2.2–5.4	0.7–1.7	6
E	Pitched-blade fan turbine[b]	4	10	4–18	8.5–54	2.1–4.5	2.8–4.0	1.0–1.2	6–8
F	Flat paddle[c]	4	8.3–10	4–24	8.5–96	2.1–4.5	2.3–4.0	0.8–1.3	2
G	Shrouded turbine	4	10	4	8.5–13	2.1–3.3	2.8–3.0	1.0	6
H	Shrouded turbine	Stator ring		4	13	3.3	3.0	1.0	6

[a] Blades mounted on edge of disk. Blade length 25% of over-all impeller diameter. Blade length 1.5 times blade width.
[b] Flat blades extending from hub. Each blade set at 45° angle. Blade length 4.0 times blade width.
[c] Blade length 50% of over-all impeller diameter. Blade length 3 times blade width.

is a long transition region in which the curve flattens and, in some cases, goes through a minimum at a Reynolds number around 200–300 (See Fig. 5 also).

(c) At Reynolds numbers above 10^4, the curves are flat for baffled systems, while those for unbaffled systems continue to fall off slightly. Thus, in a given system, when the Reynolds number is below 10 the power will vary as

$$P \propto D^3 N^2 \mu \tag{9}$$

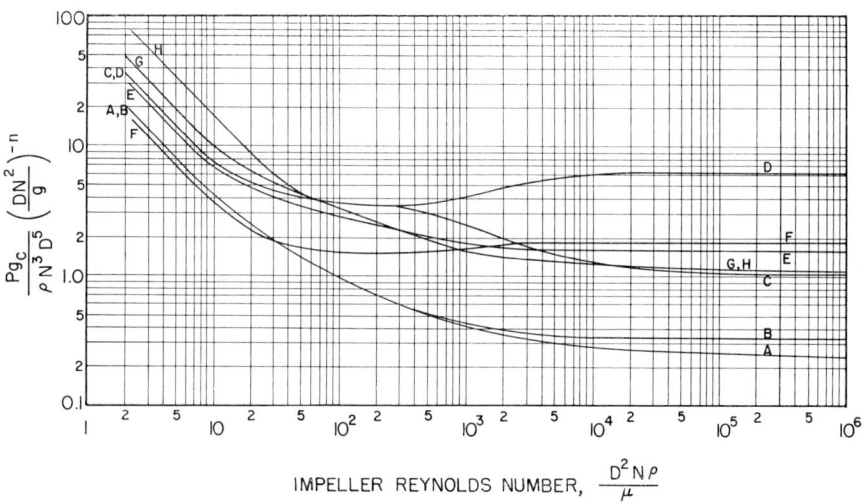

Fig. 6. Correlation of power requirements for different systems. See Table III for identification of the curves. From Rushton et al. (R13).

while at Reynolds numbers above 10^4, the relation in baffled systems will be approximately

$$P \propto D^5 N^3 \rho \tag{10}$$

It is evident that this method of correlation is not adequate to represent, by a single curve, the data from more than one kind of system; the relation between the dimensionless groups is different for each system tested. In terms of numerical values, the power required at the same value of Reynolds number for two different agitators may vary by a factor of 10 or more. Thus these correlation curves should be employed with caution, and preferably applied to mixing systems physically similar to those studied.

Rushton and co-workers drew some conclusions about the effect of several of the linear dimensions of a mixing system other than impeller diameter D. They found little effect of linear dimensions at Reynolds numbers below 200; for higher Reynolds numbers, they reported the following:

(a) In baffled tanks, the ratio of tank diameter (T) to impeller diameter has essentially no effect. This conclusion is based on tests with propellers and on a larger number with flat-blade turbines (Mixing Equipment Company type, 6 blades, 14- to 48-in. diameter) with a 96-in. diameter tank, with constant values for speed, liquid depth, and impeller height. For propellers in unbaffled tanks, the power required at a Reynolds number of 100,000 varied as $(T/D)^{0.91}$ over a T/D range from 2.7 to 4.5.

(b) On the basis of limited data, the liquid depth was found to have no effect.

These workers also discussed the influence of some aspects of impeller design, with the following conclusions:

(a) Propeller pitch (S). In general, a change in this variable changes the entire correlation curve. For a 20-in. diameter marine propeller running at 420 r.p.m. in a baffled tank, they found the power to vary as $(S/D)^{1.7}$. (S/D is unity for "square-pitch" propellers.)

(b) Number of blades (B). For Mixing Equipment Company flat-blade turbines, taking the 6-blade turbine as the reference condition, the power requirement varies as $(B/6)^{0.8}$ for turbines with 3–6 blades and as $(B/6)^{0.7}$ for turbines with 6–12 blades.

(c) Blade length (L). For Mixing Equipment Company turbines, with a constant blade-length to blade-width ratio of 1.25, the power varies as $(L/D)^{1.5}$. A major part of this effect can probably be attributed to the blade width.

O'Connell and Mack (O1) also gave some data on the effect of impeller design. They show that for 2-, 4-, and 6-blade simple turbines with flat unpitched blades, the following dimensionless relation correlated the data at Reynolds numbers greater than 10^4:

$$\frac{Pg_c}{\rho N^3 D^5} = 9.7 B^{0.5} \left(\frac{W}{D}\right)^w \tag{11}$$

where B is the number of blades, W is the blade width, and the other terms are as previously defined. The value of exponent w varied with B as

$$w = 1.33 B^{-0.1} \tag{12}$$

Here the influence of number of blades is somewhat less than given by Rushton for Mixing Equipment Company turbines. At Reynolds numbers below 30, O'Connell and Mack found

$$\frac{Pg_c}{\rho N^3 D^5} = 90 B^{0.33} \left(\frac{D^2 N \rho}{\mu}\right)^{-1.0} \left(\frac{W}{D}\right)^{w'} \tag{13}$$

where
$$w' = 0.64 B^{-0.28} \tag{14}$$

The major effects of baffles on the behavior of mixing systems has already been discussed. Mack and Kroll (M2) show that as baffle width, length, or number is increased, a maximum in power consumption is reached with three or four baffles, each $\frac{1}{10}$ to $\frac{1}{12}$ the tank diameter in width, a condition which they called the "fully baffled" state. With such baffling, power consumption was not changed significantly if the baffles were angled slightly or moved away from the wall. Rushton et al. (R13) are in general agreement with this result, but state that the power requirements increase slightly (approximately 10% with increase in baffle width from $\frac{1}{10}$ to $\frac{1}{6}$ of tank diameter).

While the general treatment of non-Newtonian liquids is beyond the scope of this review, it may be noted that limited work in this area has given results similar to those described for Newtonian liquids. Plots of power number group against Reynolds number can be used if the Reynolds number is modified to include a parameter characterizing the non-Newtonian behavior. Metzner reviews the subject of non-Newtonian flow in a recent publication (M8).

2. Discussion

A general approach to the problem of power requirements, based upon the use of dimensionless groups, provides a powerful technique for dealing with so complex a situation in a rational manner. There are, however, certain limitations to the technique; from the engineering point of view, it has not been possible to obtain a general data correlation which is independent of specific physical design features of the mixing apparatus. It is of interest to discuss some of the possible reasons for this. The mean direction and amount of flow discharged by the impeller varies primarily with its type, speed, size, and with the liquid properties. The turbulence in the discharge stream at the impeller also depends on all these factors; currents moving at different velocities also can contribute to turbulence formation. In the general case, the nature of the flow regime depends on almost every property and dimension of the fluid and of the agitating system. The power requirements to maintain the flow regime will have similar dependence; since power consumption occurs by several different gross mechanisms (small eddies at the impeller blades, flows "jetting" into the liquid mass, impact and deflection at the walls and baffles, etc.), this dependence may well be more complex than the relatively simple and obvious dimensionless groups which have been used would suggest. Relatively recently, several workers (A4, T1) have begun to consider quantitatively this problem of how the energy put into the system is dissipated.

It is of interest to draw some generalizations from the data and correlations which have been presented. First, at low values of the Reynolds group (below 10), there is the dependence of power consumption on fluid viscosity and not on density [Eq. (9)], which is typical of laminar flow. Some of the visual studies of Taylor (T1) described in Section II confirm this. With increasing Reynolds number, power shows a decreasing dependence on viscosity and an increasing dependence on density until it is essentially independent of viscosity and directly proportional to density [Eq. (10)], indicating a change in the predominant mode of energy dissipation to a type occurring through turbulent fluctuations. The "transition region" is generally long (roughly 100-fold longer, in terms of Reynolds number, than the transition region in pipe flow), not very well-defined, and variable from system to system. In this region, then, the vessel may have both laminar and turbulent regions (confirmed in photographic studies by Taylor) whose detailed characteristics are, to repeat, complex functions of most of the variables. The fact that the correlation curves for different systems flatten out at widely different power number levels indicate that there may be considerable differences in structure of the flow and the turbulence in each system.

The data which have been published (or which are available by consultation with the suppliers of mixing equipment) cover a relatively large number of commonly employed mixing systems, and now appear to be sufficient for most design uses. They are relatively ineffective for providing insight into the basic nature of the mixing systems, and there is little reason to believe that further data on total power input would be of any further value for this purpose.

B. Performance Studies: "Mixing Time"

The need for suitable parameters to characterize the performance of a mixing system remains a major problem. For one-liquid-phase systems, the problem is even more difficult than in multiphase systems, since the latter can be identified by a rate factor or by a change in physical distribution of one phase. In the limited work published, mixing systems having one liquid phase have been described by different "mixing times," each defined in terms of a particular study. In general, the "mixing time" is an interval that begins either when a small amount of tracer material is added to a large amount of uniform material under agitation, or when agitation is started in a carefully prepared but nonuniform system; this interval ends when some specified degree of over-all uniformity is achieved, the measure of which depends on the particular experimental technique which is adopted.

The use of free jets for achieving mixing in a vessel without agitators is most often found in one-liquid-phase systems, and their performance also can be identified with a "mixing time." The different techniques for such measurement will be described below.

1. *Rotating Agitators*

A remarkable effective pioneering study in this field was reported by Wood *et al.* in 1922 (W10). These workers used a 5-ft. diameter wooden vat, with a large 2-blade 45°-pitch paddle close to the bottom, without baffles. While the agitator was turning they introduced a pulse of strong salt solution near the bottom of the tank, at the start of their "mixing time" interval. Through small tubes leading from different points inside the tank, they continuously bled out small streams which they passed through conductivity cells. These cells were connected so that the sum of their resistances could be observed. When the change with time of this total resistance reached some arbitrarily selected low value, these workers considered it the end of their "mixing time" period. These "mixing times" decreased almost linearly with increasing agitator speed, from about 360 sec. at 8–9 r.p.m. to about 60 sec. at 18–20 r.p.m., and they decreased more slowly at higher speeds. The changes in conductivity with time, at different locations inside the tank, led to qualitative conclusions about the flow regime in the tank: first, there was a current moving upward at the wall; second, there was a nonuniformity in any one horizontal plane, which the authors attributed to a tangential or swirling flow having little motion in the radial direction.

Kramers *et al.* (K8), in 1953, measured some "mixing times" in small vessels (12.5- and 25-in. diameter) with 3-blade propellers and flat-blade turbines. The vessel had two widely separated conductivity probes in it. The "mixing time" interval started when a small amount of KCl solution was added to the agitated liquid and ended when the conductivity difference between the two probes settled down to an arbitrarily selected degree of uniformity. This "mixing time" depended on the amount of salt added and on the probe locations, which therefore were held constant for the bulk of the work. Impeller Reynolds numbers were in the range 4.5×10^4 to 2×10^6, and the observed "mixing times" ranged from about 5 to 20 sec. In any given geometrically-similar system, "mixing time" was inversely proportional to impeller speed and independent of the vessel size; thus,

$$N\theta = \text{constant} \tag{15}$$

where N is rotations per second, and θ is "mixing time" in seconds. For baffled vessels, with a propeller or turbine, the constant was of the order of 70–200. Without baffles, the propeller location had a significant effect on the "mixing time"; in general, a flow pattern with a vortex gave a long mixing time. Some limited tests of baffle arrangement indicated that faster mixing with propellers is obtained with baffles right at the wall, while with turbines it may be somewhat better to have a small clearance between the baffle and the wall.

A different method for determining a "mixing time" was used by van de Vusse (V1), who mixed miscible solutions of slightly different specific gravities. Passing a light through a nonhomogeneous mixture of such liquids makes it possible to see shadow patterns caused by refractive-index gradients. Van de Vusse filled the mixing vessel with a lower gravity liquid carefully floated on the top of a higher gravity liquid. The "mixing time" period started when the agitator was turned on. A beam of light was passed through the vessel, and the shadows were observed on a screen. The time when these disappeared was taken as the end of the "mixing time" interval; this end point could be observed with a reproducibility within around 5%, when the flow was turbulent.

This work employed rather small vessels (6–8 in.), usually without baffles, and turbine, paddle, and propeller agitators. The observed mixing times ranged from 5 to 30 seconds. For Reynolds numbers ($D^2N\rho/\mu$) of about 10^3–10^4, the following proportionality was found.

$$\theta \propto \left(\frac{\Delta\rho}{\rho}\right)^{0.3} \frac{H^{1.3}T^2}{N^{1.6}D^{2.6}W} \tag{16}$$

where θ is "mixing time," $\Delta\rho$ is the difference in densities between the liquids being mixed, ρ is the mixed density, H is liquid height, T is vessel diameter, N is rotational speed, and D is impeller diameter. This proportionality obviously does not apply to the case where $\Delta\rho$ is zero. The experimental values of $\Delta\rho/\rho$ are not detailed, but are presumed to be on the order of 0.02. This relation is also limited to impeller-to-tank diameter ratios of one-third or less, which is often the case in practice.

From some of the data in this study, it was possible to estimate the influence of impeller type on "mixing time." Table IV shows the estimated relative "mixing times," with all other factors held constant. It is seen that impeller type has at most about a fourfold influence on the "mixing times" measured in this study.

TABLE IV

EFFECT OF IMPELLER TYPE ON "MIXING TIME" [a]

Impeller type	Relative "mixing time"
Propeller, square pitch	1
Turbine, 12 curved blades, no pitch	0.4
Turbine, 2–3 flat blades, 30°–90° pitch	1.5–0.4

[a] From van de Vusse (V1).

Van de Vusse also made some limited investigations of other factors. For an unbaffled tank, van de Vusse's data indicate a possible variation of 40% in "mixing time" with propeller location. Limited data (from an unbaffled vessel) on the influence of turbine design indicate that the "mixing time" decreases as the proportion of radial to axial or tangential flow is increased.

Fox and Gex (F9) have studied the rate of mixing, over a wide range of operating variables, in unbaffled cylindrical vessels from 0.5 to 5 ft. in diameter, with a marine-type propeller agitator. Their "mixing time" was determined visually, as follows: a batch of liquid was made alkaline with a known amount of NaOH; phenolphthalein was added to give a deep red color; into this batch, under agitation, an equivalent amount of HCl was added. The elapsed time until the "very last wisp of red color which can be visually observed" disappeared was taken as the "mixing time." A lesser amount of work was done in a 14-ft. tank, where the "mixing time" was determined by adding a small amount of saturated oil to a large quiet batch of unsaturated oil, starting the agitator, and noting the time when samples bled from several wall ports showed a value of iodine number which was constant within the possible error of the determination. In addition to tank diameter, as just noted, Fox and Gex varied the operating conditions over the following ranges:

Viscosity, cp.	0.5–400
Liquid depth, ft.	0.5–14
Ratio, liquid depth/tank diameter	0.25–2
Rotational speed, r.p.m.	200–2200
Propeller diameter, in.	1–22
Ratio, propeller diameter to tank diameter	0.07–0.33

The value of the "mixing time" varied considerably with the location of the propeller in the tank. The location was adjusted in each test to give the minimum value of the "mixing time," which was always on the order of 10–100 sec. This minimum "mixing time" was reported and correlated in the relation:

$$\theta = K_1 \frac{H^{1/2}T}{N^{2/3}D^{4/3}g^{1/6}} \left(\frac{D^2N\rho}{\mu}\right)^{a_1} \quad (17)$$

where the constant and the Reynolds number exponent depend on the value of the Reynolds number, as follows:

Reynolds No.	K_1	a_1
Greater than 10^4	160	$-1/6$
Between 10^2 and 10^4	78,000	$-5/6$

In Eq. (17), θ is the "mixing time," T is tank diameter, H is liquid height, N is rotational speed, D is propeller diameter, g is the gravitational constant, ρ is density and μ is viscosity. This relation gave a fair correlation of the data, with some points showing deviations of up to 60%. In view of the dependence of their correlation on Reynolds number, these workers described the region of Reynolds number above 10^4 as "turbulent" and that between 10^2 and 10^4 as "incipient laminar."

Oldshue et al. (O6) have done work on mixing rates in systems different from the above, but similar to those used industrially for large-scale blending operations. They used unbaffled vessels with side-entering propeller mixers located relatively close to the bottom. They started with two stratified layers of hot and cold water, and the "mixing time" extended to the moment when thermocouples at various points in the tank reached an effectively uniform value. Water, or water-salt solutions, were used in all runs. Vessel diameters were 30, 54, and 240 in. The "mixing times" ranged from one minute to six hours, and conformed to the following proportionality.

$$\theta \propto \frac{1}{P}\left(\frac{\Delta\rho}{\rho'}\right)^{0.9}\left(\frac{T}{D}\right)^{2.3} \quad (18)$$

where P is the power supplied to the propeller, $\Delta\rho$ is the density difference between the hot and cold layers, ρ' is the density of the cold layer, and T/D is the tank to propeller diameter ratio. The range of $\Delta\rho/\rho'$ was from 0.003 to 0.07; T/D varied from 9 to 33; liquid height/tank diameter ratio was unity. It was found, as in other work, that the propeller position was critical, and the position which gave the shortest "mixing time" was used. In this position the mixer shaft was horizontal, making an angle of 7° to 10° to the left of the vertical plane of symmetry, with the propeller turning clockwise when viewed from the motor end of the shaft.

The above relation is the only one so far reported which shows so large an effect of power input. The temperature-time records for the different thermocouples provided a qualitative description of the mixing mechanism. Motion begins quickly in the lower layer of liquid; the interface between layers gradually rises, with fluid "worn away" from the upper layer being almost immediately uniformly dispersed into the lower layer. Mixing is complete when the interface reaches the top of the liquid in the tank. Similar work (R10) in a tank 65 ft. in diameter, used for mixing petroleum products, also showed this kind of mixing pattern.

Nagata et al. (N2) studied the rate of mixing of high viscosity liquids in 3.9-, 7.9-, and 11.8-in. diameter unbaffled vessels. They used a variety of agitators, including a ribbon-type mixer, and a paddle whose blade height was the same as its diameter (80% of vessel diameter). "Mixing times" were determined by observing the time for reaction of iodine in the tank with an added solution of sodium thiosulfate. For essentially similar geometric systems, with liquid viscosities from 1 to 40,500 cp. the ribbon-mixer results were correlated by the relation:

$$\theta = 33/N \tag{19}$$

where θ is "mixing time" in seconds, and N is revolutions per second. The broad-paddle mixer was slower, and the "mixing times" with it depended on the tank diameter also, as follows:

$$\theta = \frac{55}{D^{0.6} N^{1.4}} \tag{20}$$

where D is tank diameter in feet. This correlation covered data with liquids in the 900–4300 cp. range, for rotational speeds from 0.5 to 4.0 revolutions per second.

In other studies of "mixing time" Mohle and Waeser (M11) introduced a heat pulse into a vessel and observed the time for two thermocouples in the liquid to indicate the same temperature, but reported only very limited data. Beerbower et al. (B3) added a small amount of radioactive iodine-32, and used Geiger counters to measure the time required to reach a uniform dispersion, for several grease- and asphalt-blending kettles of special design.

2. Free Jets

The earliest data on the use of free jets for mixing appear to have been reported by Fossett and Prosser (F7) in 1949, and by Fossett (F6) in 1951. Their studies dealt with blending of tetraethyl lead into gasoline in large unbaffled storage tanks (up to 144 ft. in diameter), and with mixing of a variety of other petroleum products. Jet nozzle diameters ranged from 1 to 3.5 in. Data were also taken in a smaller (5-ft.) tank where the nozzle diameter was about 0.1 in. In the commercial equipment, "mixing time" was the time to reach "uniformity" as indicated by analysis of grab samples from different levels in the tank. In the small-scale work, an electrolyte was added, and "uniformity" was indicated by constant readings on two conductivity probes in the tank. The relatively small amount of additive in each case was injected into a stream withdrawn from the tank and recirculated through the jet.

In the small-scale work, the injection of additive covered about half the total period of recirculation required to achieve "uniformity." Under these conditions, the total recirculation time could be approximated by:

$$\theta = 8 \frac{T^2}{D_j u_j} \tag{21}$$

where θ is in seconds, T is tank diameter (feet), D_j is jet nozzle diameter (feet), and u_j is the linear flow velocity (feet per second) from the nozzle. However, T was not actually varied in the small scale work. "Mixing times" ranged from 3 to 25 min. The above relation was found to be independent of the nozzle Reynolds number $(D_j u_j \rho/\mu)$ for values from 4,500–80,000.

The commercial-scale tests were carried out on actual installations, and no systematic study of variables was made; the second component was added over much more than half the total recirculation period, and in some cases the mixture was satisfactorily uniform as soon as all the additive had been injected. Total recirculation times varied from 1 to 24 hr. An interesting result from both small- and commercial-scale tests was that from 6% to 50% of the total volume in the tank had to be recirculated through the jet to achieve satisfactory uniformity.

These workers also presented certain design criteria. They recommended that the jet center-line be directed upwards at an angle β from the horizontal such that

$$\tan \beta = \frac{\text{Nozzle depth below surface}}{0.67 \, T} \tag{22}$$

If the jet nozzle velocity is too low, and the liquid coming from it is much different in density from the liquid in the tank, it is possible to get stratification instead of good mixing. The minimum velocity can be estimated from fluid properties and vessel geometry, by the relation

$$(u_j^2)_{\min} = K_j g H_j \left(\frac{1}{\sin^2 \gamma}\right)\left(\frac{\Delta \rho}{\rho}\right) \tag{23}$$

where H_j is the nozzle depth below the liquid surface, g is the gravitational acceleration, γ is an angle equal to $\beta + 5°$ [β defined in Eq. (22)], $\Delta \rho$ is the difference in density between the fluid from the jet nozzle and the fluid in the tank, and ρ is the density of the heavier fluid, which is assumed to be coming from the jet nozzle. The value of the constant K_j depends on the value of $\Delta\rho/\rho$ as follows:

$\Delta\rho/\rho$	K_j
0.01	46
0.02	36
0.06	30

Additional data on jet mixing have been presented by Fox and Gex (F9), whose work on mixing with propellers was described above. They studied jet mixing in the same vessels used for the propeller tests, using the same visual color-disappearance techniques for determining a "mixing time." The ranges of the variables were:

Viscosity, cp.	0.5–400
Tank diameter, ft.	0.5–14
Liquid depth, ft.	0.5–15
Ratio, liquid depth/tank diameter	0.25–2
Jet velocity, ft./sec.	1–50
Jet diameter, in.	0.06–1.5
Ratio, jet diameter/tank diameter	0.005–0.02

The term "jet" above refers to values at the discharge nozzle. As with propellers, it was found that the relation between "mixing time" and the other variables depended on a Reynolds number, which in this case was evaluated at the nozzle discharge:

$$\theta = K_2 \frac{H^{1/2} T}{(D_j u_j)^{4/6} g^{1/6}} \left(\frac{D_j u_j \rho}{\mu}\right)^{a_2} \tag{24}$$

where the constants have the values

Reynolds No.	K_2	a_2
Greater than 2000	118	$-1/6$
Between 200 and 2000	8×10^5	$-8/6$

and θ is the "mixing time," H is liquid depth, T is tank diameter, u_j is velocity of jet discharge, D_j is diameter of jet discharge, and g is the gravitational constant. The values of θ in the study ranged between about 0.5 and 30 min. The jet nozzle location was found not to be critical provided that it did not cause a pronounced swirl in the tank and did not feed directly towards the recirculation pump suction port. The results of both Fossett [Eq. (21)] and Fox and Gex [Eq. (24)] show that the effect of jet nozzle size and jet velocity can be expressed in terms of the product $D_j u_j$, and that the "mixing time" will decrease as this product is increased. Meny and Velykis (M7) describe an educator mixer, developed primarily for use in large tanks, which draws fluid from an internal perforated standpipe. The engineering details for the design of this mixing arrangement are presented very completely in the reference.

Information gathered during aerodynamic investigations of gas jets should prove applicable to the problem of mixing liquids in a tank. Folsom and Ferguson (F4), in 1949, using this approach with the data available at that time, formulated the following relation for the amount of fluid induced into a turbulent jet from the surrounding fluid.

$$Q_i = Q_j \left(\frac{0.234x}{D_j} - 1 \right) \quad (25)$$

where Q_i is the total volumetric rate of fluid induction between the nozzle and a point located at x feet along the jet axis from the nozzle, Q_j is the volumetric rate of flow from the jet nozzle, and D_j is the nozzle diameter in feet. Later workers (F5) confirm the general assumption of similarity of behavior of gas and liquid jets, but recommend a value of 0.33 for the coefficient of x/D_j. No experimental test of this relation in a conventional mixing tank has been reported; the presence of recirculating flows and other factors make the situation hydrodynamically more complicated than the situations from which the relation is derived. Folsom and Ferguson conclude that for a circular tank of one radius depth, the effective length for inducing flow is about 75% of the tank diameter; thus, for a 100-ft. diameter tank and a 3-in. diameter jet nozzle, the flow inducted by the turbulent jet will be approximately 100 times the flow out of the jet nozzle itself.

The blending of liquids by blowing air, steam, or other gases through them is a procedure which still finds application. Kauffman (K4) states empirically that superficial air velocities of from 0.7 to 3.1 ft./min. will give states of agitation ranging from "moderate" to "violent" in a liquid depth of 9 ft. Dunkley and Perry (D3) and Perry and Dunkley (P4) have found the blending performance to depend in a complicated way on tank and air inlet geometry, depth of liquid, and liquid viscosity; while satisfactory blending could be accomplished by air agitation, they concluded that mechanical agitation gave consistently faster performance.

3. *Scale-Up*

It is of interest to examine the performance studies discussed above, from the point of view of scale-up. It will be assumed here that the liquid properties are fixed (densities and viscosities are determined by the chemical processing requirements of the operation), and the concern of the engineer is the interrelation between "mixing time," agitator speed, and system size. These relations are summarized in Table V.

The influence of agitator speed in a vessel with a fixed liquid volume is seen from the first column, which confirms the generality that higher speed results in faster mixing. For top-entering agitators, the "mixing time" varies roughly as the inverse first power of the speed, the exponents on N varying from -0.8 to -1.6. The work of Oldshue et al. on side-entering propellers shows a different relation, with the "mixing time" inversely proportional to the cube of the speed.

Next, consider the effect of changes in vessel size for a geometrically-similar series of vessels (recalling that in such a series the impeller type is constant). The second column of Table V shows considerable variation

TABLE V
Scale Relations for One-Liquid-Phase Systems
(Liquid Properties Constant)

	One vessel	Geometrically-similar series of vessels	
	"Mixing time" varies with speed of rotation as	"Mixing time" varies as	For constant "mixing time," power per unit volume varies as
Top-entering agitators:			
Kramers et al. (K8)	N^{-1}	N^{-1}	$V^{0.67}$
van de Vusse (V1)	$N^{-1.6}$	$N^{-1.6}D^{-0.3}$	$V^{0.48}$
Fox and Gex (F9)			
$N_{Re} > 10^4$	$N^{-0.83}$	$N^{-0.83}D^{-0.17}$	$V^{0.46}$
$N_{Re} < 10^4$	$N^{-1.5}$	$N^{-1.5}D^{-1.5}$	—
Side-entering propellers:			
Oldshue et al. (O6)	N^{-3}	$N^{-3}D^{-5}$	V^{-1}

N is rotational speed. V is volume of liquid. D is impeller diameter, and is in direct constant proportion to any other dimension in a geometrically similar system.

in the nature of this size effect. Although there is no very strong basis for recommending any particular one of these results, it should be recalled that the Fox and Gex data are based on a wider range of sizes than any of the other studies.

A very common scale-up characterization of mixing vessels, widely used in the literature, is the effect that increasing size in a geometrically-similar system has on the power per unit volume required to achieve the same mixing performance. By use of Eq. (10) for the power input at Reynolds numbers above 10^4, the relations in the last column of Table V are obtained. These show that the power per gallon required to keep the "mixing time" constant as the batch size is increased goes up with the volume of the batch when top-entering agitators are used. The different studies show the power per unit volume varying with batch size raised to a power ranging from 0.46 to 0.67. On the other hand, the work of Oldshue *et al.* on side-entering propellers in large tanks shows a different result, with power per unit volume decreasing as volume is increased. In any case, all the studies show that the concept of equal power per unit volume to give equal performance is not valid in these investigations.

For systems which are not geometrically similar, the relations of van de Vusse [Eq. (16)] and Fox and Gex [Eq. (17)] for top-entering agitators indicate different effects for variations in liquid depth or tank diameter. There is no a priori reason to choose one or the other, except to note again that the work of Fox and Gex covered the wider range of variables.

The scale-up of tanks mixed by free jets is still more limited by lack of data. In general, the "mixing time" can be decreased by increases in $D_j u_j$ (product of jet diameter and velocity), while increases in tank volume will increase the "mixing time"; the specific relations are indicated by Eqs. (21) or (24). The work of Fox and Gex shows that if $D_j u_j$ is constant, the "mixing time" increases as the square root of the batch volume; in a given tank, the "mixing time" varies as $(D_j u_j)^{-5/6}$ or $(D_j u_j)^{-2}$, depending on the jet Reynolds number.

It must be observed again that the "mixing times" depend on the particular experimental techniques used to evaluate them. It is probably the different definitions of "mixing time" that are responsible for the differences in scale relations seen in Table V, although the limited range of some of the studies can also be a factor. The engineer faced with the problem of scale-up must decide if any of the studies outlined above are based on a "mixing time" definition which is especially appropriate to the situation with which he is concerned.

4. *Discussion*

In the foregoing investigations, the significance of the "mixing time" used in each case is not altogether clear. The reported "mixing times" depend on the particular experimental techniques used to determine them; there is no way to relate these observations to a common quantitative measurement of uniformity. However, each method does measure some effective average rate for achieving a kind of homogenization, and, in a general sense, the shorter the "mixing time," the better the performance. The relations obtained are probably more reliable in the comparative, than in the absolute, sense; the same physical variables have a somewhat similar influence in the different studies.

The work of Fox and Gex [Eqs. (17) and (24)] does represent a systematic approach to the problem, and their definition of mixedness (time when the "last wisp" of red color disappeared) is probably a conservative one. If the rash assumption is made that the results of different workers can be compared, the data of van de Vusse indicate that the results of Fox and Gex with propeller mixers would probably predict results for other agitator types within about a fourfold factor. While they are far from precise in a general way, these relations should be useful for rough estimates of mixing time.

IV. Gas-Liquid Systems

The contacting of gases with liquids, in mixing vessels agitated by rotating impellers, is a relatively common operation encompassing such applications as high-pressure hydrogenation, biochemical fermentations, and liquid-phase oxidations. Most of the literature available on this type of mixing is concerned with restricted ranges of operation in particular chemical systems.

A. Basic Studies

The over-all behavior of a gas-liquid agitated system will depend on basic phenomena of mass transfer, bubble dynamics, and the fluid-dynamic regime in the mixing vessel. A general discussion of mass transfer is beyond the scope of this review, and the book by Sherwood and Pigford (S5) is recommended as a guide to that subject.

A growing body of published information on bubbles has been concerned with such areas as the flow field around bubbles, and of bubble formation at orifices. However, this information has not yet been related to the problem of gas-liquid contacting, and hence will also be excluded from this review. The interested reader is referred to the recent annual reviews on "Fluid Dynamics" in *Industrial and Engineering Chemistry* for summaries of the work in this field.

With more specific reference to agitated systems, information is lacking on the effects of gas bubbles on basic properties like mean flow pattern, impeller discharge rate, or turbulence characteristics. The observations presented in Section II, based mainly on one-liquid-phase data, must therefore be considered as the best available approximations for the flow regimes in gas-liquid agitated systems. There have been a few papers of somewhat basic nature with direct application to these systems and these will be discussed in the remainder of this section.

Foust *et al.* (F8) studied the contacting of air and water in cylindrical baffled vessels agitated with Mixing Equipment Company dispersers, which had arrowhead-shaped blades mounted at the edge of a flat disk. Measurements were made with a 6-blade 4-in. disperser in a 1-ft. diameter tank, and with a 10-blade 20-in. disperser in an 8-ft. diameter tank. Air in each case was introduced from a single opening below and close to the center of the impeller. The superficial air flow velocity (volumetric flow divided by cross-sectional area of tank) ranged from 1 to 5 ft./min. Power input to the agitator was measured, as well as the increase in volume of the vessel contents due to the presence of the air bubbles. This increase of volume, called the gas "holdup," varied from 2 to 10% of the ungassed liquid volume. An "average gas contact time per foot of liquid depth" (θ_c) was calculated as:

$$\theta_c = \frac{\text{Volume of gas holdup}}{(\text{Volume gas per second})(\text{Liquid depth})} \tag{26}$$

"Contact-time" values varying from about 0.4 to about 2.1 sec./ft. of depth were observed. This "contact time" is a simple average. Actually, it seems likely that the residence time of any one gas molecule in the vessel could vary widely from the mean, because the gas bubbles follow relatively tortuous paths from gas inlet to liquid surface, and also may coalesce to a varying extent.

The experimental data in this study were correlated in the dimensional form:

$$\theta_c = K_3 \left[\frac{(HP.)}{VU_o'} \right]^{0.47} \tag{27}$$

where

θ_c = contact time, seconds per foot of liquid depth
$(HP.)$ = Horsepower delivered to the impeller
V = volume of air-free water, ft.^{3}
U_o' = superficial air velocity, ft./sec.
K_3 = a constant

The values of K_3 depended on liquid depth and impeller immersion (distance from surface to impeller) as follows:

Liquid depth (tank diameters)	Impeller immersion (tank diameters)	K_3
1.0	0.67	1.65
1.0	0.50	1.43
0.75	0.67	1.53
0.75	0.38	1.26

Equation (27) correlated the data from the 1-ft. diameter tank to within about 30%, and was somewhat less satisfactory for limited data taken in the 8-ft. vessel. The relation between θ_c and the operating variables is even more complex than indicated because the horsepower input varies with the gas velocity as well as with the other operating variables. Furthermore, according to these workers, unpublished data of theirs show that gas holdup is strongly affected by the nature of the liquid being gassed, which seems reasonable. This effect would lead to different values of K_3 and perhaps also of the exponent. Nevertheless, Eq. (27) should give order-of-magnitude estimates of gas holdup in this class of agitated system.

Bartholomew et al. (B2) present data on air holdup in a streptomycin broth agitated in a 6-in. diameter baffled fermentor equipped with two flat-blade turbines on a common shaft. Air holdups up to 14% of ungassed liquid volume were observed, for gas rates up to 1.2 ft./min. superficial velocity. Holdup was found to increase relatively rapidly with gas rate at low gas velocities and less rapidly at higher gas rates, due to the formation of larger, faster-rising bubbles at higher gas rates. It was also found that at any fixed gas rate the holdup volume increased with agitator speed and was greater for a sintered sparger than for a constricted-pipe gas inlet.

Fundamental understanding of gas-liquid mass transfer requires knowledge of the interfacial area between the phases. Vermeulen et al. (V2) reported a study of the average interfacial area between two liquids, and between gas and liquid, which are contacted in an agitated baffled vessel. In their gas-liquid work, various gases were beaten into several different liquids, and the average interfacial area in a volume of mixture was measured by means of a photoelectric probe which monitored about 1 cc. of mixture. Measurements were made in a fully enclosed and fully baffled tank, 10 in. in diameter and 9 in. high, agitated by a stirrer with four flat blades. The ratio of blade width to impeller diameter ranged from 0.25 to 0.65, and tank-to-impeller diameter ratio was low, about 1.5. Impeller speeds varied from about 100 to about 400 r.p.m. The photoelectric probe was not small relative to the vessel size, and quite probably affected the flow pattern in its vicinity. There was also evidence of variation in interfacial area with position in the tank. The extent of this is not given, but all values reported were for the probe at one fixed location. The observed bubble diameters were usually 0.1–0.5 cm., and interfacial areas on the order of 1 cm.^{2}/cc. for air in water were found. The following correlation was presented:

$$A = \frac{1400 N^{1.5} D \rho_c^{0.5} \mu_d^{0.75} \phi}{\sigma \mu_c^{0.25} F(\phi)} \qquad (28)$$

where
- A = interfacial area per unit volume of mixture
- N = impeller rotational speed
- D = impeller diameter
- ρ_c = continuous (liquid) phase density
- μ_d = dispersed (gas) phase viscosity
- μ_c = continuous phase viscosity
- σ = interfacial surface tension
- ϕ = volume fraction of gas phase
- $F(\phi)$ = a function which varied with ϕ approximately as follows:

$$F(\phi) = 2.5\phi + 0.75 \tag{29}$$

Any consistent set of dimensions may be used. This equation gives only a fair correlation of the data, with a scatter of about 50% around it. It is also indicated that trace impurities affect the interface in a random and unknown manner. The large-diameter impellers used for this study, as well as the fact that the gas was beaten in rather than introduced through a pipe or sparger, makes this agitation system different in two important particulars from those normally encountered.

In a later investigation, Calderbank (C1) measured gas holdup and interfacial area in baffled vessels, agitated by a flat-blade turbine, with continuous air injection beneath the impeller. A considerable variation of local values of holdup and interfacial area with location in the tank was observed. Air holdup was in the range 1–8%; bubble sizes were 0.2–0.5 cm.; interfacial areas were in the range of about 0.2–1.0 cm.^{2}/cc.

The effect of liquid shear on mass transfer from bubbles is reported upon in a thesis by Tereschkevitz (T3), who measured oxygen absorption from bubbles rising through water in the annulus between a rotating drum and a stationary cylinder. Absorption coefficients on the order of 0.01–0.03 cm./sec. were found, based on a concentration driving force. It was concluded that the mechanism of absorption in circulating bubbles under shear is that of unsteady-state diffusion as postulated by the penetration theory (S5). A finding directly applicable to agitated vessels was that, even for high average rates of shear (velocity gradient across annulus between drum and cylinder), the mass transfer *per unit of bubble surface* was not greatly affected by changes in shear rate. For example, the mass-transfer coefficient for absorption of oxygen by distilled water at an average shear rate of 200 sec.$^{-1}$ (much higher than an average rate estimated for a conventional gas-liquid mixing vessel) was only 40% greater than the coefficient under no-shear conditions. The maximum effect of shear was observed for absorption by a relatively viscous (170 cp.) aqueous solution of a carboxypolymethylene thickening agent containing 15 p.p.m. of surfactant (pentapropylbenzene sulfonate); here the transfer coefficient under an average shear of 230 sec.$^{-1}$ was about 3.5 times the no-shear value.

Another conclusion of this study was that the presence of the surface active agent could reduce the transfer coefficient per unit area to values as low as 30% of those measured in the absence of this agent. While the effect of a surfactant in any specific case is related in a complex manner to the bubble sizes, the flow pattern around the bubbles, and the structure of the interface itself, this result is an indication of the large influence that the merest traces of material may exert on gas-liquid mass transfer systems.

B. Power Requirements

Some power measurements on gas-liquid agitated systems have been made in conjunction with investigations of mass-transfer properties. The power requirements for fixed conditions of gas rate varied as the cube of the impeller speed, in experiments by Bartholomew et al. (B2) with a 6-in. diameter baffled fermentor and by Wegrich and Schurter (W1) with a 2000-gal. baffled fermentor; this dependence is the same as for the turbulent agitation of gas-free liquids. On the other hand, Elsworth et al. (E1) using a 13-in. diameter baffled vessel, 7-in. liquid depth, and 4-, 5-, and 6-in. diameter vaned-disk impellers, found that power input varied with impeller speed raised to a power varying from 1.7 to 4.1, and with impeller diameter raised to a power varying from 2.6 to 5.5. In another study John (J1) reports that power input depends on impeller speed raised to a power between 2.5 and 3.0, depending on the mixer geometry and the nature of the material being agitated.

Consider an impeller rotating at a fixed speed in a liquid. If a stream of gas is introduced beneath the impeller, there will be a decrease in the amount of power required by it. Several investigators have reported their data in terms of the ratio of agitator power with gas flow to the power required with no gas flow, all other conditions being constant. Oyama and Endoh (O11) correlated this ratio with a dimensionless group Q_g/ND^3, where Q_g is the volumetric gas flow rate, N is impeller rotational speed, and D is impeller diameter. They found that the correlation depended primarily on impeller design. Kalinske (K1) also used these dimensionless groups. Calderbank (C1) measured power requirements for a flat-blade disk-turbine mixing air into water, ethanol, or glycol, and correlated his data by use of these groups. A typical correlation of the power ratio with the group Q_g/ND^3 is given in Fig. 7. While the specific nature of such a correlation curve depends on the particular impeller design and system geometry, the curve in Fig. 7 should be useful for estimating purposes

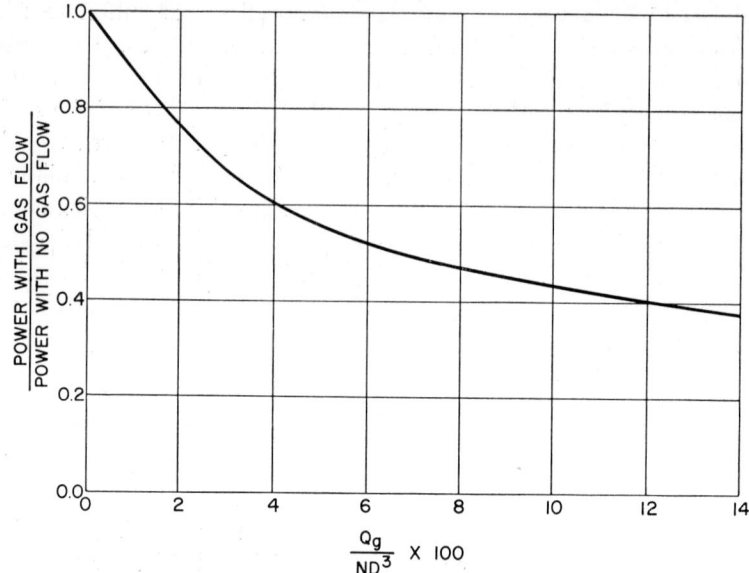

Fig. 7. Effect of gassing on power requirements.

since it agrees within about 0.1 with the power ratios reported by various workers (C7, C1, F8, K1, O11).

The data on the power reduction due to gas flow show that if power input is designed for a system by the correlations for no-gas conditions, there will certainly be ample power for the gassed state. Conversely, a drive that is closely designed to provide only a certain power input at a given gas rate may be badly undersized with respect to operation at the same speed for lower gas rates. It is also of interest to note that the magnitude of the decrease of power consumption with gassing indicates that the added gas does more than decrease the over-all average density of the liquid medium in the tank; a gas flow giving 10% holdup may correspond to a power reduction of as much as 60% or more.

A few scattered data (N3, W1) indicate that when gas is introduced beneath the bottom turbine of a multiple-turbine system the fractional decrease in total power requirement is much smaller than given in Fig. 7. It appears that gassing affects the power requirements through changes in the flow regime in the vicinity of the impeller being sparged, and not so much by changes around any other impellers present.

C. Performance Studies

The investigations of performance of gas-liquid systems in agitated vessels have been concerned, for the most part, with determining over-all average gas-to-liquid mass-transfer rates in chemically-reacting systems. The results of these studies apply almost exclusively to the particular systems in which they were carried out, but nevertheless show the kinds of effects which may be observed.

Air oxidation of sodium sulfite solutions, containing a trace of cupric ion as catalyst, was first studied by Cooper et al. (C7). These workers found the reaction rate to be essentially independent of sulfite and sulfate concentrations and therefore especially convenient for their study. Most of their runs were made using a vaned-disk agitator (a flat disk with sixteen radial vanes on its lower face) in baffled vessels 6–17 in. in diameter. Tank-to-impeller diameter ratio was 2.5, and the impeller was 0.3 tank diameter off the bottom, in every case. Impeller speeds were 60–900 r.p.m. A few measurements were made with a simple flat paddle, with a total length one-fourth the vessel diameter, in 9.5- and 96-in. diameter tanks. Air was introduced from a single open-end pipe just below the center of the agitator. Superficial air velocities were 0.3–11 ft./min.

The rate of oxygen absorption was calculated by measuring the sulfite concentration in the liquid at the beginning and end of a run. Then, by assuming irreversible absorption (no oxygen back-pressure from the liquid), an absorption coefficient (K_v) was calculated as (lb.-moles oxygen absorbed)/(ft.^{3} of solution) (log-mean oxygen partial pressure, atm.) (hr.). K_v varied widely with operating conditions, ranging from 0.0006 to 0.13, while absorption efficiencies varied from less than 1% (without agitation) up to 42%. Power input to the agitators were in the range of about 0.04–12 HP./1000 gal. It was found that the absorption coefficient increased with increases in power input and/or gas velocity. For this vaned-disk agitator, with liquid depth equal to tank diameter, the following dimensional equation correlated the results fairly well for power inputs above about 0.3 HP. per 1000 gallons:

$$K_v = 7.8 \times 10^{-3} U_o^{0.67}(\text{hp.})^{0.95} \qquad (30)$$

where U_o is the superficial gas velocity in feet per minute and (hp.) is the horsepower input per 1000 gallons. For U_o values greater than about 6.5 ft./min. (a "loading" point), this relation no longer held. Visual observation of this "loading" condition showed large bubbles beginning to pass around the impeller and rise quickly through the liquid.

Data for a paddle agitator were correlated by a similar equation, except that the exponent on (hp.) was 0.53. Limited data with the vaned disk agitator in the 6-in. diameter vessel showed that the absorption coefficient increased with liquid depth as follows *if* the power input per volume was kept constant:

Ratio, liquid depth to tank diameter	Relative value of K_v at constant power/volume
0.5	0.6
1.0	1.0
2.0	1.4
4.0	1.7

The oxygen-transfer coefficient in the air oxidation of sodium sulfite solutions has been used to characterize gas-liquid agitation systems by a number of other workers. Rushton *et al.* (R14) investigated the effects of multiple impellers on gas-liquid contacting by use of this reaction. With fully baffled tanks, 6 and 12 in. in diameter, air was introduced underneath the lowest impeller through an open pipe in the small tank or through a sparger ring in the large tank. Mixing Equipment Company flat-blade turbines were used. Tests were made with various liquid heights, and with one, two, or three turbines at various positions on the axial shaft. Values of the absorption coefficient were found in the range 0.01–0.1 lb.-moles oxygen/(hr.) (ft.^{3} of liquid) (atm.), for superficial gas velocities 1.5–6 ft./min., with power inputs 0.9–8.4 HP. per 1000 gal. These values are on the same order as those of Cooper *et al.* cited above.

The number and spacing of turbines had a complex influence on the transfer coefficients, depending upon the air flow rate, power input, and probably upon factors of system geometry. With liquid depth/tank diameter ratios below 2.5, it was not possible to change the coefficient more than 10% by the use of multiple turbines. With a liquid depth four times the tank diameter, and at high air flows and power levels, multiple turbines might give a 25% improvement in the coefficient but, if not "properly" positioned, they can also give a 50% decrease. Photographic studies cited were said to indicate radical changes of flow pattern with turbine spacing. For single and multiple turbines, at constant gas rate, the transfer coefficient was found to vary with the 0.8 power of power input per unit volume [to be compared with the 0.95 exponent reported by Cooper *et al.*, Eq. (30)].

The fermentation industry, stimulated by the large demand for antibiotics, has made major application of gas-liquid contacting in agitated vessels, and has contributed to numerous publications which have appeared describing specific biological systems. Finn (F2) has written a thorough review on the subject of aeration as applied in biochemical processing, showing that the performance of the system (in terms of yield, antibiotic activity, etc.) is definitely affected by impeller design and speed.

Bartholomew et al. (B2) give an extended analysis of oxygen transfer in the production of penicillin and streptomycin. They found that over-all oxygen transfer depends on liquid-phase properties and on the nature of the oxygen-consuming reactions, as well as on the properties of the gas phase and interface. Agitation not only increases the interfacial area, but also may lower the resistance to mass transfer in the fermentation broth, which is actually a complex non-Newtonian suspension. These workers used air rates of 0.2–1.2 ft./min. and observed over-all oxygen-transfer coefficients up to 0.015 lb.-moles oxygen/(ft.^{3}) (atm.) (hr.). The relation between mass transfer coefficient and operating conditions was quite complex. The presence of a surface-active agent was observed to lower the over-all oxygen transfer. In a later study (K2), the same investigators discussed the scale-up of fermentation processes over a range from 5 liters to 15,000 gal. In terms of antibiotic production, they could scale up fairly well by using the correlation of Cooper et al., even though there are differences of chemistry, physical properties, and mixer geometry. With multiple impellers, it appeared that the sparged impeller contributed most to the absorption of oxygen; hence they recommended introducing air to each impeller separately when multiple arrangements are used.

Chain et al. (C4) studied aeration by use of the sulfite-oxidation system in a 6-liter reaction vessel. Using flat-blade turbines, with baffles, with a single-point sparger, they found an increase in oxygen absorption as the power input was increased; the form of their relation was more complicated than the one reported by Cooper et al. As noted earlier, Elsworth et al. (E1), working with the sulfite system in a baffled vessel, found a complex relation between the impeller speed and diameter, air rate, and power input to the agitator. The oxygen absorption coefficient, considered separately from the power input, was found to be independent of the air flow-rate and proportional to the 2.4 power of the impeller speed. Friedman and Lightfoot (F12) observed similar results with a 6-in. vessel and simple flat-blade impellers; in a range of air velocities of 9.5–86.6 ft./hr., they found the oxygen transfer coefficient essentially independent of gas rate, as long as the coefficient was greater than 0.002 lb.-moles oxygen/(ft.^{3}) (hr.) (atm.). In other studies of sulfite oxidation, Finn (F2) has reviewed the extensive literature on the subject and Schultz and Gaden (S3) and Carpani and Roxburgh (C2) have conducted experimental work. These studies show that the individual steps which make up the over-all process may be of considerable complexity. Since sodium sulfite solutions are also very simple physically, especially as compared to biochemical reaction media, caution is advisable in interpreting the results of this test system.

Johnson *et al.* (J3) suggest the use of the hydrogenation reaction of α-methylstyrene with a suspended palladium-alumina catalyst as an alternative test system to establish the effect of agitation variables on liquid-phase mass-transfer coefficients. They found the over-all hydrogen transfer coefficient to vary in a complex manner with agitator speed, and to increase with the 0.6 power of the superficial gas velocity up to a point beyond which the transfer showed no further change with gas velocity.

The effect of spargers on system performance has also been considered. Bartholomew *et al.* report a considerable increase in oxygen diffusion rate and in antibiotic yield when using a sintered sparger as compared with a constricted pipe. Chain *et al.*, in the work cited above, found a porous sparger to be more effective than a ring sparger, the difference decreasing with increasing agitation. A ring sparger and an open pipe gave roughly equivalent results.

Chain *et al.* also report some results from operation without baffles or a sparger; under conditions where a vortex was produced, enough air was drawn in through the vortex to give oxygen uptake rates equal to those observed with baffled operation and a submerged air inlet. However, Kneule (K6), in a very brief report, finds that aeration under the usual baffled conditions is preferable to vortex aeration.

Shu (S6) has taken advantage of the dependence of fermentor performance on agitator speed, by setting up a "feedback" control loop in which the agitator speed is varied automatically so as to cause the broth oxygen uptake to vary with time in a predetermined manner.

Some data on the agitation in laboratory rocking and shaking autoclaves are given in publications by Hoffman *et al.* (H12) and by Snyder *et al.* (S7). Studies of the agitation in shaken flasks have also been reported (A6, R3).

D. Discussion

Certain generalizations can be drawn from these investigations of gas-liquid systems. It is clear that the impeller power requirement increases with rotational speed. It also decreases with rate of gas throughput, as indicated in Fig. 7, although quantitative differences in the variation indicate the sensitivity of these systems to geometric parameters and liquid properties. With respect to oxygen transfer rates in air-liquid agitated systems, if the air rate is kept constant, the mass-transfer coefficient increases when power input to the impeller is increased. In terms of primary operating variables, an increase of impeller speed or impeller diameter will increase the rate of mass transfer at a given air rate. Some workers describe a marked increase of mass transfer with increasing air rate, if the power input is kept constant by increasing impeller speed.

But if impeller speed and impeller diameter are fixed, an increase in air rate will have only a slight effect on the oxygen transfer coefficient. For efficient operation it is necessary to avoid the "loading" condition described by Cooper et al.; i.e., operating under conditions of such high gas rate (or, in some cases, power input) that the system performance is no longer strongly dependent on these variables. As yet, there are no general correlations for determining the "loading" point.

In multiple impeller systems, it appears that conditions in the region around the impeller nearest to the point of gas injection have the greatest influence on system performance. In view of the work of Tereshkevitz, which showed a large influence of bubble size (as compared with liquid shear) on mass transfer, a tentative conclusion is that the average bubble size is determined mostly by the conditions near the gassed impeller. It follows that primary emphasis in process development and scale-up should be placed upon selection of the impeller and the gas-inlet conditions; variations in the over-all system geometry, with the possible exception of liquid depth, may be of secondary importance.

The air oxidation of aqueous sodium sulfite solutions has had wide use as a test system for evaluating the performance of agitated gas-liquid contactors. As has been noted, gas-liquid processes may involve rate-controlling steps far different from those of the sulfite oxidation system. Green (G2) found it was not useful for application to a hydrogenation problem. Nevertheless, since it is quite convenient to carry out experimentally, sulfite oxidation should always be considered as a possible test system, subject to actual demonstration, in cases where a substitute is needed for the actual process under consideration. For example, Karow et al. (K2) and others have applied sulfite oxidation successfully to fermentation situations.

In some cases, surface-active agents can reduce significantly the rate of mass transfer. The relationships involved are complex and poorly understood. Since only trace quantities may be sufficient to exert a marked influence, surface-active contamination must always be considered as a possible hidden variable in these systems.

V. Liquid-Liquid Systems

Liquid-liquid systems with mechanical agitation are widely employed by the chemical industry to carry out chemical reactions like nitration, alkylation, and polymerization; to prepare emulsions; and to perform a variety of extraction, washing, and other operations involving transfer of material from one liquid phase to another. The literature on these systems is limited, and the many different liquids used by different investigators make it difficult to compare the published studies. As in the case of any heterogeneous system, the performance of many liquid-liquid agitated systems is related to the fundamental process of mass transfer. Basic mass-transfer phenomena are discussed in detail by Sherwood and Pigford (S5). The engineering design of extraction operations is covered by Treybal (T4).

A. Basic Studies

The properties of agitated liquid-liquid systems are determined basically by the physical laws that govern such phenomena as circulation inside liquid drops or the break-up of one immiscible liquid added to another in a particular flow regime. Published work on these problems (B1, C6, H1, K7) ranges from purely theoretical studies to experimental measurements in systems relatively simple from a fluid-dynamic viewpoint. For agitated systems containing several liquid phases, no studies have been made of mean flow patterns or flow velocities; as with gas-liquid systems, it is necessary to use the descriptive work on single-phase systems (Section II) as the basis for consideration of gross flow patterns in liquid-liquid systems.

Two studies have been concerned with measurement of the interfacial area obtained by agitation of liquid-liquid systems. Each of these investigations relied on the use of a photoelectric probe which measured the light transmission of the two-phase dispersion. Vermeulen and co-workers (V2) made measurements in two geometrically similar, baffled vessels of 10- and 20-in. diameter. They used a very simple four-blade paddle-like stirrer, with a tank-to-impeller diameter ratio of about 1.5, and a 0.25 blade-width/impeller-diameter ratio. The impeller was located midway between the top and bottom of the vessel, which had a cover and was run full. Impeller speeds varied from about 100 to 400 r.p.m. A wide variety of liquids was employed. Volume fractions of dispersed phase varied from 10% to 40%. The mean droplet diameters observed ranged from 0.003 to 0.1 cm. The results were correlated with a mean deviation of about 20% by an empirical equation relating the specific interfacial area near the impeller to several system and operating variables as follows:

$$A = 72 \left(\frac{D^3 N^2 \rho_e}{\sigma}\right)^{0.6} \left(\frac{1}{D}\right)\left(\frac{\phi}{F(\phi)}\right) \tag{31}$$

where
 A = interfacial area per unit volume of the mixture
 N = impeller rotational speed
 D = impeller diameter
 ρ_e = an "effective mean density" [see Eq. (32) below]
 σ = interfacial surface tension
 ϕ = volume-fraction of dispersed phase

$F(\phi)$ is the function whose value is well estimated by Eq. (29). Any consistent set of units may be used. The "effective mean density" ρ_e is defined by:

$$\rho_e = 0.6\rho_d + 0.4\rho_c \tag{32}$$

where ρ_d and ρ_c are the densities of the dispersed and continuous phases, respectively. It was also found that the mean area was essentially unaffected by viscosity of either phase, or by density difference of the phases, although these quantities were varied over wide ranges.

Fick et al. (F1) found that Eq. (31) applied over a range of 0.13–0.65 in impeller width-to-diameter ratio. In addition, measurements without baffles showed that for other conditions constant the interfacial area in the unbaffled system was about 0.6 of the area in the baffled vessel.

Rodger et al. (R6, T6) measured the interfacial area produced by agitation of water with one of a large number of organic liquids, in a fully baffled cylindrical vessel with liquid depth equal to vessel diameter. They used equal volumes of each phase. In all the dispersions studied, water was the continuous phase. The impeller was a Mixing Equipment Company six-flat-blade turbine located at the interface of the two liquids. Most of the measurements were made in a 5.6-in. diameter vessel, and some were also made in 11.4- and 17.4-in. vessels. Interfacial areas were roughly on the order of 20–90 cm.²/cm.³ The following highly complicated empirical expression was presented as correlating the results, with an average deviation of 6.3% and all points except three (out of 244) within 20%.

$$DA = 42 \left(\frac{D^3 N^2 \rho_c}{\sigma}\right)^{0.36} \left(\frac{D}{T}\right)^k \left(\frac{\nu_d}{\nu_c}\right)^{1/5} \left(\frac{\theta_s}{\theta_{s0}}\right)^{1/6} \exp\left(\frac{3.6\Delta\rho}{\rho_c}\right) \tag{33}$$

where
- ν_d = kinematic viscosity of the dispersed phase
- ν_c = kinematic viscosity of the continuous phase
- θ_s = emulsion settling time
- θ_{s0} = "standard" settling time
- $\Delta\rho$ = difference in densities
- T = tank diameter

and the other symbols are defined as for Eq. (31). The exponent k on the impeller-to-tank diameter ratio varied from 0.75–1.4, depending on the nature of the liquids studied.

These workers were not able to correlate the values of k or of settling time with any more general properties. To free the correlation from the restriction imposed by the presence of factors which could be determined only by specific experiments, the following relation was given:

$$A = 79 \left(\frac{D^3 N^2 \rho_c}{\sigma}\right)^{0.26} \left(\frac{1}{T}\right) \left(\frac{\nu_d}{\nu_c}\right)^{1/5} \exp\left(\frac{3.6 \Delta \rho}{\rho_c}\right) \tag{34}$$

where the symbols are the same as above. With this relation they could correlate about 80% of the data (for settling times between 0.2 and 3 min.) with an average deviation of 12%, and all but seven points (of this group) within 30%.

The "settling time" in this work appears to be a measure of the surface-active contamination present in the particular system being studied. Vermeulen et al. also noted considerable variability in their results which they ascribed to surface-active agents. Neither they nor Rodger et al. were able to observe changes in interfacial surface tension to account for the apparent changes in interface contamination. It would appear that equilibrium drop size in liquid-liquid systems is sensitive to remarkably small amounts of interface contaminations.

Rodger et al. used a dimensionless Weber group ($D^3 N^2 \rho_c/\sigma$) in their correlation. This group was also used by Vermeulen et al. and Fick et al., although they used an "effective mean density" [Eq. (32)] instead of the continuous phase density. A Weber group generally appears in any theoretical or dimensional-analysis consideration of fluid dynamic systems in which interfacial tension effects are significant.

It is seen that Eq. (31) shows an inverse relation of interfacial area to impeller diameter for constant Weber numbers while Eq. (34) shows an inverse relation to tank diameter. Rodger et al. show a small effect due to viscosity and density, while the other workers do not. The differences may be due to differences in technique or equipment, but the Rodger equation does fit their data better.

In view of this fit, and of the fact that Rodger et al. used an impeller type that is in common use, Eq. (34) seems to be preferable for engineering estimates. However, the strong effects of variations in impeller design or system geometry which were not studied, and the possible influence of the relatively large probe on the flow, and the absence of measurements at other volume fractions or with other continuous phases, all must be considered as factors which indicate the need of caution in any application of Eq. (34).

The problem of coalescence in these systems, which can in some cases cause considerable point-to-point variation in interfacial area within a vessel, has not been studied and is poorly understood. The difference in location of measuring elements suggests that more coalescence may have occurred in Rodger's systems.

The prediction of which of two immiscible liquids will be the dispersed phase in an agitated system is a problem which has been discussed only in qualitative terms. Thus, Rodger *et al.*, in the work cited above, noted that while with most systems at low energy input the stable form was oil-in-water (they made their interfacial area measurements on this type), it was possible to cause some systems to change suddenly to water-in-oil dispersion by increasing the agitator energy input. They also stated that the desired dispersion form could generally be obtained if the impeller was located so that, at rest at the start of the run, it was completely in the phase which was desired to be continuous. Treybal (T5) points out that these generalizations may not hold for continuous operations. He also points out that for liquid volume ratios $\frac{1}{3}$ or smaller, the liquid present in the smaller amount will probably be the dispersed phase.

B. Power Requirements

Most of the published power measurements on liquid-liquid agitation systems have been made in connection with performance studies that will be discussed in the next section.

Miller and Mann (M10) measured power consumption in several different oil-water systems, in 6-, 12-, and 18-in. diameter unbaffled vessels. A relatively simple torque-table was used for the power measurements. Several different simple flat-blade, paddle-type impellers were studied as well as some with pitched blades. A shallow-pitch propeller and special shrouded turbine were also used. There was undoubtedly some degree of vortexing present in all the runs, but apparently it was not severe enough to uncover the impeller in any case. It was found that for any impeller the power required to agitate any given two-phase system varied with impeller speed in the same manner as for one-liquid-phase systems. Further, for any given impeller, its one-liquid-phase correlation of power number with Reynolds number (of the type described in Section III) also correlated the two-liquid-phase power data, provided that the volume-average mixture density was used and that the following empirical relation was used to calculate an effective mean viscosity for the two-phase system:

$$\mu_e = \mu_c{}^{\phi_c}\mu_d{}^{\phi_d} \tag{35}$$

where μ_c and μ_d are the viscosities of each phase, respectively, and ϕ_c and ϕ_d are the volume fractions of each phase, respectively. Later studies by Olney and Carlson (O7) using baffled vessels and by Wingard *et al.* (W8) in both baffled and unbaffled vessels showed that the power requirements for any given geometric system and agitator type could be correlated by the usual power-number Reynolds-number relations, if these numbers were based on the average density and on average viscosity as defined above. A similar conclusion was reached by Fick *et al.* (F1), but with use of a different average viscosity.

Laity and Treybal (L1) report on experiments with a variety of two-phase systems in a covered vessel which was always run full, so that there was no air-liquid interface at the surface of the agitated material. Under these circumstances no vortex was present, even in the case of operation without baffles. Mixing Equipment Company flat-blade disk-turbines were used in 12- and 18-in. diameter vessels whose heights were about 1.07 times their diameters. Impeller diameter was one-third of tank diameter in each case. For operation without baffles, using only one liquid phase, the usual form of power-number Reynolds-number correlation fit the data, giving a correlation curve similar to that given in Fig. 6 for disk-turbines in unbaffled vessels. In this case, however, the Froude number did not have to be used in the correlation because of the absence of a vortex. For two-phase mixtures, Laity and Treybal could correlate the power consumption results for unbaffled operation by means of the same power number-Reynolds number correlation as for one-phase systems provided the following equations were used to calculate the effective mean viscosity of the mixture: For water more than 40% by volume:

$$\mu_e' = \frac{\mu_w}{\phi_w}\left[1 + \frac{6\phi_o\mu_o}{\mu_w + \mu_o}\right] \quad (36)$$

For water less than 40% by volume:

$$\mu_e'' = \frac{\mu_o}{\phi_o}\left[1 - \frac{1.5\phi_w\mu_w}{\mu_w + \mu_o}\right] \quad (37)$$

The subscripts "o" and "w" refer to the oil and water phases, respectively.

Laity and Treybal also measured power consumption with four baffles in the vessel, each baffle being 1/6 of the tank diameter in width. With single liquids, they found the same power number-Reynolds number correlation as given in Section III for normal baffled operations with the presence of an air-liquid interface at the surface. Again, with two-phase systems, the same relation correlated the data provided that an effective mean viscosity calculated as follows was used:

$$\mu_e''' = \frac{\mu_c}{\phi_c}\left[1 + \frac{1.5\phi_d\mu_d}{\mu_c + \mu_d}\right] \quad (38)$$

The subscripts "c" and "d" refer to the continuous and dispersed phases, respectively. Equation (38) is a relation suggested by Vermeulen *et al.* (V2), and Eqs. (36) and (37) are modifications of it. Laity and Treybal found that use of the simpler Eq. (35) for effective mean viscosity was not satisfactory for correlation of their data.

It appears that the power requirements for liquid-liquid agitated systems can generally be dealt with in the same manner as for the case of one-liquid-phase systems. The appropriate effective mean viscosity of the two-phase mixture must be used as indicated above. Similarly, the average density must be used. In the power studies cited, a simple volume-average was used:

$$\rho_a = \phi_c \rho_c + \phi_d \rho_d \tag{39}$$

The results cited above are limited by the ranges of systems and variables studied. Even within these ranges, occasional deviations in behavior were observed which remain unexplained. In any event, the many problems and limitations noted in Section III in connection with the prediction of power requirements for one-liquid-phase systems apply equally well to liquid-liquid systems.

C. Performance Studies

1. *Mixing Index*

Miller and Mann (M10) studied the performance of a liquid-liquid agitation system from the point of view of the "degree of mixing" and "mixing index" concepts developed by Hixson and Tenney (H10) in their work on solid-liquid systems. Thus, Miller and Mann used 14-mm. diameter glass sample tubes ("thieves") which grabbed samples of 4.5 or 8 cc. from different locations in an unbaffled vessel while the agitator was running. The volume-fraction of each phase in the grab sample was observed by allowing the sample to settle. In general, the volume-fraction of one of the phases was smaller than the average volume fraction of that phase in the vessel as a whole; the ratio of this observed volume fraction to the over-all average value determined the "percentage mixed" for the sample. This quantity then could vary from zero (only one phase present) to 100% (sample same composition as over-all vessel contents). The "mixing index" for the whole system was defined as the arithmetic average of the "percentage mixed" values for a number of grab samples.

This "mixing index" is only a crude measure of the performance of the agitator system as a producer of homogeneity on a very coarse scale. It gives no information on interfacial area or drop size, on the distribution of inhomogeneities in the vessel, or on the nature of the flow patterns in the vessel. It is difficult to assign any real value to the "mixing index," except to the extent that a high degree of large-scale uniformity is itself an objective of the mixing operation. For their unbaffled vessels, Miller and Mann concluded that the major factor affecting the value of the "mixing index" was power input to the impeller. The variables of impeller type, liquid and impeller levels, and liquid properties had irregular but secondary effects. Curves of "mixing index" versus power input per volume generally showed a maximum "index" at power inputs of about 1–2 HP./1000 gal. Limited data indicated that scale-up between the 6-, 12-, and 18-in. tanks, in terms of constant "mixing index," could be accomplished by keeping equal values of power input per unit volume.

Wingard et al. (W8) also used the "mixing index" as a measure of performance, but took grab-samples from only one point in the vessel. They also observed the change in "mixing index" with time. Their conclusions of general interest were that the rate of increase in "mixing index" increased with agitator speed and that the presence of baffles also increases the rate of mixing (as measured by the "index").

2. Extraction Rates

Continuous-flow agitated liquid-liquid systems are commonly employed for extraction applications. Oldshue and Rushton (O5) and Karr and Scheibel (K3) studied the performance of compartmented columns agitated by many impellers on a common shaft. Flynn and Treybal (F3) studied continuous extraction in a system consisting of one agitated vessel.

The latter work (F3) involved extraction of benzoic acid from either toluene or kerosene, into water. Two geometrically similar systems were used, with vessels 6 and 12 in. in diameter, vessel height equal to diameter, four baffles $\frac{1}{6}$ of vessel diameter, and a 6-flat-blade disk-turbine with a diameter one-third that of the vessel. Murphree-type stage efficiencies were calculated, assuming cocurrent flows. The stage efficiency was considerable (45–90% in some cases) even without agitation. The stage efficiency increased from the no-agitation value when the impeller was rotated; the additional efficiency due to agitation was compared to the maximum increase in efficiency that could have been obtained. For each system of liquids, at a fixed phase-ratio in the feed, the extra efficiency due to agitation depended on the agitator energy consumption per unit volume of liquid flowing through the vessel. For fixed flow-rate and power input, the stage efficiency tended to increase with the ratio of kerosene or toluene to water, but this effect was not consistent over the entire range of power inputs used.

Overcashier and co-workers (O8) also studied agitated-vessel performance as related to liquid-liquid extraction. They used a 14.75-in.

diameter vessel with a liquid depth of about 19 in. The system was operated in continuous flow, with the vessel full, so that no vortexing occurred when baffles were absent. Feed streams were introduced at the wall at the bottom of the vessel and the mixture was withdrawn at the top. The liquids were water and kerosene, and the solute was n-butylamine. A variety of impellers were studied, including propellers, flat-blade turbines, and spiral-blade turbines, with diameters of 4–10 in. Exit concentrations were measured by running for 10 min. or so, and collecting the outlet stream for the last 4 min. of the run to minimize the effect of random fluctuations. The performance of the system was expressed in terms of an extraction-stage efficiency calculated on the basis that the system was considered countercurrent; efficiencies from about 0.4 to almost 1.0 were observed. The results were presented in the form of curves of efficiency as a function of power input to the impeller for each system.

Somewhat surprisingly, the power required to reach a given level of efficiency was always less for the unbaffled than for the baffled vessel. In unbaffled operations the type of impeller had relatively small effect; for each type the optimum size was about 40% of tank diameter, and gave about 0.95 theoretical stage at 1–2 HP. per 1000 gal. For baffled operation the optimum sizes were larger, and the radial-flow impellers were better than propellers. These workers suggested that the better performance of the unbaffled system is due to better utilization of the mass-transfer driving force, because of lessened end-to-end back mixing. The efficiency data for the various impellers, operated at 1 min. residence time, one-half liquid-depth impeller submergence, and a kerosene-water volume ratio of 1.57, fit the relation:

$$1 - E = \frac{3.18 \times 10^{15}}{(N_{\text{Re}})^{3.2}(N_P)^{1.37}} \left(\frac{D}{T}\right)^c \tag{40}$$

where E is the fractional stage efficiency; D and T are impeller and tank diameter, respectively; N_{Re} and N_P are impeller Reynolds and power numbers based on a mean density calculated by Eq. (39) and a mean viscosity calculated by Eq. (35); the exponent c is zero for operation with baffles, and 1.6 without baffles.

Other data showed that longer residence time gave higher efficiency for the same power input; but the total energy input per gallon processed was increased by increased residence time. Baffling the spiral turbine with either wall baffles or a stator ring resulted in the same power-efficiency relation for that impeller. In the absence of baffles, variations in impeller submergence had no effect. For baffled operation a submergence of one-third liquid depth was somewhat better than three-quarter liquid depth submergence, but the effect was not large. A few measurements indicated that feeding to the impeller rather than at the wall might improve the efficiency for a given power input.

In connection with these observations of Overcashier *et al.* with respect to power required for a given stage efficiency, Fick *et al.* (F1) also found that, although interfacial area was less when baffles were used, the power requirements for the unbaffled case were also much smaller. In their system, these effects were almost exactly counterbalanced; with all other conditions constant, they found approximately the same interfacial area for a given power input in either a baffled or unbaffled system. This result lends credence to the possibility proposed by Overcashier *et al.* that an effect such as end-to-end back mixing could become determining in their particular system. Nevertheless, as these workers recognized, it would be unwise to generalize their results without great caution.

Treybal (T5) has reported that the volume-fraction of dispersed phase actually present in an agitated vessel may be considerably less than the volume-fraction in the feed, especially in an upflow system when the dispersed phase is the lighter one. For baffled vessels and flat-blade turbines, with cocurrent upward flow, the volume-fraction of dispersed phase in the vessel is about 20% of the volume-fraction in the feed at energy inputs less than 100 ft.-lb./ft.^{3} of feed. This ratio rises rapidly with increased power above this level and is between 80% and 100% at energy inputs above 400 ft.-lb./ft.^{3} of feed. This effect is still incompletely understood.

In the same article Treybal utilizes information on mass transfer from solids and heat transfer to spheres, to arrive at a procedure for making rough estimates of extraction stage efficiencies for baffled vessels agitated with flat-blade turbine impellers. Results from this calculation procedure are in general agreement with the experimental results of Flynn and Treybal and of Overcashier *et al.*

VI. Solid-Liquid Systems

Most of the early experimental investigations were performance studies concerned with the homogeneity of solid-liquid suspensions or with the rate of solution of solid particles as related to various operating characteristics. This class of systems has not been studied to any extent from a fundamental standpoint; the following sections will indicate the large regions of interest which remain to be investigated.

A. Basic Studies

It is not known how the presence of solid particles affects the fluid dynamic regime in an agitated vessel; hence, it is necessary to assume that solid-liquid systems behave similarly to one-liquid-phase systems as described in Section II. Obviously this assumption will be best for systems in which the solids are small, are not very different in density from the liquid, are present in low concentration, and do not affect the Newtonian characteristics of the liquid.

A complete understanding of solid-liquid agitated systems will require basic information about the behavior of solid particles in a variety of fluid-flow fields. In this context, Stokes' law applies to the behavior of a single spherical particle in an infinite laminar flow field. Other relations have been developed to account for cases where other particles are present, where there is a wall effect, etc. (P3). Considerable work on the more complicated problem of behavior of solid particles in a turbulent liquid flow-field is now being done by workers in the fields of meteorology and civil engineering, as well as in chemical engineering (F11). Turbulent flow situations lead to especially complex relations when the particle is large enough so there is a large relative velocity between it and the surrounding fluid or when there are enough particles present so that they interfere with each other. Considerable work in this area remains to be done before results applicable for engineering use are available.

In connection with solid-liquid systems agitated so as to achieve interphase mass transfer or heterogeneous chemical reaction it may be noted that various workers have begun to consider the combined fluid dynamic, mass-transfer, and chemical kinetic problem in which a fluid moves past a solid with which it reacts chemically. The paper by Acrivos and Chambré (A1) is an example of this approach.

B. Power Requirements

Early data on impeller power requirements (H9) indicated a proportionality between power input and roughly the cube of impeller speed for a sand-water mixture. Data taken by Mack and Marriner (M4), using flat-blade turbines in baffled tanks, also have shown the power requirements to vary with the cube of impeller speed. These sketchy results seem to suggest the use of the one-liquid-phase power correlations of Section III, possibly with some appropriate density and viscosity corrections as in the case of liquid-liquid systems. If the opportunity exists, actual measurements on the system of interest should be made.

C. Performance Studies

1. *Suspension of Solids*

White, Sumerford and co-workers (W3, W6), in 1932–33, reported on sand-water suspensions produced by a flat paddle in an unbaffled tank. A wide range (more than twentyfold) of local sand concentration was found in the vessel; the concentration pattern varied with sand size, impeller size and location, and impeller speed. They found that sand concentration (at any given location) increased with speed up to a certain point, then leveled off. This "critical point" was suggested as a criterion for the effectiveness of the agitation. Other studies of sand in water were made by Hixson and Tenney (H10) with a four-pitched-blade turbine in an 18-in. diameter unbaffled tank. Here a 0.5-in. diameter grab-sampler tube was used to measure the sand concentrations at a number of points in the vessel, and from these data a "mixing index" was calculated as the arithmetic average of "percentage mixed" values for each sample location. The latter values, based on the component present in smaller concentration than in the vessel as a whole, and expressed as the ratio of that component's local concentration to its over-all average value, thus varied between 0 and 100. The "mixing index" was found to increase with impeller speed until it leveled off at a value of roughly 90. For constant speed, the "mixing index" increased with liquid viscosity. It seems clear that the "mixing index" is only a crude measure of the very-large-scale homogeneity in the vessel.

Hirsekorn and Miller (H2) made visual qualitative observations of the suspension of solids by paddle agitation in very viscous liquids (to about 50,000 cp.). For low impeller Reynolds numbers (about 10) in geometrically similar systems (6-, 12-, 18-in. vessels) the major factor in effecting particle suspension appeared to be power input per unit volume. In any given case the power required for complete suspension of all the particles was affected by system geometry and the settling velocity of the solids. No detailed correlation of the observations was presented.

Additional work on the suspension of solid particles in agitated systems has been reported by Oyama and Endoh (O13), on suspension of sands and resin particles, in baffled vessels 5.5, 6.7, and 10.8 in. in diameter. These authors used a 3.6-in. vaned disk, and 2.6- and 3.6-in. flat-blade turbines. A light beam passed through the vessel onto a photoelectric tube was used to monitor the particle concentration in a horizontal plane at a height about $\frac{2}{3}$ of the total liquid height. At low speeds the particles tended to congregate near the vessel bottom. As the speed increased, the particle concentration at the test level increased up to a "critical" value of impeller speed at which the particle concentration became essentially constant. This "critical" speed was the same for different particle loadings, but all the solids concentration values used were quite low, on the order of 1% by weight. At the condition of "critical" impeller speed, the following relation was found between the properties of the materials being agitated and the impeller power consumption per mass of agitated liquid:

$$\left(\frac{P_c g_c}{M}\right)^{2/3} = \frac{g d_p^{1/3}(\rho_s - \rho)}{16\rho} \quad (41)$$

where P_c is the power to the impeller when it is turning at the "critical" speed, M is the weight of liquid in the vessel, d_p is particle diameter, ρ_s is solid density, ρ is liquid density, g is the acceleration of gravity, and g_c is the usual conversion factor.

Pavlushenko et al. (P2) studied the suspension of screened fractions of sand and iron ore in a variety of liquids, with a 1-ft. diameter unbaffled vessel filled to a depth of one foot. Square-pitch three-blade propellers of 3-, 4-, and 5-in. diameter were used, and most of the observations were made with a 1 to 4 weight ratio of solids to liquid. Thief samples were taken at various levels in the vessel. In some cases, the contents did not become uniform at any impeller speed; in other cases the contents became uniform at some impeller speed and remained so at higher speeds; in a third type of behavior, the upper part of the vessel reached the over-all vessel average and then exceeded it as impeller speed was increased. Using the observations from the second and third types of behavior, a "critical" speed was defined as the lowest impeller speed at which the solids concentration at each level, or in the "upper layers" of the liquid, was equal to the over-all average solids concentration. This "critical" speed N_c in revolutions per second had the following relation to the operating variables:

$$N_c = 0.105 \frac{g^{0.6} \rho_s^{0.8} d_p^{0.4} T^{1.9}}{\mu^{0.2} \rho^{0.6} D^{2.5}} \quad (42)$$

Tank diameter (T) and acceleration of gravity (g) are in this equation as a result of dimensional analysis. Tank diameter was not varied during the experiments. The values for ρ and μ are those of the liquid medium. A series of tests with one solid showed that this correlation applied over a range of solid-to-liquid ratios of $\frac{1}{5}$ to $\frac{1}{2}$. The "critical" speed decreased with solids concentration below a $\frac{1}{5}$ weight ratio. A vortex was present during these experiments, and Eq. (42) did not apply if the vortex was deep enough to reach the impeller.

Zwietering (Z1) studied the suspension of solids in a series of baffled vessels, varying in diameter from 6 to 24 in. Paddles, six-flat-blade disk-turbines, vaned disks, and propeller-like impellers with flat blades were used. Screened fractions of sand and sodium chloride were mixed with a variety of liquids. Solids concentration was varied from 0.5 to 20% by weight. In this study, a "critical" impeller speed was defined as the lowest impeller speed with which there were no solids resting on the bottom of the vessel. It should be noted that the vessel contents were not necessarily homogeneous at this particular "critical" speed. It was found that for a particular impeller operating at a particular ratio of tank diameter to impeller height, the following equation correlated the observations:

$$N_c = \psi \left(\frac{T}{D}\right)^t \frac{g^{0.45}(\rho_s - \rho)^{0.45}\mu^{0.1}d_p^{0.2}(100R)^{0.13}}{D^{0.85}\rho^{0.55}} \quad (43)$$

In this equation, R is the weight ratio of solid to liquid, ψ and t have values which depend on impeller type and relative height, and the other symbols are as defined above. The values of ψ were in the range 1.0–2.0, centered around a value of 1.5. The exponent t had a value of about 1.4.

Considerable variation is encountered in the conclusions reported on suspension of solids. A major reason for this is that the various "critical" speeds and "mixing indices" used as performance criteria are not, in fact, equivalent measures of system performance. Other sources of variation are the major differences in geometry and in solids concentrations and densities. For design purposes, the results of a particular investigation should be used only after due consideration of the significance of the particular performance parameter in relation to the process under consideration.

The pulp-water systems encountered in the agitation of paper stock represent a highly specialized case of solid-liquid agitation, discussed in a few papers (C10, K5, O4), which will not be reviewed here. Lamont (L2) discusses the operation of pachuca tanks, used for ore-leaching operations, in which agitation results from air introduced at the tank bottom.

2. Mass Transfer

The rate of solution of solid particles in a solid-liquid agitated system has been studied by a number of investigators. In early work, Murphree (M13) presented basic equations which succeeding workers have used with some modifications. Basically, the rate of solution is proportional to the product of interfacial surface area and of concentration difference between saturation conditions and bulk-liquid conditions, with the proportionality constant an over-all mass-transfer coefficient related in some way to the materials and the agitation in the system. A number of workers (H4, H8, H11, W7) integrated this rate equation for a variety of cases where the particles may keep the same shape as they dissolve, or where only a small amount of solids dissolve, or where the solution changes in volume as solids dissolve, etc. They also showed experimentally that their results for rate of solution follow the expected functional form, and also that the mass-transfer coefficient varies with changes in impeller speed or agitator system geometry.

Hixson and Wilkens (H11) proposed a dimensionless-group relation, involving the mass-transfer coefficient and a Reynolds number for the system. Hixson and Baum (H4, H6) amplified this idea and extended the range of experimental measurements. They used unbaffled vessels with diameters of 6–24 in., a four-pitched-blade turbine, and a marine propeller entering at a 60° angle with the vertical. Different solid and liquid materials were studied with particle sizes on the order of 0.25 in. in diameter. Solution mass-transfer coefficients on the order of 0.005 g./(sec.) (cm.² of solid surface) (unit concentration driving force in gm./cm.³) were observed. For each geometrically similar system, using the turbine or the propeller agitator, the mass-transfer coefficient could be correlated with impeller Reynolds number and with Schmidt number by an equation of the following form, analogous to the Gilliland-Sherwood (P3, S5) equation:

$$\frac{K_s T}{\mathfrak{D}} = \Phi(N_{Re})^r (N_{Sc})^s \tag{44}$$

where
- K_s = mass-transfer coefficient
- T = tank diameter
- $\mathfrak{D}$ = diffusivity of solute in the liquid
- N_{Re} = impeller Reynolds number, $D^2 N \rho / \mu$
- N_{Sc} = Schmidt number, $\mu / \rho \mathfrak{D}$

Different values of the constant Φ and exponent r were used for different system geometries and operating conditions. Thus, for the turbine agitators at a Reynolds number less than 7400, Φ was 0.00058 and r was 1.4. For the turbine agitators at a Reynolds number above 7400, Φ was 0.62 and r was 0.62. For the propellers, over a range of Reynolds numbers from 3300 to 330,000, Φ was 0.0043, and r was 1.0. In every case, the Schmidt number exponent s was 0.5. Although these equations fit the data fairly well, there is a scatter of the individual points. Hence, for any particular solid-liquid combination at constant temperature in a geometrically similar series of vessels, the results could also be correlated by a direct proportionality between the mass-transfer coefficient and the impeller tip speed.

Mack and Marriner (M4) studied the solution of benzoic acid particles in two baffled tanks (9.7- and 16-in. diameter) using flat-blade turbines. They found that the rate of solution increased with roughly the 0.3 power of the power number ($Pg_c/N^3D^5\rho$) and the 0.6 power of the impeller Reynolds number. The solution rate was not affected by particle size in the range of approximately 0.02–0.3 in.

Oyama and Endoh (O12) studied the solution of sugar in water in 6.7- and 10.8-in. baffled vessels using paddles and flat-blade turbines. They report a mass-transfer coefficient which was proportional to the cube root of the particle diameter and to the cube root of the impeller power consumption per unit mass of agitated liquid.

Humphrey and Van Ness (H15) also studied the rate of solution of suspended crystals ($Na_2S_2O_3 \cdot 5H_2O$) in a continuous system, feeding salt and water and withdrawing solution. The mixing vessel was baffled, 1 ft. in diameter, and fitted with either a three-blade marine propeller or a six-flat-blade disk-turbine. The ratio of tank to impeller diameter was 3.0. The mass-transfer coefficients were correlated by a relation of the same form as Eq. (44), but with different values of Φ and r which varied with impeller type. Again, the Schmidt number exponent was 0.5 for both the propeller and the turbine. For the propeller, Φ was 0.47 and r was 0.58, while for the turbine Φ was 0.022 and r was 0.87. The ranges of the Reynolds numbers covered were 63,000–330,000 for the propeller, and 31,000–89,000 for the turbine. No trend was observed for the mass-transfer coefficient with varying particle size, in turbine runs with three size fractions in the 4-mesh to 12-mesh screen size range.

Johnson and Huang (J2) measured the rate of solution in an agitated vessel by a different technique, casting a ring of solid solute flush into the bottom of the tank. A small vessel (6-in. diameter) was used, fitted with baffles which could be removed, and agitated by a 3-in. diameter six-flat-blade disk-turbine. A variety of solid-solvent systems was used. For baffled operation, they also could correlate their observations by use of Eq. (44). In this case Φ was 0.092 and r was 0.71. The Schmidt number exponent was again found to be 0.50, and in this work the Schmidt number was varied over a wide range (208–14,800). The Reynolds numbers ranged from 2300–49,000. It is of interest that the final correlation of this rate is not very different from that for suspended crystals even though the physical situation of the dissolving material is quite different. This may be a result of a high degree of uniformity throughout these vessels.

The variation of the values of the constant and of the Reynolds number exponent in Eq. (44) from system to system shows that the over-all situation is more complex than is allowed for in the form of that equation. In the first place, it seems unlikely that one simple relation will cover laminar, transition, and all the possible turbulent flow states. In addition, a mixing vessel will generally have regions in it which have widely varying flow characteristics, influenced by the vessel geometry. Further, it seems reasonable to expect a complex relation between particle size, density, and loading and the flow properties in mixing vessels. These factors mitigate against the use of a solution-rate factor as a general characterization of agitation effectiveness, since the performance of a given agitator system may depend on the geometry and operating variables in an entirely different manner when it is used for operations other than dissolving.

VII. Heat Transfer

The approach to problems of heat transfer in agitated systems has been entirely analogous to that employed for heat transfer to fluids in pipes or in other physical situations. In studies of agitated systems, the general correlation technique has been to use a Nusselt number, an impeller Reynolds number, a Prandtl number, and a viscosity ratio in the usual heat-transfer relation:

$$\frac{hT}{k} = \Phi'(N_{Re})^{r'}(N_{Pr})^{p}(\mu/\mu_s)^{v} \qquad (45)$$

where
- h = surface heat transfer coefficient
- k = liquid thermal conductivity
- T = tank diameter
- N_{Re} = impeller Reynolds number, $D^2 N \rho/\mu$
- N_{Pr} = Prandtl number, $c'\mu/k$, with c' = heat capacity
- μ/μ_s = ratio of the viscosity of the bulk liquid to the liquid viscosity at the temperature of the heat transfer surface.

Some workers have modified this equation to include certain characteristic dimension ratios of the agitation system.

The results of the more extensive experimental investigations are outlined in Table VI. Except for scattered early data (P6, R2), the work of Chilton et al. (C5) was the first study to use the conventional groups for correlating heat transfer to jackets or coils. In some cases the temperatures at different points in the liquid varied by as much as 5°F. They used a variety of liquids, and found values of h in the range of 20–2000 B.T.U./(hr.)(ft.²)(°F.), depending on the conditions used.

TABLE VI
Heat Transfer to Agitated Systems

Investigators	Heat transfer surface	System geometry	Correlation	Range of Reynolds number
Chilton et al. (C5)	Jacket	1-ft. diameter vessel—no baffles. 2-blade flat paddle, 0.6-ft. diameter, near bottom.	$\frac{hT}{k} = 0.36(N_{Re})^{2/3}(N_{Pr})^{1/3}(\mu/\mu_s)^{0.14}$	$300-3 \times 10^5$
	Coil	0.8-ft. diameter coil of 0.5-in. tubing. Few tests on 5-ft. diameter vessel.	$\frac{hT}{k} = 0.87(N_{Re})^{0.62}(N_{Pr})^{1/3}(\mu/\mu_s)^{0.14}$	$300-3 \times 10^5$
Brown et al. (B9)	Jacket	5-ft. vessel with hemisphere bottom. No baffles. Propeller, $T/D = 2.5$. U-shape anchor, 1- and 5-in. clearance.	$\frac{hT}{k} = 0.55(N_{Re})^{0.67}(N_{Pr})^{0.25}(\mu/\mu_s)^{0.14}$	$5 \times 10^3 - 4 \times 10^4$
Rushton et al. (R15)	Vertical tubes	4-ft. vessel—4 ft. water depth. Mixing Equipment Co. flat blade turbines: 16 in., 6 blades, heating. 16 in., 6 blades, cooling. 12 in., 4 blades, heating. 12 in., 4 blades, cooling.	$h = 0.00285 N_{Re}$ $h = 0.00265 N_{Re}$ $h = 0.00235 (N_{Re})^{0.7}$ $h = 0.00220 (N_{Re})^{0.7}$	$1 \times 10^5 - 5 \times 10^5$
Cummings and West (C11)	Jacket	30-in. diameter vessel—no baffles. 12-in. impellers: curved-blade (2 on same shaft in some runs) and pitched-blade turbines.	$\frac{hT}{k} = 0.40(N_{Re})^{2/3}(N_{Pr})^{1/3}(\mu/\mu_s)^{0.14}$	$300-6 \times 10^5$
	Coil	2-ft. diameter coil of 1-in. diameter tubing.	$\frac{hT}{k} = 1.01(N_{Re})^{0.62}(N_{Pr})^{1/3}(\mu/\mu_s)^{0.14}$	$200-6 \times 10^5$
Kraussold (K9)	Jacket	40-in. diameter vessel—no baffles. Paddle, 20-in. diameter, 28.4 in. wide.	$\frac{hT}{k} = 0.36(N_{Re})^{2/3}(N_{Pr})^{1/3}(\mu/\mu_s)^{0.14}$	$1 \times 10^3 - 6 \times 10^5$
	Coil	App. 36-in. diameter coil of 1-in. tubing.	$\frac{hT}{k} = 0.87(N_{Re})^{0.62}(N_{Pr})^{1/3}(\mu/\mu_s)^{0.14}$	$4 \times 10^4 - 1 \times 10^5$
Dunlap and Rushton (D4)	Vertical tubes	2- and 4-ft. vessels. Mixing Equipment Co. flat blade turbine, $T/D = 3$-6.	$\frac{hd_e}{k} = 0.09(N_{Re})^{0.65}(N_{Pr})^{0.3}\left(\frac{D}{T}\right)^{0.33}(\mu/\mu_l)^{0.40}$	$10^3 - 10^6$
Oldshue and Gretton (O3)	Coil	4-ft. vessel with baffles. Mixing Equipment Co. flat blade turbines: 12–28-in. diameter. 3⁄4–28-in. diameter coil with 7⁄8- and 1.75-in. tubing.	$\frac{hd_e}{k} = 0.17(N_{Re})^{0.67}(N_{Pr})^{0.37}\left(\frac{D}{T}\right)^{0.1}\left(\frac{d_e}{T}\right)^{0.5}(\mu/\mu_s)^{\nu}$ $\nu = 0.7\mu^{-0.2}$	$400-1.5 \times 10^6$
Uhl (U1)	Jacket	2-ft. diameter vessel. Impellers: 14-inch paddle.	$\frac{hT}{k} = 0.42(N_{Re})^{2/3}(N_{Pr})^{1/3}(\mu/\mu_s)^{0.24}$	20–4000
		12-in. pitched-blade turbine.	$\frac{hT}{k} = 0.54(N_{Re})^{2/3}(N_{Pr})^{1/3}(\mu/\mu_s)^{0.24}$	20–200
		U-shape anchor, 7⁄8-in. clearance.	$\frac{hT}{k} = 1.0(N_{Re})^{1/2}(N_{Pr})^{1/3}(\mu/\mu_s)^{0.18}$	30–300
		Same as above.	$\frac{hT}{k} = 0.38(N_{Re})^{2/3}(N_{Pr})^{1/3}(\mu/\mu_s)^{0.18}$	300–4000
Pursell (P8)	Jacket	2-ft. diameter vessel. Paddles, 6–19-in. diameter. 3 in. wide.	$\frac{hT}{k} = 0.112(N_{Re})^{0.75}(N_{Pr})^{0.44}\left(\frac{T}{D}\right)^{0.4}\left(\frac{W}{D}\right)^{0.13}(\mu/\mu_s)^{0.25}$	$600-5.4 \times 10^5$

Rushton et al. (R15) gave the first data on the use of vertical coils to act as baffles as well as heat transfer surfaces. Using Mixing Equipment Company flat-blade turbines, they noted that the heat transfer was best when the turbine was midway between tank bottom and water surface, based on tests with the liquid depth equal to tank diameter. The entries in Table VI are for this condition. Values of heat transfer coefficients on the order of 300–1200 B.T.U./(hr.) (ft.^{2}) (°F.) were observed. Only one liquid was used and only a small Reynolds-number range was covered.

Cummings and West (C11) extended the earlier work of Chilton et al. to a larger vessel and some different impellers. They obtained essentially the same correlations and also showed that their correlation for helical coil heat transfer was adequate to correlate the data of Rushton et al. (R15) for vertical baffle-coils. With a few scattered data, Cummings and West also indicate that changing the direction of rotation of their pitched-blade turbine had no significant effect on the heat-transfer coefficients. Data by Kraussold (K9), from a vessel in which a paddle with a very high ratio of blade width to diameter was used, also confirm the correlations of Chilton et al.

Dunlap and Rushton (D4) extended the investigation of heat transfer from vertical tube-baffles in agitated vessels. Their final correlation used a Nusselt group based on the tube diameter and a viscosity ratio based on viscosity evaluated at the mean film temperature. The correlation also was expanded to include the effect of the ratio of impeller diameter to tank diameter. Oldshue and Gretton (O3) made an extensive study of heat transfer through helical coils in baffled systems. They also correlated the coefficient for heat transfer in a Nusselt group involving coil-tube diameter rather than vessel diameter. The correlation, somewhat more complicated than that of earlier workers, shows an effect of both impeller diameter and coil diameter besides those indicated by inclusion of these variables in the Reynolds and Nusselt groups, respectively. Although these diameters are expressed in terms of their ratios to tank diameter, the latter dimension was not varied in this study. The effect of viscosity ratio is considerably different from that of earlier investigators of agitated-vessel or pipe-flow heat transfer, and probably reflects a difference in flow patterns. Oldshue and Gretton also noted that variations in the position of the baffles (at the vessel wall, 1 in. off the wall, or inside the coil) had little effect on the heat transfer. There was also some indication that a wider tube spacing in the coil lowered the heat-transfer coefficient by a small amount. Thus, for the 1.75-in.-tubing coil, changing the turn spacing from 2 to 4 tube diameters reduced the heat transfer coefficient by 4% for water (0.4 cp. viscosity) and by 12% for oil (50 cp. viscosity).

Uhl (U1) extended the data for jacketed vessels into the low Reynolds number ranges commonly encountered with viscous materials. As seen in Table VI, his correlations follow the same general form as those of other workers except for a difference in the viscosity-ratio exponent and, for the anchor agitator, the change in Reynolds number exponent when the value of the Reynolds number is below 300. Again, an effect of agitator type is reflected in different values of the correlation parameters. Uhl also made some runs with baffles, using the paddle agitator, and found that the baffles had no appreciable effect in this range. On the basis of power measurements, Uhl concluded that power requirements would not have much influence on the choice of impeller type for heat transfer applications.

Pursell (P8) continued the study of paddle agitators, and developed a correlation given in Table VI. This equation fits his data in addition to the data obtained by previous investigators using paddle agitators. It will be noted that he has included terms to account for the effect of blade width W and the ratio of tank diameter to impeller diameter.

In a different kind of study, Hixson and Baum (H5) measured the heat-transfer coefficients between liquid and solid phases of the same material in unbaffled agitated vessels by putting relatively large frozen pieces (up to about 3 in.) into the liquid and observing the rate of melting. They correlated their results using the same groups as discussed previously:

$$\frac{hT}{k} = 0.83(N_{Re})^{0.63}(N_{Pr})^{0.5} \tag{46}$$

This correlation was found to be less valid at the end of the melting process, which suggests that particle size should be taken into account.

Pramuk and Westwater (P7) have shown that agitation increases the heat transfer for nucleate, transition, and film boiling. They point out that boiling itself creates a high degree of liquid agitation, hence extra mechanical agitation has to be fairly intense if its influence is to be observed. Under some conditions of agitation, the boiling heat flux can be increased by amounts of 25%–100% over the no-agitation values.

In summary, because of the uncertain effect of system geometry, it is recommended that for purposes of estimation or design Table VI should be used to select a correlation which was developed from tests on a system as similar as possible to the one under consideration. For jacketed vessels using paddle agitators the correlation of Pursell is indicated. For other jacketed vessels or for the case of helical coils relatively close to the vessel wall, the correlations of Cummings and West are indicated. For helical coils in baffled vessels with flat-blade turbines the equation of Oldshue and Gretton is recommended. For jacket heat transfer using anchor agitators in viscous materials, the correlation of Uhl is recommended.

VIII. Scale-Up of Heterogeneous Systems

A. Introduction

The subject of scale-up of agitated systems containing more than one phase has been deferred to this section because of the essential similarities in this regard among gas-liquid, liquid-liquid, and solid-liquid systems. Much of the material in Section II, dealing with the general characteristics of mixing operations, is of interest in this connection. Also, subsection III, B, concerned with one-liquid-phase systems, contains background material applicable to the scale-up of more complex systems. Studies of mixing systems have employed various types of performance parameters; their exact significances are not clear, and there are variations in their relations to various characteristics of the mixing systems. A natural result of this situation, with respect to scale-up, has been the wide variation in scaling relations for one-liquid-phase systems, as given in Table V. Heterogeneous agitated systems are complicated by the same factors which act in the single-liquid-phase case, in addition to others resulting from the presence of different phases. It is not surprising that scale-up of these systems cannot be reduced to simple and universally-applicable rules.

B. General Considerations

In broad terms scale-up is an engineering technique for translating performance in a "small" system to performance in a "large" system. A useful review of the formal material available on theories of models, similitude, dimensional analysis, etc., written from the chemical engineering point of view, is available in a recent book (J4). The practical applications of these theories involve the use of dimensionless groups, such as Reynolds number, in correlations which describe the performance of a system in terms equally applicable to "large" or "small" systems. This method of scale-up is familiar to all engineers.

Some particular aspects of mixing systems should be considered in relation to the general principles of model theory. First, if a general correlation is to be used for scale-up, it must be certain that the variables used in the correlation are the only ones acting in the given situation. Because of the complexity of mixing systems, we often lack knowledge of the proper characteristic variables or variable-groups to use in a "universal" correlation. For example, as discussed in Section III, it is not possible to present a dimensionless-group correlation which would account for the effect of impeller type on power requirements. There are very many possible groupings of variables that can be conjectured, and only actual tests can show if any particular group is of value.

Another complication in the use of generalized dimensionless correlations for scale-up of mixing systems lies in the difficulty of establishing an adequate performance parameter. In some cases there may be several different parameters, like conversion and purity, for example, or particle size and catalytic activity; the correlations between the different parameters and the agitation system properties may not be the same, and this may make the scale-up more difficult and more arbitrary.

Further complications arise in applying principles of similitude to heterogeneous systems, because of the impossibility of achieving absolute similitude. The physical characteristics of the material being agitated are not subject to independent variation, as is the size of the mixing vessel. In fact, in the engineering approach to scale-up, it is usually a prime objective to retain constant physical properties while the scale of the operation is changed. Since the performance of many heterogeneous agitated systems may depend on a relation between some characteristic length of the vessel-impeller system on the one hand, and some physical property like crystal particle size or droplet or bubble diameter on the other, great care must be used in "small-scale" studies of these systems.

C. Scale-Up Techniques

There are several different scale-up procedures which have been published by various workers in this field. One general recommendation is that *geometrical similarity* should be maintained. It will be remembered that geometrical similarity means that impeller type, vessel shape, baffling, etc., do not change with change in system capacity or, in other words, for any two geometrically-similar agitation systems there will be a constant ratio between any corresponding lengths. It may be frequently impossible, or at least economically undesirable, to satisfy this criterion of similarity in every respect. Experience and/or experiment must be relied on to establish what deviations might be allowed. The most important characteristics whose geometric similarity should be preserved appear to be impeller type, vessel shape, baffle arrangements, and, possibly less rigorously in some cases, tank-to-impeller diameter ratio, impeller height, impeller width/diameter ratio, and liquid depth. At any rate, the rest of this section will discuss other scale-up criteria on the basis that geometrical similarity between the "small" and "large" operation will be maintained.

A commonly-mentioned scale-up parameter is the *power input per unit volume* of material to be mixed, often expressed as HP./1000 gal. Hirsekorn and Miller (H2) and Miller and Mann (M10) recommend this based on relatively limited experiments (Sections V and VI); Chaddock (C3) recommends it although he gives no supporting data; one mixing equipment supplier (P5) recommends power per volume as a conservative scale-up parameter for "difficult" cases. The best-documented use of power per volume as a scaling parameter has been in gas-liquid systems carrying out fermentations or sulfite oxidations; in Eq. (30) power per unit volume at a fixed air velocity is a main factor affecting the value of the over-all average oxygen transfer coefficient. In liquid-liquid extraction, under certain conditions, Flynn and Treybal (F3) relate extraction efficiency for continuous operation to the power input per unit volume of feed stream. In some studies of solids suspension and rates of solution (O12, O13), the power input per unit mass of agitated liquid was related to the system performance.

It is difficult to predict from fundamental considerations that constant power per unit volume should be a generally significant scale-up criterion. In fact, as Rushton (R8) shows, use of equal power per volume for scaling can result in serious error in many cases. Thus, the successful application of this concept to certain operations must be regarded as a somewhat fortuitous result of the specific interactions present in those particular cases.

Another more general scale-up criterion is that of *impeller Reynolds number*, $D^2N\rho/\mu$. This approach, strongly recommended by Rushton (R8), is based on the observation that in many mixing operations the performance of the system can be successfully described by relations like Eq. (44) or (45), involving the familiar Nusselt or Sherwood numbers and the impeller Reynolds number. For geometrically-similar systems and constant material properties, these relations reduce to the forms

$$hD = \Gamma(D^2N)^r \quad (47)$$

$$K_sD = \Gamma'(D^2N)^{r'} \quad (48)$$

where D is the impeller diameter, N its rotational speed, h is a heat transfer coefficient, K_s is a mass transfer coefficient, and Γ, Γ', r, and r' are constants. It has been seen in the preceding chapters that equations of this form serve to correlate much of the published experimental data on mixing systems. Furthermore, if the assumptions are made that the impeller diameter (D) is truly the characteristic length parameter, that the impeller tip speed (ND) is truly the characteristic velocity parameter, and that no gravitational, surface, or other forces are important, then the principles of similitude would predict that equations of the above types could be used for correlation and scale-up.

The significance of these equations for scale-up is straightforward. When geometric similarity is maintained, the value of hD or K_sD in the "large" system will be the same as in the "small" one, if the impeller Reynolds number (proportional to ND^2) is the same. More generally, if the values of the constants Γ and r (or Γ' and r') are determined by experiment on the "small" scale, the equations can be used to predict the value of h or K_s on any other scale (providing the "other" scale operations are in the same range of Reynolds numbers as the "small" scale experiments). Also, since the power input for the "turbulent" systems to which the above equations generally apply (impeller Reynolds number above 10^3 to 10^4) can be expressed (for baffled systems, in particular) by

$$P \propto N^3 D^5 \qquad (49)$$

it is possible by simple algebraic manipulation to show, in terms of r (or r'), the relations between power required for the "large" and "small" operations for any selected ratios of the h (or K_s) values and D values. It should be noted that equal power per unit volume (proportional to P/D^3) will correspond to equal values of h (or K_s) only in the special case where r (or r') is equal to 0.75; published data show that this is not often the case.

Use of Reynolds number alone as a scale-up function, as just described, requires that the system performance be expressed in terms of a simple parameter like h or K_s. Also, the several assumptions implied by Eqs. (47) and (48) must be verified. Especially with heterogeneous systems, forces like interfacial tension or particle drag may influence performance enough to require correlation in terms of other groups. The engineer must rely on previously reported studies or on his own experiments to decide if this is the case. If equations like those above, or more complicated ones, are established, scale-up is readily accomplished. The major problem is not usually in the scale-up proper, but rather in obtaining a true relationship between the desired performance and the controlling variables or variable-groups. Experimentation for this purpose is discussed in the next section.

IX. Experimentation with Agitated Systems

A. General

The large body of published literature on properties of agitated systems contains valuable information with respect to the kinds of experimentation which ought to be carried out with these systems. In this section we will assemble some of this information about the general nature of experimental programs that will utilize properly the published experience, for objectives such as the design of a commercial plant, or the improvement of an existing process.

Consider now the general problem: a process is being developed, or possibly improved, and one or more steps involve operations which will be carried out in an agitated vessel. These may involve chemical reactions and heterogeneous systems or may appear relatively simple. It is now the duty of the engineer to perform a series of experiments (in fact or on paper) to define the relationships between performance and the several variables of the system. These relationships must be valid on the eventual scale of the commercial process as well as on the scale of the experimentation. In order for the experimental program to satisfy these objectives, it is suggested that the following three criteria be satisfied:

1. The performance parameters for the process must be established.
2. Similarity should be maintained as much as possible.
3. The nature and influence of the controlling variables or variable-groups must be determined.

These criteria will be discussed below; it will be clear that they are not absolute but rather a guide.

B. Performance Parameters

Any given mixing operation is carried out in order to accomplish some particular action. A primary object of a good experimental program is to establish clearly the nature of one or more parameters by which the degree of performance of this action can be expressed. The problem is sometimes simple; in the case of chemical reaction without any by-products, product yield is a reliable performance parameter. In more complex cases, both the amount of desired product and the extent and/or nature of other substances produced must be considered in assessing the system performance. Often physical considerations, like particle size of the material produced, must also be considered.

In dealing with complex systems, the experimenter must understand the nature of the materials being processed and also the underlying nature of the process requirements. Thus, it might turn out that several apparently different product parameters may all depend on a single property; drop size, for example. Expressing performance in terms of an interfacial mass-transfer coefficient, or in terms of one or two apparent reaction-rate constants, might simplify certain situations. If the experimenter is fortunate, he might discover an "equivalent process," easy to experiment with, whose results provide a suitable performance parameter for the actual system of interest. An example of this is the use of the oxygen mass-transfer coefficient in the oxidation of sodium sulfite as a performance parameter to describe the behavior of various fermentor systems for antibiotic production (Section IV).

There are no simple rules for the establishment of performance parameters. Preferably they should be as few in number as will be adequate to describe the system fully. In the general case, more than one parameter will be of concern and no single "process result" will be sufficient to characterize the operation.

C. Similarity

The preceding sections have clearly demonstrated the importance of maintaining similarity during any change of the scale of operation in a mixing system. In theory, this similarity has several aspects; geometric, kinematic, dynamic, etc., all of which should be satisfied. In practice it is frequently true (especially with chemically reacting systems) that the experimenter has not determined, at the outset, what variables or variable-groups will provide similarity on different scales of operation. There is, however, one kind of similarity which is relatively easy to maintain, and which the published studies in this field have suggested to be very important, namely, geometric similarity.

The provision of geometric similarity requires some knowledge of the nature of the full-scale agitation system. Even when there is considerable freedom in design of the large unit, it is likely in most cases that it will have a relatively conventional form. An interesting rule-of-thumb guide for estimating the probable nature of a mixing system is given in an article by Lyons and Parker (L4). Here, a selection chart defines the recommended impeller type and position, tank-to-impeller diameter ratio, and tank height/diameter ratio for nine different types of service. An important item to be considered in choosing the agitation system geometry is the general nature of the flow pattern desired; as was discussed in detail in Section II, the impeller type is of major importance in setting the general flow patterns.

In some cases it may be necessary to experiment with a variety of arrangements, preferably early in the development program, in order to fix the system geometry.

The problem of "scale-down" may occur also, in which the geometry of the full-scale system is fixed by reasons beyond the experimenter's control. In such cases it is equally important to maintain geometric similarity on the small scale, even if the system would appear to benefit by design changes.

Physically the maintenance of geometric similarity should be possible for all but the very smallest-scale experiments. Some reactor manufacturers make a line of geometrically-similar vessels with various capacities. Small accurately-scaled impellers of various types are available from mixer manufacturers. Their use is far more desirable than the use of locally fabricated equipment of a nonstandard design. Baffled vessels of conventional design, small enough to work with a few liters of material, can be readily fabricated from metal or plastic. "Resin" flasks, into which a "cage" holding baffles may easily be inserted, are useful as small glass reactor vessels. The interested experimenter should have little difficulty in devising means to meet the requirement of geometrical similarity.

D. Controlling Variables

When the point in a development program is reached at which the nature of the performance parameters and the agitation system geometry (at least with respect to impeller type) are fairly well established, it then becomes necessary to work out the relation between performance and the operating variables. For purposes of scale-up, it is desirable to know this relation in terms of dimensionless groups, equally applicable to large or small operations. The results of similar work that may have been published can be helpful here.

For almost every case, operation without a vortex is desirable, both because performance is generally better and because the absence of the vortex often eliminates a variable-group otherwise necessary to account for gravitational forces. As discussed in Sections II and III, operation with baffles, with particular propeller locations, or with completely-filled closed tanks will eliminate vortexing.

A useful generalization noted in the previous section is the widespread applicability of impeller Reynolds number for correlating performance data from different-scale operations in geometrically similar systems. In some heterogeneous systems, it may be necessary to modify the definitions of density and viscosity for use in this Reynolds number, and to introduce groups like the Weber number to account for interfacial forces (see Section V). The main point is that it requires experiment to establish finally the form of the controlling groups.

In the normal program, the physical properties of the materials do not change with the scale of the operation. In a given vessel then, the major agitation variables are the agitator speed N and the impeller diameter D. In the application of dimensionless groups to agitated systems, the characteristic velocity term has usually been ND (proportional to the impeller tip speed) and the characteristic length term has usually been the diameter D. Thus, even though the dimensionless group concept may be of great assistance, the experimental problem may still reduce to a measurement of the effects of variations in N and D. In some cases the ratio of impeller diameter to tank diameter will be important. If this is so, it may be necessary to experiment with different vessel diameters as well as with different impellers.

It is of interest here to discuss ideas proposed by Rushton and Oldshue (R12), on how to interpret observations of the influence of impeller diameter. Many mixing systems operate in an impeller Reynolds number range such that the power input P is proportional to N^3D^5:

$$P \propto N^3D^5 \qquad (49)$$

By analogy to hydraulics, this power input is also proportional to the product of the volumetric impeller discharge flow Q and a head $\mathcal{H}$. Thus,

$$P \propto Q\mathcal{H} \qquad (50)$$

The discharge is the product of a velocity (proportional to ND) and an area (proportional to D^2):

$$Q \propto ND^3 \qquad (51)$$

While this is dimensionally correct, there is not a great deal of published data to support it. We have seen that the mean fluid velocity near the impeller tip is related to, but less than, the impeller-tip velocity itself (Section II). Eqs. (49) and (50) together show that

$$Q\mathcal{H} \propto N^3D^5 \qquad (52)$$

Substitution for Q from Eq. (51) gives

$$\mathcal{H} \propto N^2D^2 \qquad (53)$$

Now, if power input is kept constant by appropriately changing N at the same time that D is varied, manipulation of these relations shows that

$$\left(\frac{Q}{\mathcal{H}}\right)_P \propto D^{8/3} \qquad (54)$$

Thus, Rushton and Oldshue suggest that varying the impeller diameter while keeping the power input constant is a method of evaluating the relative effects of flow and turbulence on the process.

They also suggest that the effect of power input itself be observed, before variations in impeller diameter are tried. As shown by Eq. (49), varying power for a fixed diameter is equivalent to varying N. In fact, if D is constant and N is changed, even though the power varies it can be seen from Eqs. (51) and (53) that

$$\left(\frac{Q}{\mathcal{H}}\right)_D \propto \frac{1}{N} \tag{55}$$

and so it could be said that increasing the agitator speed in a given vessel corresponds to increasing the proportion of the power input dissipated as turbulence.

These ideas of impeller flow, head and power input as related to operating variables have some merit for a qualitative description of the effects of the operating variables on the process. However, it requires extensive experience, and usually actual experiments, to decide whether a system performance is favored by a particular combination of "flow" and "head." (Rushton and Oldshue (R12) note that high values of $Q/\mathcal{H}$ are preferred for blending and solid suspension, low ratios for liquid-liquid and gas-liquid operations.) This approach still requires the systematic study of impeller speed and diameter as process variables.

For most experiments, there seems to be no particular advantage in restricting the changes in system variables so that system performances can always be compared at equal values of power input; in many cases there is not a clear relation between power input and performance. The major controllable variables are such factors as impeller and vessel diameter, and impeller speed; it is the relation between these and system performance which must be discovered. Eventually, the power requirement becomes a factor in the usual engineering balance of cost versus level of performance. The experience of manufacturers of mixing equipment can be helpful when it is desired to know the power requirements for a system operating under specified conditions of size, impeller speed, etc.

The preceding paragraphs have not considered the possible influences of geometric factors like impeller height, liquid depth, etc. As a starting point it seems reasonable to use the widely adopted "square" batch, with liquid height equal to vessel diameter, and to set the impeller one impeller-diameter above the vessel bottom. Later, one should check experimentally to see if any of these factors is influencing the process performance.

Finally, it should be noted that external variables like gas rate, volume-ratio of different phases, rate of throughput with continuous processes, etc., may need to be considered in a particular process. Other processing variables, like temperature, reactant ratios, etc., may not be directly connected with the nature of the agitation, but will often have major significance in determining the performance of the process and may thus interact with some of the agitation variables.

E. Discussion

The preceding sections have indicated the requirement for a considerable degree of complexity in a thorough development program involving an agitated system. The extent of the actual program in any particular case will be the result of compromise between cost, available time, nature of the dependence between performance and system variables, and the many other factors which influence the course of any development program. The number of performance parameters used to describe the system may have to be restricted to one or two. Previously published studies may help to eliminate various factors from consideration as variables in the physical design of the system or from consideration as factors in correlations of system performance. Experience, either with the particular process or with agitated systems in general, may be useful for reducing the extent of the required experimentation.

It might be useful in cases where development programs involve systems of roughly similar properties to consider the standardization of most of the geometric properties of the agitation system. This could provide a way to apply previous experience to a given process with a greater degree of confidence than would be the case if the previous experience were gained with a widely-different mixer geometry.

Finally, it is important that an experimental system be designed so that the impeller and impeller speed, and possibly other items, can be conveniently changed. This may complicate the mechanical engineering problem and increase the first cost, but in the long run it will be advantageous.

Acknowledgment

The writer prepared this review as a member of the Engineering Research Section, Central Research Division, American Cyanamid Company, whose permission to publish this work is gratefully acknowledged.

Nomenclature

Any consistent set of units may be used. For the few equations which are not dimensionless, the appropriate units are given at the place where they are introduced in the text.

a	Constant in Eq. (8) and Table II	c	Exponent in Eq. (40)
a_1	Exponent in Eq. (17)	c'	Heat capacity
a_2	Exponent in Eq. (24)	d_p	Particle diameter
A	Interfacial area per volume of two-phase mixture	d_c	Heat transfer tube diameter
		$\mathfrak{D}$	Molecular diffusivity
b	Constant in Eq. (8) and Table II	D	Impeller diameter
B	Number of blades on an impeller	D_j	Jet diameter at nozzle

E	Fractional stage efficiency for extraction	S	Propeller pitch
g	Acceleration of gravity	t	Exponent in Eq. (43)
g_c	Conversion factor for consistency of units	T	Tank diameter
		u_j	Velocity of jet at the nozzle
h	Heat transfer coefficient	U_o	Superficial gas velocity
(hp.)	Agitator horsepower input per 1000 gallons of mixture	v	Exponent in Eq. (45)
		V	Volume of liquid in tank
$\mathcal{K}$	Impeller discharge head	w	Exponent in Eq. (11)
H	Height of liquid in a vessel	w'	Exponent in Eq. (13)
H_j	Height of liquid above jet nozzle	W	Width of impeller blade
(HP.)	Horsepower delivered to the impeller	x	Distance from jet nozzle
		$\propto$	Symbol for "is proportional to"
k	Thermal conductivity	β	Inclination angle for jet nozzle
K	Constant in Eq. (7)	γ	Angle equal to $\beta + 5$ degrees
K_j	Constant in Eq. (23)	Γ	Constant in Eq. (47)
K_v	Absorption coefficient per volume of liquid	Γ'	Constant in Eq. (48)
		Δ	Symbol for "difference"
K_s	Mass-transfer coefficient for dissolving	θ	"Mixing time" in a one-liquid-phase system
K_1	Constant in Eq. (17)	θ_c	Gas contact time per unit height of liquid
K_2	Constant in Eq. (24)		
K_3	Constant in Eq. (27)	θ_s	Emulsion settling time
L	Length of impeller blade	θ_{s0}	"Standard" settling time in Eq. (33)
m	Exponent in Eq. (7)		
M	Mass of liquid in tank	μ	Viscosity
n	Exponent in Eq. (7)	μ_c	Viscosity of the continuous phase
N	Impeller speed, revolutions per unit time	μ_d	Viscosity of the dispersed phase
		μ_e	Effective mean viscosity defined by Eq. (35)
N_c	Critical impeller speed for solids suspension	μ_e'	Effective mean viscosity defined by Eq. (36)
N_P	Power number, $Pg_c/\rho N^3 D^5$	μ_e''	Effective mean viscosity defined by Eq. (37)
N_{Re}	Impeller Reynolds number, $D^2 N \rho / \mu$		
		μ_e'''	Effective mean viscosity defined by Eq. (38)
N_{Pr}	Prandtl number, $c'\mu/k$		
N_{Sc}	Schmidt number, $\mu/\rho \mathfrak{D}$	μ_f	Viscosity at a mean film temperature
p	Exponent in Eq. (45)		
P	Power input to impeller	μ_o	Viscosity of oil in an oil-water dispersion
P_c	Critical power input for solids suspension		
		μ_s	Viscosity at the heat transfer surface temperature
Q	Volumetric discharge rate from impeller		
		μ_w	Viscosity of water in an oil-water dispersion
Q_g	Volumetric rate of gas flow		
Q_i	Volumetric rate of fluid induction by jet	ν_c	Kinematic viscosity of the continuous phase
Q_j	Volumetric rate of jet flow at the nozzle	ν_d	Kinematic viscosity of the dispersed phase
r	Exponent in Eq. (44)	ρ	Density
r'	Exponent in Eq. (45)	ρ'	Density of cold fluid in Eq. (18)
R	Ratio by weight of solids to liquid	ρ_a	Average density defined by Eq. (39)
s	Exponent in Eq. (44)		

ρ_c Density of the continuous phase
ρ_d Density of the dispersed phase
ρ_e Effective mean density defined by Eq. (32)
ρ_s Density of solid particles
σ Interfacial tension
ϕ Volume fraction
ϕ_c Volume fraction of the continuous phase
ϕ_d Volume fraction of the dispersed phase
ϕ_o Volume fraction of oil in an oil-water dispersion
ϕ_w Volume fraction of water in an oil-water dispersion
Φ Constant in Eq. (44)
Φ' Constant in Eq. (45)
ψ Constant in Eq. (43)

References

A1. Acrivos, A., and Chambré, P. L., *Ind. Eng. Chem.* **49**, 1025 (1957).
A2. Aiba, S., *Chem. Eng. (Tokyo)* **20**, 280 (1956).
A3. Aiba, S., *Chem. Eng. (Tokyo)* **20**, 288 (1956).
A4. Aiba, S., *Chem. Eng. (Tokyo)* **21**, 139 (1957).
A5. Alexander, L. G., Baron, T., and Comings, E. W., *Univ. Illinois Eng. Expt. Sta. Bull.* **No. 413**, 1953.
A6. Auro, M. A., Hodge, H. M., and Roth, N. G., *Ind. Eng. Chem.* **49**, 1237 (1957).
B1. Baranaev, M. K., Teverovskii, E. N., and Tregubova, E. L., *Doklady Akad. Nauk S.S.S.R.* **66**, 821 (1949).
B2. Bartholomew, W. H., Karow, E. O., Sfat, M. R., and Wilhelm, R. H., *Ind. Eng. Chem.* **42**, 1801, 1810 (1950).
B3. Beerbower, A., Forster, E. O., Kolfenbach, J. J., and Vesterdal, H. G., *Ind. Eng. Chem.* **49**, 1075 (1957).
B4. Bissell, E. S., *Ind. Eng. Chem.* **30**, 493 (1938).
B5. Bissell, E. S., *Ind. Eng. Chem.* **36**, 497 (1944).
B6. Bissell, E. S., Everett, H. J., and Rushton, J. H., *Chem. Eng. Progr.* **43**, 649 (1947).
B7. Bissell, E. S., Miller, F. D., and Everett, H. J., *Ind. Eng. Chem.* **37**, 426 (1945).
B8. Bissell, E. S., Hesse, H. C., Everett, H. J., and Rushton, J. H., *Chem. Eng. Progr.* **43**, 649 (1947).
B9. Brown, R. W., Scott, M. A., and Toyne, C., *Trans. Inst. Chem. Engrs. (London)* **25**, 181 (1947).
B10. Brumagin, I. S., *Chem. & Met. Eng.* **53**, No. 4, 110 (1946).
C1. Calderbank, P. H., *Trans. Inst. Chem. Engrs. (London)* **36**, 443 (1958).
C2. Carpani, R. E., and Roxburgh, J. M., *Can. J. Chem. Eng.* **36**, 73 (1958).
C3. Chaddock, R. E., *Chem. & Met. Eng.* **53**, No. 11, 151 (1946).
C4. Chain, E. B., Paladino, S., Callow, D. S., Ugolini, F., and Van der Sluis, J., *Bull. World Health Organization* **6**, 73 (1952).
C5. Chilton, T. H., Drew, T. B., and Jebens, R. H., *Ind. Eng. Chem.* **36**, 510 (1944).
C6. Clay, P. H., *Proc. Acad. Sci. Amsterdam* **43**, 852, 979 (1940).
C7. Cooper, C. M., Fernstrom, G. A., and Miller, S. A., *Ind. Eng. Chem.* **36**, 504 (1944).
C8. Corrsin, S., *A.I.Ch.E. Journal* **3**, 329 (1957).
C9. Corrsin, S., *Phys. of Fluids* **1**, 42 (1958).
C10. Couture, J. W., *Pulp Paper Mag. Can.* **46**, 765 (1945).
C11. Cummings, G. S., and West, A. S., *Ind. Eng. Chem.* **42**, 2303 (1950).
D1. Danckwerts, P. V., *Appl. Sci. Research* **A3**, 279 (1952).
D2. Danckwerts, P. V., *Research (London)* **6**, 355 (1953).
D3. Dunkley, W. L., and Perry, R. L., *J. Dairy Sci.* **40**, 1165 (1957).

MIXING AND AGITATION

D4. Dunlap, I. R., and Rushton, J. H., *Chem. Eng. Progr.* **49**, *Symposium Ser. No. 5*, 137 (1953).
E1. Elsworth, R., Williams, V., and Harris-Smith, R., *J. Appl. Chem.* (*London*) **7**, 261 (1957).
F1. Fick, J. L., Rea, H. E., and Vermeulen, T., Univ. California Radiation Lab. Rept. No. 2545 (1954).
F2. Finn, R. K., *Bacteriol. Revs.* **18**, 254 (1954).
F3. Flynn, A. W., and Treybal, R. E., *A.I.Ch.E. Journal* **1**, 324 (1955).
F4. Folsom, R. G., and Ferguson, C. K., *Trans. A.S.M.E.* **71**, 73 (1949).
F5. Forstall, W., and Gaylord, E. W., *J. Appl. Mechanics* **22**, 161 (1955).
F6. Fossett, H., *Trans. Inst. Chem. Engrs.* (*London*) **29**, 322 (1951).
F7. Fossett, H., and Prosser, L. E., *Proc. Inst. Mech. Engrs.* (*London*) **160**, 224 (1949).
F8. Foust, H. C., Mack, D. E., and Rushton, J. H., *Ind. Eng. Chem.* **36**, 517 (1944).
F9. Fox, E. A., and Gex, V. E., *A.I.Ch.E. Journal* **2**, 539 (1956).
F10. Friedland, W. C., Peterson, M. H., and Sylvester, J. C., *Ind. Eng. Chem.* **48**, 2180 (1956).
F11. Friedlander, S. K., *A.I.Ch.E. Journal* **3**, 381 (1957).
F12. Friedman, A. M., and Lightfoot, E. N., Jr., *Ind. Eng. Chem.* **49**, 1227 (1957).
G1. Graybeal, P. E., and Bechtel, R. J., *Ind. Eng. Chem.* **49**, No. 3, 42A (1957).
G2. Green, S. J., *Trans. Inst. Chem. Engrs.* (*London*) **31**, 327 (1953).
G3. Gunness, R. C., and Baker, J. G., *Ind. Eng. Chem.* **30**, 497 (1938).
H1. Hinze, J. O., *A.I.Ch.E. Journal* **1**, 289 (1955).
H2. Hirsekorn, F. S., and Miller, S. A., *Chem. Eng. Progr.* **49**, 459 (1953).
H3. Hixson, A. W., *Ind. Eng. Chem.* **36**, 488 (1944).
H4. Hixson, A. W., and Baum, S. J., *Ind. Eng. Chem.* **33**, 478 (1941).
H5. Hixson, A. W., and Baum, S. J., *Ind. Eng. Chem.* **33**, 1433 (1941).
H6. Hixson, A. W., and Baum, S. J., *Ind. Eng. Chem.* **34**, 120 (1942).
H7. Hixson, A. W., and Baum, S. J., *Ind. Eng. Chem.* **34**, 194 (1942).
H8. Hixson, A. W., and Crowell, S. H., *Ind. Eng. Chem.* **23**, 923, 1002, 1160 (1931).
H9. Hixson, A. W., and Luedecke, V. D., *Ind. Eng. Chem.* **29**, 927 (1937).
H10. Hixson, A. W., and Tenney, A. H., *Trans. A.I.Ch.E.* **31**, 113 (1935).
H11. Hixson, A. W., and Wilkins, G. A., *Ind. Eng. Chem.* **25**, 1196 (1933).
H12. Hoffman, A. N., Montgomery, J. B., and Moore, J. K., *Ind. Eng. Chem.* **41**, 1683 (1949).
H13. Hooker, T., *Chem. Eng. Progr.* **44**, 833 (1948).
H14. Hughes, R. R., *Ind. Eng. Chem.* **49**, 947 (1957).
H15. Humphrey, D. W., and Van Ness, H. C., *A.I.Ch.E. Journal* **3**, 283 (1957).
H16. Hunsaker, J. C., and Rightmire, B. G., "Engineering Applications of Fluid Mechanics." McGraw-Hill, New York, 1947.
J1. John, G., *Chem.-Ing.-Tech.* **30**, 529 (1958).
J2. Johnson, A. I., and Huang, G.-J., *A.I.Ch.E. Journal* **2**, 412 (1956).
J3. Johnson, D. L., Saito, H., Polejes, J., and Hougen, O. H., *A.I.Ch.E. Journal* **3**, 411 (1957).
J4. Johnstone, R. E., and Thring, M. W., "Pilot Plants, Models, and Scale-Up Methods in Chemical Engineering." McGraw-Hill, New York, 1957.
K1. Kalinske, A. A., *Sewage and Ind. Wastes* **27**, 572 (1955).
K2. Karow, E. O., Bartholomew, W. H., and Sfat, M. R., *J. Agr. Food Chem.* **1**, 302 (1953).

K3. Karr, A. E., and Scheibel, E. G., *Chem. Eng. Progr.* **50**, *Symposium Ser. No. 10,* 73 (1954).
K4. Kauffman, H. L., *Chem. & Met. Eng.* **37**, 178 (1930).
K5. Keon, J. J., *Pulp Paper Mag. Can.* **47**, 188 (1946).
K6. Kneule, F., *Chem.-Ing.-Tech.* **30**, 529 (1958).
K7. Kolmogoroff, A. N., *Doklady Akad. Nauk S.S.S.R.* **66**, 825 (1949).
K8. Kramers, H., Baars, G. M., and Knoll, W. H., *Chem. Eng. Sci.* **2**, 35 (1953).
K9. Kraussold, H., *Chem.-Ing.-Tech.* **23**, 177 (1951).
K10. Krzywoblocki, M. Z., *Jet Propulsion* **26**, 760 (1956).
L1. Laity, D. S., and Treybal, R. E., *A.I.Ch.E. Journal* **3**, 176 (1957).
L2. Lamont, A. G. W., *Can. J. Chem. Eng.* **36**, 153 (1958).
L3. Lyons, E. J., *Chem. Eng. Progr.* **44**, 341 (1948).
L4. Lyons, E. J., and Parker, N. H., *Chem. Eng. Progr.* **50**, 629 (1954).
M1. Mack, D. E., *Chem. Eng.* **58**, No. 3, 109 (1951).
M2. Mack, D. E., and Kroll, A. E., *Chem. Eng. Progr.* **44**, 189 (1948).
M3. Mack, D. E., and Uhl, V. W., *Chem. Eng.* **54**, No. 10, 115 (1947).
M4. Mack, E. M., and Marriner, R. E., *Chem. Eng. Progr.* **45**, 545 (1949).
M5. MacLean, G., and Lyons, E. J., *Ind. Eng. Chem.* **30**, 489 (1938).
M6. Martin, J. J., *Trans. A.I.Ch.E.* **42**, 777 (1946).
M7. Meny, R. B., and Velykis, R. B., *Oil Gas J.* **55**, No. 43, 88 (1957).
M8. Metzner, A. B., *Advances in Chem. Eng.* **1**, 77 (1956).
M9. Miller, F. D., and Rushton, J. H., *Ind. Eng. Chem.* **36**, 499 (1944).
M10. Miller, S. A., and Mann, C. A., *Trans. A.I.Ch.E.* **40**, 709 (1944).
M11. Möhle, W., and Waeser, B., *Chem.-Ing.-Tech.* **24**, 494 (1952).
M12. Morton, A. A., and Redman, L. M., *Ind. Eng. Chem.* **40**, 1190 (1948).
M13. Murphree, E. V., *Ind. Eng. Chem.* **15**, 148 (1923).
N1. Nagata, S., and Yokoyama, T., *Chem. Eng. (Tokyo)* **20**, 272 (1956).
N2. Nagata, S., Yanagimoto, M., and Yokoyama, T., *Chem. Eng. (Tokyo)* **21**, 278 (1957).
N3. Nelson, H. A., Mason, W. D., and Elferdink, T. H., *Ind. Eng. Chem.* **48**, 2183 (1956).
N4. Newitt, D. M., Shipp, G. C., and Black, C. R., *Trans. Inst. Chem. Engrs. (London)* **29**, 278 (1951).
O1. O'Connell, F. R., and Mack, D. E., *Chem. Eng. Progr.* **46**, 358 (1950).
O2. Oldshue, J. Y., *Ind. Eng. Chem.* **48**, 2194 (1956).
O3. Oldshue, J. Y., and Gretton, A. T., *Chem. Eng. Progr.* **50**, 615 (1954).
O4. Oldshue, J. Y., and Gretton, A. T., *Tappi* **39**, 378 (1956).
O5. Oldshue, J. Y., and Rushton, J. H., *Chem. Eng. Progr.* **48**, 297 (1952).
O6. Oldshue, J. Y., Hirschland, H. E., and Gretton, A. T., *Chem. Eng. Progr.* **52**, 481 (1956).
O7. Olney, R. B., and Carlson, C. J., *Chem. Eng. Progr.* **43**, 473 (1947).
O8. Overcashier, R. H., Kingsley, H. A., and Olney, R. B., *A.I.Ch.E. Journal* **2**, 529 (1956).
O9. Oyama, Y., and Aiba, S., *Chem. Eng. (Tokyo)* **15**, 367 (1951).
O10. Oyama, Y., and Aiba, S., *J. Sci. Research Inst. (Tokyo)* **46**, 172 (1952).
O11. Oyama, Y., and Endoh, K., *Chem. Eng. (Tokyo)* **19**, 2 (1955).
O12. Oyama, Y., and Endoh, K., *Chem. Eng. (Tokyo)* **20**, 576 (1956).
O13. Oyama, Y., and Endoh, K., *Chem. Eng. (Toyko)* **20**, 666 (1956).
P1. Pai, S., "Fluid Dynamics of Jets." Van Nostrand, New York, 1954.

P2. Pavlushenko, I. S., Kostin, N. M., and Matveev, S. F., *Zhur. Priklad. Khim.* **30**, 1160 (1957).
P3. Perry, J. H., ed., "Chemical Engineers' Handbook." McGraw-Hill, New York, 1950.
P4. Perry, R. L., and Dunkley, W. L., *J. Dairy Sci.* **40**, 1152 (1957).
P5. Philadelphia Gear Works, Inc., Mixer Catalog A-27. Philadelphia, Pennsylvania.
P6. Pierce, D. E., and Terry, P. B., *Chem. Met. Eng.* **30**, 872 (1924).
P7. Pramuk, F. S., and Westwater, J. W., *Chem. Eng. Progr.* **52**, *Symposium Ser. No. 18,* 79 (1956).
P8. Pursell, H., M.S. Thesis, Newark College of Engineering, Newark, New Jersey, 1954.
R1. Reavell, B. N., *Trans. Inst. Chem. Engrs. (London)* **29**, 301 (1951).
R2. Rhodes, F. H., *Ind. Eng. Chem.* **26**, 944 (1934).
R3. Rhodes, R. P., and Gaden, E. L., Jr., *Ind. Eng. Chem.* **49**, 1233 (1957).
R4. Riegel, E. R., "Chemical Machinery." Reinhold, New York, 1944.
R5. Roberts, A. G., *British Coal Utilisation Research Association Bulletin* **20**, No. 5, 189 (1956).
R6. Rodger, W. H., Trice, V. G., and Rushton, J. H., *Chem. Eng. Progr.* **52**, 515 (1956).
R7. Rushton, J. H., *Ind. Eng. Chem.* **37**, 422 (1945).
R8. Rushton, J. H., *Chem. Eng. Progr.* **47**, 485 (1951).
R9. Rushton, J. H., *Chem. Eng. Progr.* **50**, 587 (1954).
R10. Rushton, J. H., *Petrol. Refiner* **33**, No. 8, 101 (1954).
R11. Rushton, J. H., and Mahoney, L. H., *J. Metals* **6**, 1199 (1954).
R12. Rushton, J. H., and Oldshue, J. Y., *Chem. Eng. Progr.* **49**, 161, 267 (1953).
R13. Rushton, J. H., Costich, E. W., and Everett, H. J., *Chem. Eng. Progr.* **46**, 395, 467 (1950).
R14. Rushton, J. H., Gallagher, J. B., and Oldshue, J. Y., *Chem. Eng. Progr.* **52**, 319 (1956).
R15. Rushton, J. H., Lichtman, R. S., and Mahoney, L. H., *Ind. Eng. Chem.* **40**, 1082 (1948).
R16. Rushton, J. H., Mack, D. E., and Everett, H. J., *Trans. A.I.Ch.E.* **42**, 441 (1946).
S1. Sachs, J. P., Ph.D. Thesis, Illinois Inst. Technol., Chicago, 1952.
S2. Sachs, J. P., and Rushton, J. H., *Chem. Eng. Progr.* **50**, 597 (1954).
S3. Schultz, J. S., and Gaden, E. L., Jr., *Ind. Eng. Chem.* **48**, 2209 (1956).
S4. Shanahan, C. E. A., and Cooke, F., *J. Appl. Chem. (London)* **7**, 645 (1957).
S5. Sherwood, T. K., and Pigford, R. L., "Absorption and Extraction." McGraw-Hill, New York, 1952.
S6. Shu, P., *Ind. Eng. Chem.* **48**, 2204 (1956).
S7. Snyder, J. R., Hagerty, P. F., and Molstad, M. C., *Ind. Eng. Chem.* **49**, 689 (1957).
S8. Stoops, C. E., and Lovell, C. L., *Ind. Eng. Chem.* **35**, 845 (1943).
T1. Taylor, J. S., M.S. Thesis, University of Delaware, Newark, Delaware, 1955.
T2. Tennant, B. W., M.S. Thesis, Illinois Inst. Technol., Chicago, 1952.
T3. Tereshkevitch, W., Sc.D. Thesis, Massachusetts Inst. Technol., Cambridge, Massachusetts, 1956.
T4. Treybal, R. E., "Liquid Extraction." McGraw-Hill, New York, 1951.

T5. Treybal, R. E., *A.I.Ch.E. Journal* **4**, 202 (1958).
T6. Trice, V. G., and Rodger, W. A., *A.I.Ch.E. Journal* **2**, 205 (1956).
U1. Uhl, V. W., *Chem. Eng. Progr.* **51**, *Symposium No. 17*, 93 (1955).
V1. van de Vusse, J. G., *Chem. Eng. Sci.* **4**, 178, 209 (1955).
V2. Vermeulen, T., Williams, C. M., and Langlois, G. E., *Chem. Eng. Progr.* **51**, 85 (1955).
W1. Wegrich, O. G., and Shurter, R. A., *Ind. Eng. Chem.* **45**, 1153 (1953).
W2. Weidenbaum, S. S., *Advances in Chem. Eng.* **2**, 249 (1958).
W3. White, A. M., and Sumerford, S. D., *Ind. Eng. Chem.* **25**, 1025 (1933).
W4. White, A. M., and Sumerford, S. D., *Chem. & Met. Eng.* **43**, 370 (1936).
W5. White, A. M., Brenner, E., Phillips, G. A., and Morrison, M. S., *Trans. A.I.Ch.E.* **30**, 570, 585 (1934).
W6. White, A. M., Sumerford, S. D., Bryant, E. O., and Lukens, B. E., *Ind. Eng. Chem.* **24**, 1160 (1932).
W7. Wilhelm, R. H., Conklin, L. H., and Sauer, T. C., *Ind. Eng. Chem.* **33**, 453 (1941).
W8. Wingard, R. E., Vinyard, M. N., and Craine, C. M., Jr., *Alabama Polytech. Inst. Eng. Expt. Sta. Bull. No. 17*, 3 (1952).
W9. Winter, E. F., and Deterding, J. H., *Brit. J. Appl. Phys.* **7**, 247 (1956).
W10. Wood, J. C., Whittemore, E. R., and Badger, W. L., *Trans. A.I.Ch.E.* **14**, 435 (1922).
Y1. Yamamoto, K., and Kawahigasi, Z., *Chem. Eng. (Tokyo)* **20**, 686 (1965).
Z1. Zwietering, T. N., *Chem. Eng. Sci.* **8**, 244 (1958).

DESIGN OF PACKED CATALYTIC REACTORS

John Beek

Shell Development Company, Emeryville, California

I. Introduction	204
II. Reduction of Chemical and Rate Equations to an Independent Set	205
A. Definition of Stoichiometric Matrix and Submatrices	205
B. Relations between Conversions and Concentrations	207
C. Calculation of Additive Properties	209
D. Introduction of Virtual Conversions	209
E. Illustration	210
F. Exceptional Cases	210
III. Equations Describing Simultaneous Reaction and Transport Processes	211
A. Statement of Assumptions	211
B. Derivation of Conservation Equations	214
C. Adiabatic Reactor	222
D. The One-Dimensional Approximation	222
E. Calculation of the Pressure	223
IV. Estimation of Transport Properties	224
A. Introductory Discussion	224
B. Velocity Profile	225
C. Eddy Diffusivity	227
D. Other Contributions to Effective Thermal Conductivity	229
E. Heat-Transfer Coefficient at the Wall	232
F. Friction Factor	234
V. Numerical Solution of Equations	235
A. Automatic Computer Assumed	235
B. One-Dimensional Approximation	235
C. Sources of Error	236
D. Stability of Partial Difference Equations	240
E. Introduction of r^2 as Radial Variable	240
F. Explicit Partial Difference Equations	241
G. Implicit Partial Difference Equations	244
H. Adiabatic Reactor	249
I. Illustrations	250
VI. Stability of Packed Tubular Reactors	257
A. Statement of Problem	257
B. Description of Barkelew's Criterion	258
VII. Scale Models of Packed Tubular Reactors	259
A. General Discussion	259
B. Derivation of Requirements for the General Case	261
C. Selection of Important Quantities	262

D. Limitations	265
E. The One-Dimensional Approximation	265
F. Example	267
Nomenclature	268
References	269

I. Introduction

It is understood by practicing chemical engineers that the design of complicated pieces of equipment has not been reduced to a straightforward operation: there is no formula into which we can substitute the required conditions to give explicitly the corresponding design parameters. The design process involves, rather, the calculation of the conditions that would be realized in the equipment with the design parameters fixed at various sets of values, and then selecting a suitable set of parameters for the design.

It is the purpose of this chapter to discuss presently known methods for predicting the performance of nonisothermal continuous catalytic reactors, and to point out some of the problems that remain to be solved before a complete description of such reactors can be worked out. Most attention will be given to packed catalytic reactors of the heat-exchanger type, in which a major requirement is that enough heat be transferred to control the temperature within permissible limits. This choice is justified by the observation that adiabatic catalytic reactors can be treated almost as special cases of packed tubular reactors. There will be no discussion of reactors in which velocities are high enough to make kinetic energy important, or in which the flow pattern is determined critically by acceleration effects.

In order to provide a logical framework for the discussion, an example will be worked out in enough detail to illustrate the methods used. There is no question of providing a recipe for designing nonisothermal reactors; methods of working that are useful will be presented, and their application to a more or less typical problem will be described.

It will be supposed that the kinetics of all the reactions that are going on and the thermodynamical and molecular transport properties of all the substances present are known, and that it is desired to find out how the composition of the effluent from a reactor depends on the conditions that are imposed. The conditions that must be fixed are the composition, pressure, temperature, and flow rate of the reactant mixture, the dimensions of the reactor and of the catalyst pellets, and enough properties of the heat-transfer medium to determine a relation between the temperature of the tube wall and the heat flux through it.

Only systems in a quasi-steady state will be considered; that is, the discussion will not include transient states in which the change is fast enough to give the heat capacity of the system any importance, but will apply to those systems in which a progressive deterioration of the catalyst makes changes that become significant only after several hours. Some attention will be given to a cooperative property of the system that Amundson has called parametric sensitivity, but this has reference to the magnitude of differences between two steady states corresponding to some difference in the conditions.

The discussion will also be restricted to reactors in which the range of temperature is too wide to permit the use of an average temperature to characterize the whole catalyst bed. No sharp line can be drawn following this restriction, but it clearly includes any case in which the rate of reaction cannot be described satisfactorily as a linear function of temperature over the whole range covered by the reactor. The only way to find out in doubtful cases whether the variation of temperature is significant is to make the calculation taking the variation into account.

With the above restriction, an explicit representation of the output of a reactor as a tabulated function of the input conditions is out of the question. In most cases, the output must be calculated by numerical methods. This means that the conditions within the reactor must be calculated stepwise, starting with the known conditions at the inlet. This process may be thought of as finding a numerical solution for a set of differential equations, or, following the suggestion of Deans and Lapidus (D3), as solving a set of difference equations, moving along the reactor by steps of a particle diameter. The important feature that all such processes have in common is that, in order to find out what comes out of the reactor, it is necessary to find out what the conditions are everywhere in the reactor. The first problem is to formulate the equations that are used to get this information. Toward this end, the equations that describe conditions within the reactor are formulated, suggestions are made for evaluating the transport properties that appear in the equations, and then methods for solving the equations are given. Finally, approximate methods are presented for evaluating the parametric sensitivity of reactors, and for changing the scale of a reactor without changing the course of the reaction.

II. Reduction of Chemical and Rate Equations to an Independent Set

A. DEFINITION OF THE STOICHIOMETRIC MATRIX AND SUBMATRICES

1. *The Complete Matrix*

It is not unusual to encounter a reacting system in which a given set of products may be obtained by more than one reaction path. In a simple case there is no difficulty in writing down the stoichiometric relations, but in complicated situations, such as the catalytic reforming of naphthas, it is useful to have a systematic scheme for relating the concentrations of the various substances to the concentrations of a minimum number of key substances. The scheme employed here will take advantage of the compact language of matrix mathematics (A1, H4). An algebraic treatment of multiple-reaction problems has been given by Brötz (B8a). The reader should note that a knowledge of matrix methods is not essential for understanding the later sections of this article.

The quantities involved in such a scheme are derived from the stoichiometric coefficients that appear in the chemical equations, the equations being written in such a form that the coefficients tell the number of moles of the substance in question produced by one mole of the reaction.

A familiar example of multiple reaction paths is the catalytic oxidation of ethylene to ethylene oxide, where the by-products water and carbon dioxide are produced both by direct oxidation of ethylene and by further oxidation of ethylene oxide. For this example, the equations may be written in the form

$$0 = -C_2H_4 - \tfrac{1}{2} O_2 + C_2H_4O$$
$$0 = -C_2H_4 - 3 O_2 + 2 CO_2 - 2 H_2O$$
$$0 = - \tfrac{5}{2} O_2 - C_2H_4O + 2 CO_2 + 2 H_2O$$

The array of coefficients on the right-hand side of the equations is manipulated as a matrix, which may be denoted as **S**, the matrix of stoichiometric coefficients. In the example,

$$\mathbf{S} = \begin{pmatrix} -1 & -\tfrac{1}{2} & 1 & 0 & 0 \\ -1 & -3 & 0 & 2 & 2 \\ 0 & -\tfrac{5}{2} & -1 & 2 & 2 \end{pmatrix} \qquad (2\text{-}1)^*$$

If any inert substances are present, they are represented in the matrix by columns of zeros. In this and the other matrices used in the calculation, the substances and the quantities that pertain to them are arranged by columns, while the reactions and the quantities that pertain to them are arranged by rows. For example, the set of heat capacities, represented as $\mathbf{C}_p$, is a row matrix, while the set of reaction rates **R** is represented as a column matrix.

2. The Reduced Matrix

The number of independent reactions is just the rank of the matrix **S**, that is, the rank of the largest nonsingular square submatrix of **S**. In the example above, the second row is the sum of the other two, so that the rank cannot be 3. On the other hand, the four elements in the upper left-hand corner obviously form a nonsingular submatrix, so that the rank is 2. A nonsingular square submatrix **s** having the rank of **S**, is selected

---

* Throughout this review the first digit of a numbered equation refers to the section numbered with the corresponding Roman numeral.

as a basis for the calculation. The choice amounts to selecting a set of key substances and a corresponding set of key reactions, in terms of which the composition of the system is to be represented. There is some advantage in choosing this matrix from columns referring to substances that have easily determined concentrations. In the example, a good choice would be the first two rows of the columns for ethylene oxide and carbon dioxide; here there is the additional advantage that the submatrix is diagonal, having the form

$$\mathbf{s} = \begin{pmatrix} 1 & 0 \\ 0 & 2 \end{pmatrix} \tag{2-2}$$

It is desirable at this point to rearrange the matrix $\mathbf{S}$ so that the submatrix $\mathbf{s}$ appears in the upper left-hand corner. Then $\mathbf{S}$ is written in the form

$$\mathbf{S} = \begin{pmatrix} 1 & 0 & -1 & -\frac{1}{2} & 0 \\ 0 & 2 & -1 & -3 & 2 \\ -1 & 2 & 0 & -\frac{5}{2} & 2 \end{pmatrix} \tag{2-3}$$

as though the chemical equations were

$$0 = \quad C_2H_4O \qquad\qquad C_2H_4 - \tfrac{1}{2} O_2$$
$$0 = \qquad\qquad 2\ CO_2 - C_2H_4 - 3\ O_2 + 2\ H_2O$$
$$0 = -C_2H_4O + 2\ CO_2 \qquad\qquad - \tfrac{5}{2} O_2 + 2\ H_2O$$

3. *Other Submatrices*

Two other submatrices of $\mathbf{S}$ are needed in the calculation: they are $\mathbf{S}_R$, made up of the rows of $\mathbf{S}$ that appear in $\mathbf{s}$, and $\mathbf{S}_c$, made up of the columns of $\mathbf{S}$ that appear in $\mathbf{s}$. In the example, they are

$$\mathbf{S}_R = \begin{pmatrix} 1 & 0 & -1 & -\frac{1}{2} & 0 \\ 0 & 2 & -1 & -3 & 2 \end{pmatrix} \tag{2-4}$$

and

$$\mathbf{S}_c = \begin{pmatrix} 1 & 0 \\ 0 & 2 \\ -1 & 2 \end{pmatrix} \tag{2-5}$$

The fact that

$$\mathbf{S} = \mathbf{S}_c \mathbf{s}^{-1} \mathbf{S}_R \tag{2-6}$$

will be made use of later.

B. Relations between Conversions and Concentrations

1. *Fundamental Relation*

It is the relation between the concentrations and the conversions that gives meaning to the stoichiometric coefficients. The relation is

$$\mathbf{Y} = \mathbf{Y}^0 + \mathbf{X}^T\mathbf{S} \tag{2-7}$$

or, for a single reaction,

$$Y = Y^0 + XS$$

We note that $\mathbf{X}^T$ is the *transpose* of $\mathbf{X}$, in which rows and columns are interchanged. Equation (2-7), when put into words, says that the number of moles of each substance present in a certain mass of reacting mixture is the number of moles there initially plus whatever amount has been formed by all the reactions that have occurred. The conversion X is defined as the number of moles of the corresponding reaction, as written, that have occurred in a mass of reactant equal to the average molecular weight of the initial mixture, and the concentration Y is the number of moles of the substance in question in this same mass; Y^0, the initial value of Y, is just the initial mole fraction.

2. *Composition in Terms of Concentrations of Key Components*

Now the key components, those that correspond to the columns of $\mathbf{s}$, are singled out for special attention. The set of the concentrations of these components is denoted as $\mathbf{y}$, and a part of Eq. (2-7) is written in the form

$$\mathbf{y} = \mathbf{y}^0 + \mathbf{X}^T\mathbf{S}_c \tag{2-8}$$

The formula for calculating $\mathbf{Y}$ from $\mathbf{y}$ is obtained by writing Eq. (2-7) with $\mathbf{S}_c\mathbf{s}^{-1}\mathbf{S}_R$ substituted for $\mathbf{S}$. The equation becomes

$$\begin{aligned}\mathbf{Y} &= \mathbf{Y}^0 + \mathbf{X}^T\mathbf{S}_c\mathbf{s}^{-1}\mathbf{S}_R \\ &= \mathbf{Y}^0 + (\mathbf{y} - \mathbf{y}^0)\mathbf{s}^{-1}\mathbf{S}_R \\ &= (\mathbf{Y}^0 - \mathbf{y}^0\mathbf{s}^{-1}\mathbf{S}_R) + \mathbf{y}(\mathbf{s}^{-1}\mathbf{S}_R)\end{aligned} \tag{2-9}$$

The last form of this equation is in the best form for computing, since $\mathbf{Y}$ is found from a constant plus a multiple of $\mathbf{y}$. In following the course of a set of reactions, the two matrices in parentheses would be calculated once for all, and then used as constants. Equation (2-9) gives a systematic procedure for finding the concentrations of all components from the concentrations of the key components, which are followed through the reactor by using Eq. (2-8) either directly or indirectly.

3. *Calculation of Mole Fractions*

In order to calculate the mole fraction of a component, the total number of moles in the reference mass must be known. This total number of moles is the sum of all the elements of $\mathbf{Y}$. Sometimes it is more convenient to calculate this sum with a subsidiary equation than to sum the concentrations directly. The column matrix formed by summing each row of $\mathbf{S}_R$ is denoted by $\boldsymbol{\nu}$; it gives the net number of moles formed by the key reactions. Then the desired relation is obtained by summing rows in the postfactors of Eq. (2-9), giving the equation

$$\Sigma Y = (1 - \mathbf{y}^0 \mathbf{s}^{-1} \mathbf{\nu}) + \mathbf{y}(\mathbf{s}^{-1} \mathbf{\nu}) \tag{2-10}$$

C. CALCULATION OF AN ADDITIVE PROPERTY

Any additive property of the mixture can be obtained directly from the formula for **Y**. Let **P** be a row matrix, the elements of which are the molar values of the property for the several components. Then $\overline{P}$, the additive property of the mixture, is found from the equation

$$\overline{P} = \mathbf{Y} \mathbf{P}^T$$
$$= (\mathbf{Y}^0 - \mathbf{y}^0 \mathbf{s}^{-1} \mathbf{S}_R) \mathbf{P}^T + \mathbf{y}(\mathbf{s}^{-1} \mathbf{S}_R \mathbf{P}^T) \tag{2-11}$$

It must be remembered that $\overline{P}$ calculated in this way refers to the mass of mixture equal to the initial average molecular weight, that is, to ΣY moles.

From the point of view of the computer, it is worth noting that in each of the working equations, just as many multiplications as there are independent reactions are used in calculating on element of the desired matrix.

D. INTRODUCTION OF VIRTUAL CONVERSIONS

1. *Definition of Virtual Conversions*

It is sometimes useful to construct a set of virtual conversions, related to virtual reaction rates, that can be used instead of the concentrations for describing the composition of a reacting mixture. The column matrix **x** can be defined by putting $\mathbf{x}^T \mathbf{s}$ for the term $\mathbf{X}^T \mathbf{S}_c$ that appears in Eq. (2-8), that is, for the amounts of the key components that have been formed. From the equation

$$\mathbf{x}^T \mathbf{s} = \mathbf{X}^T \mathbf{S}_c \tag{2-12}$$

it follows immediately that

$$\mathbf{x}^T = \mathbf{X}^T \mathbf{S}_c \mathbf{s}^{-1} \tag{2-13}$$

and that

$$\mathbf{x}^T \mathbf{S}_R = \mathbf{X}^T \mathbf{S}_c \mathbf{s}^{-1} \mathbf{S}_R$$
$$= \mathbf{X}^T \mathbf{S} \tag{2-14}$$

The last result shows that the set of virtual conversions can be used to calculate all the concentrations. By considering the variation of x, the relation

$$\mathcal{R} = \mathbf{R}^T \mathbf{S}_c \mathbf{s}^{-1} \tag{2-15}$$

may be seen to hold, in which $\mathcal{R}$ is the column matrix of virtual rates.

2. Formulas for Other Quantities

The formulas of Eqs. (2-9), (2-10), and (2-11) can be transformed to expressions in **x**. The results follow.

$$\mathbf{Y} = \mathbf{Y}_0 + \mathbf{x}^T \mathbf{S}_R \tag{2-16}$$

$$\Sigma Y = 1 + \mathbf{x}^T \boldsymbol{\nu} \tag{2-17}$$

$$\mathbf{P} = \mathbf{Y}^0 \mathbf{P}^T + \mathbf{x}^T \mathbf{S}_R \mathbf{P}^T \tag{2-18}$$

The concentrations of key substances and the virtual conversions are practically equivalent as variables for determining the composition of a reacting mixture.

E. Illustration

The use of these relations may be illustrated by continuing with the example of the oxidation of ethylene. The factors needed are calculated, and then used in the expressions for the desired quantities.

$$\mathbf{s}^{-1} = \begin{pmatrix} 1 & 0 \\ 0 & \frac{1}{2} \end{pmatrix} \tag{2-19}$$

$$\mathbf{s}^{-1}\mathbf{S}_R = \begin{pmatrix} 1 & 0 & -1 & -\frac{1}{2} & 0 \\ 0 & 1 & -\frac{1}{2} & -\frac{3}{2} & 1 \end{pmatrix} \tag{2-20}$$

$$\boldsymbol{\nu} = \begin{pmatrix} -\frac{1}{2} \\ 0 \end{pmatrix} \tag{2-21}$$

$$\mathbf{s}^{-1}\boldsymbol{\nu} = \begin{pmatrix} -\frac{1}{2} \\ 0 \end{pmatrix} \tag{2-22}$$

$$\Sigma Y = 1 + \tfrac{1}{2} y_1^0 + \tfrac{1}{2} y_1 \tag{2-23}$$

The concentration of oxygen is

$$Y_4 = Y_4^0 + \tfrac{1}{2} y_1^0 + \tfrac{3}{2} y_2^0 - \tfrac{1}{2} y_1 - \tfrac{3}{2} y_2 \tag{2-24}$$

and its mole fraction is this concentration divided by ΣY. In terms of the virtual conversions, similar expressions are obtained.

$$\mathbf{S}_c \mathbf{s}^{-1} = \begin{pmatrix} 1 & 0 \\ 0 & 1 \\ -1 & 1 \end{pmatrix} \tag{2-25}$$

$$\Sigma Y = 1 - \tfrac{1}{2} x_1 \tag{2-26}$$

$$\mathfrak{R}_1 = R_1 - R_3 \tag{2-27}$$

$$\mathfrak{R}_2 = R_2 + R_3 \tag{2-28}$$

F. Exceptional Cases

For lack of information about the kinetics of the reactions, it is sometimes necessary to approximate the selectivity for a desired product in the form of some function of the conversion with respect to some one reaction, giving no attention to the detailed effects of conditions on the rates of side reactions. In such a situation, with the rates of side reactions nominally determined by the conversion for one reaction, the side reactions are not independent, and only one concentration variable needs to be followed, no matter what the rank of the stoichiometric matrix is. In fact, there is formally only one reaction, with nonintegral stoichiometric coefficients that may vary with conversion.

Another situation in which the number of independent reactions is less than the rank of the formal stoichiometric matrix arises when the rates of two reactions are always in the same ratio. In this situation, the two reactions should be combined into one, and instead of the two corresponding rows of the matrix, the appropriate linear combination of them should be used.

III. Equations Describing Simultaneous Reaction and Transport Processes

A. STATEMENT OF ASSUMPTIONS

1. *Smooth Variation of Properties*

It is in order to state first the problem in rather general form, and in as complete detail as is warranted by our knowledge of the processes that are involved. In the first place, a bed of catalyst pellets is treated as though it were statistically homogeneous, its average density depending on distances from walls or other disturbances to the structure of the packing, but no account being taken of the actual detailed location of individual pellets, or of the actual shape of passages among the pellets. Accordingly, it is assumed that such quantities as the superficial mass velocity and the friction factor vary smoothly through the bed, and, as a further consequence, that the temperature, pressure and composition vary smoothly.

Since it is obvious that the smooth variation of these quantities does not extend to the surface of catalyst pellets, much less through them, it must be explained that the smoothness has significance only for a succession of points that are in some sense in equivalent positions relative to the particles of packing. We might, for example, think of a network of points that are roughly in the centers of the holes among the pellets: taking account only of these points, the temperature can be described as a fairly smooth function of position. An alternative network would be the points at the centers of catalyst pellets.

As the statistical description of the packed bed must depend on some kind of averaging to represent the properties of the bed as functions of position, the property is averaged over a representative region having the same order of magnitude as a particle of packing. For the solid, this region is just the volume of a particle. For the fluid, the region can be thought of as bounded by constrictions in the passages among the particles, or, somewhat vaguely, as being a hole in the packing. It is an essential weakness of this statistical description of random packed beds that these regions vary in size and shape, so that the average of a property is only approximately a smooth function of position.

In the traditional view, the desired smoothness is induced by considering only averages over regions that include enough particles to give statistical stability to the averages, but that are still small compared to the size of the bed. This concept is hardly satisfactory when the diameter of the tube is only 10 times the diameter of the particles, so that the whole range of radial variation is covered in 5 particle diameters. The average over a small region is subject to larger fluctuations, but at least it gives more meaning to the concept of smooth variation of properties in the packed bed. Baron (B3) has discussed the effect of spatial fluctuations on the behavior of packed reactors; it may be hoped that our knowledge of the structure of packed beds will develop enough so that this effect can be taken into account, but at present it can only be characterized as small.

2. Eddy Diffusion Dominant

Another feature of packed beds that must be discussed before material and enthalpy balances can be formulated is the apparent eddy diffusion that is observed, both transverse to the flow and parallel to the flow. It is generally agreed that the transverse diffusion results from a random walk of elements of fluid through the passages in the bed (B3), or equivalently, from mixing of converging streams (B1, R1) and that the diffusion parallel to the flow arises primarily from the stagewise mixing in the interstices of the bed (A4, K2), with some contribution from the variation of mass velocity with radial position (C1, C5). In the usual formulation of the equations, terms for the accumulation of material and of heat by both transverse and parallel diffusion have been included (see, for example, S4). The radial transport of heat in a heat-exchanger type of tubular reactor cannot, of course, be neglected, and the description of the mixing in a packed bed by a diffusion coefficient is so good that the term for the transverse diffusion can be included without further discussion.

3. Axial Diffusive Processes Neglected

There are two reasons for adopting a different attitude toward the apparent diffusion parallel to the flow. One is that in the reactors that one expects to encounter in commercial practice, the effect is small. The Peclet number for axial diffusion is based on the particle diameter, and is between 1 and 2; then the Peclet number based on the length of the bed, which gives some idea of the importance of diffusion in modifying a process going on in the bed, is between 1 and 2 times the number of particle diameters in some characteristic length, such as the length of a region in which the temperature or concentration is changing rapidly. In any case, this length has the same order of magnitude as the length of the bed, so that the significant Peclet number is at least 100. Another observation indicating the negligible magnitude of the diffusive effect is that with a Peclet number of 2, the concentration 3 particle diameters upstream from a steady source is about 1/400 what it is just downstream from the source.

The second reason goes back to the nature of the processes giving rise to the apparent diffusion. What is observed is that an impulsive signal introduced into a fluid flowing through a packed bed is dispersed as it goes downstream, or that a harmonic signal is attenuated. From the extent of the dispersion or attenuation an apparent diffusivity is calculated, and this diffusivity is useful in predicting to the first order the small effects of the dispersive process. It must be kept in mind, however, that if the dispersion arises only from the processes mentioned above, that is, from stagewise mixing and the variation of velocity, it does not lead to one of the important effects of diffusion, which is the propagation of a signal upstream. Only if there is a significant amount of backmixing, so that a signal is sent upstream, is it reasonable to represent the dispersive process as diffusion. The importance of this distinction was emphasized by Beek and Miller (B4) who pointed out that if there is backmixing, boundary conditions at both ends of the reactor must be used to define the conditions within the reactor; this is just the same as the situation that was shown by Wehner and Wilhelm (W2) to hold when a term for axial diffusion is introduced into the differential equations describing the system.

From the point of view of practical computation, the equations without axial diffusion are solved by starting at the inlet to the reactor and computing the conditions stepwise downstream, making no adjustment for what happens further downstream. When there is axial diffusion, the conditions at all points in the reactor have to be adjusted to meet boundary conditions at both ends, and the difficulty of the computation is multiplied by a large factor.

Deans and Lapidus (D3), assuming that backmixing is negligible, have recommended that, instead of trying to use diffusion to represent an essentially different dispersive effect, one should formulate a set of difference equations that describe the processes that are actually going on in the packed bed. There is no question but that this scheme gives the best representation of the behavior of a packed bed that is at present available, but this good representation is very expensive in computing time, both because it requires axial steps about equal to a particle diameter and because the simultaneous difference equations that must be solved are not linear in the unknown quantities. As will be explained below, this method is the one to use when the reactor is so short—measured in particle diameters—that the dispersive effects must be taken into account.

In summary, the argument for writing the equations describing the reactor without accounting for axial dispersion is that this effect usually has very little importance, while the extra effort required to account for it is large. The equations derived in this section will be based on the assumptions that the axial dispersion is negligible, and that the conditions within the bed are sufficiently smooth functions of position to be related by differential equations. These assumptions involve the reservation that the bed is not extremely short.

4. *Heat-Transfer Coefficient at the Wall*

Another assumption that will be made hardly needs stating, as it has come to be generally accepted. It will be assumed that there is a local resistance to heat transfer from the fluid to the wall of the reactor, giving rise to what is practically a discontinuity in the radial temperature profile at the inside surface of the wall. It has not been possible to determine the exact character of the wall effect that is observed, so that the description used here is not certainly the best one. On the basis of the experiments that have been made, however, this description is indistinguishable from an alternative one that assigns a reduced thermal conductivity to a layer adjoining the wall (see Y1 and Y3).

B. Derivation of Conservation Equations

1. *Nature of Transverse Eddy Diffusion*

a. Concentration Based on Mass

There is a sound experimental basis (B6, L1) for the conclusion that in packed tubes when the Reynolds number is above 100, the transport of both matter and heat transverse to the flow is dominated by eddy diffusive processes. To the extent that these processes dominate, all conserved entities in the fluid diffuse alike, provided the gradients of the concentrations of these entities in amount per unit mass of fluid are used as driving forces for the diffusion. In all that follows, it will be assumed that molecular diffusion is negligible in comparison with eddy diffusion. The system has the property that the flux of an entity transverse to the flow can be related to a Peclet number by the equation

$$J_x = -(Gd_p/N_{Pe})\partial c/\partial x \tag{3-1}$$

the same Peclet number serving for all conserved quantities. In this equation, J_x is the component in the x-direction of the superficial diffusive flux, G is the magnitude of the superficial mass velocity, c is the concentration of a conserved entity, and x is a coordinate perpendicular to the direction of flow. All flows and rates are based on unit area or volume of the reactor, without regard to what fraction of that area or volume is occupied by packing.

It may be noticed that the quantity Gd_p/N_{Pe} does not have the dimensions of a diffusivity, but of a diffusivity multiplied by a density. This situation arises from the fact that the amount per unit mass, rather than the amount per unit volume, is used for a concentration. It is necessary to use the mass of fluid as a basis for concentration, since the mass but not the volume is conserved in the flow.

b. Definition of Eddy-Diffusion Operator

The equations describing the variation of concentration and temperature are derived from material and enthalpy balances. The enthalpy balance is used instead of a heat balance because the enthalpy is a conserved quantity in the flow. The balances are based on the diffusive flux shown in Eq. (3-1), which will now be used in the form

$$\mathbf{J} = -E \text{ grad } c \tag{3-2}$$

This form for the equation is chosen to take advantage of the vector notation; and to make it possible to indicate a purely transverse flux as arising from a gradient of concentration that may have a nonzero component parallel to the flow. In this equation, $\mathbf{J}$ is the superficial diffusive flux, and the eddy diffusion coefficient E may be thought of either as a tensor of second rank, or, in Gibbs' nomenclature, as a dyadic. $-E$ operates on a gradient of concentration to give a flux in a direction opposite to the projection of the gradient of concentration on a plane perpendicular to the flow, with a magnitude equal to the product of Gd_p/N_{Pe} and the magnitude of the projection.

In a steady state the balance involves accumulation by convection, by diffusion, or conduction, and by reaction. The statement that the sum of these must be zero takes the form

$$-\text{div }(Gc - E \text{ grad } c - k_p \text{ grad } c) + A = 0 \tag{3-3}$$

in which A is the rate of production of the entity in question by reaction, and the term div k_p grad c is added to account for diffusive transport by processes in addition to eddy diffusion. k_p is taken as zero for the diffusion of material.

2. Material Balance

All the required material balances can be written in one equation by putting for c the set of concentrations of key components, designated by the row matrix $\mathbf{y}$. Corresponding to Eq. (2-8), the reduced matrix of local production rates may be seen to be $\mathbf{R}^T\mathbf{S}_c$. With these substitutions, the equation giving the material balances for the key substances is

$$-\operatorname{div}(\mathbf{G}\mathbf{y} - E \operatorname{grad} \mathbf{y}) + \mathbf{R}^T\mathbf{S}_c = 0 \tag{3-4}$$

Only eddy diffusion is involved in the material balance. In the steady state, the condition

$$\operatorname{div} \mathbf{G} = 0 \tag{3-5}$$

permits some simplification of Eq. (3-4) to the form

$$-\mathbf{G} \cdot \operatorname{grad} \mathbf{y} + \operatorname{div}(E \operatorname{grad} \mathbf{y}) + \mathbf{R}^T\mathbf{S}_c = 0 \tag{3-6}$$

3. Enthalpy Balance

a. General Statement

For the enthalpy balance, c becomes H, the enthalpy of unit mass of fluid, which is naturally expressed in terms of the temperature and composition of the fluid. In this case, the source term becomes zero, corresponding with the conservation of enthalpy in the flow. The heat of reaction enters when the enthalpy balance is transformed into an equation involving the temperature. The enthalpy balance is then

$$-\mathbf{G} \cdot \operatorname{grad} H + \operatorname{div}(E \operatorname{grad} H) + \operatorname{div}(k_p \operatorname{grad} T) = 0 \tag{3-7}$$

The term div $(k_p$ grad $T)$ accounts for the conduction of heat through the packing. Since the conduction operates partly in series and partly in parallel with the eddy transport, the exact relationship with the mass velocity is complicated, and has not been successfully analyzed. Argo and Smith (A3) have given the most complete discussion of the problem.

b. Relation to Temperature and Composition

The notation $\mathbf{H}$ is now introduced for the matrix of the partial molar enthalpies of the components of the fluid. Then the enthalpy is expressed in the form [see Eq. (2-8)]

$$H = (\mathbf{Y}^0 - \mathbf{y}^0\mathbf{s}^{-1}\mathbf{S}_R)\mathbf{H}^T + \mathbf{y}\mathbf{s}^{-1}\mathbf{S}_R\mathbf{H}^T \tag{3-8}$$

and the gradient is

$$\operatorname{grad} H = (\mathbf{Y}^0 - \mathbf{y}^0 \mathbf{s}^{-1}\mathbf{S}_R + \mathbf{y}\mathbf{s}^{-1}\mathbf{S}_R)\operatorname{grad} \mathbf{H}^T + (\operatorname{grad} \mathbf{y})\mathbf{s}^{-1}\mathbf{S}_R\mathbf{H}^T \quad (3\text{-}9)$$

When the partial enthalpies vary significantly with pressure and composition as well as with temperature, three terms appear in the working expression for grad H. In the derivation that follows, the partial enthalpies will be assumed to depend only on the temperature, which is equivalent to saying that they are just the molar enthalpies of the separate components. If there is an important effect of composition or pressure, it is easy to carry the extra term through the derivation.

With the elements of H depending on the temperature only, their variation can be expressed in the form

$$\operatorname{grad} \mathbf{H} = \mathbf{C}_p \operatorname{grad} T \quad (3\text{-}10)$$

with $\mathbf{C}_p$ standing for the matrix of molar heat capacities of the components. Then it follows that

$$\operatorname{grad} H = (\mathbf{Y}^0 - \mathbf{y}^0 \mathbf{s}^{-1}\mathbf{S}_R + \mathbf{y}\mathbf{s}^{-1}\mathbf{S}_R)\mathbf{C}_p{}^T \operatorname{grad} T + (\operatorname{grad} \mathbf{y})\mathbf{s}^{-1}\mathbf{S}_R\mathbf{H}^T \quad (3\text{-}11)$$

The product $(\mathbf{Y}^0 - \mathbf{y}^0 \mathbf{s}^{-1}\mathbf{S}_R + \mathbf{y}\mathbf{s}^{-1}\mathbf{S}_R)\mathbf{C}_p{}^T$ may be recognized as C_p, the heat capacity of the unit mass of mixture.

Each of the terms in grad H contributes to div $(E \operatorname{grad} H)$. On expanding, it is found that

$$\operatorname{div}(E \operatorname{grad} H) = \operatorname{div}(EC_p \operatorname{grad} T) + \operatorname{div}[E(\operatorname{grad} \mathbf{y})\mathbf{s}^{-1}\mathbf{S}_R\mathbf{H}^T] \quad (3\text{-}12)$$

The last term in this equation may be expanded to give an expression involving more directly measured quantities.

The expression is

$$\operatorname{div}[E(\operatorname{grad} \mathbf{y})\mathbf{s}^{-1}\mathbf{S}_R\mathbf{H}^T] = [\operatorname{div}(E \operatorname{grad} \mathbf{y})]\mathbf{s}^{-1}\mathbf{S}_R\mathbf{C}_p{}^T$$
$$+ (\operatorname{grad} T) \cdot (E \operatorname{grad} \mathbf{y})\mathbf{s}^{-1}\mathbf{S}_R\mathbf{C}_p{}^T \quad (3\text{-}13)$$

Making use of these relations, the enthalpy balance may be written as

$$-C_p \overline{\mathbf{G}} \cdot \operatorname{grad} T - \overline{\mathbf{G}} \cdot (\operatorname{grad} \mathbf{y})\mathbf{s}^{-1}\mathbf{S}_R\mathbf{H}^T + \operatorname{div}(EC_p \operatorname{grad} T)$$
$$+ [\operatorname{div}(E \operatorname{grad} \mathbf{y})]\mathbf{s}^{-1}\mathbf{S}_R\mathbf{H}^T + (\operatorname{grad} T) \cdot (E \operatorname{grad} \mathbf{y})\mathbf{s}^{-1}\mathbf{S}_R\mathbf{C}_p{}^T$$
$$+ \operatorname{div}(k_p \operatorname{grad} T) = 0 \quad (3\text{-}14)$$

c. Use of Material Balance to Introduce Rates

This equation may be considerably simplified and made easier to use by combining it with the material balance. Equation (3-6) is multiplied by $\mathbf{s}^{-1}\mathbf{S}_R\mathbf{H}^T$ as a postfactor, and then subtracted from Eq. (3-14). The result is

$$-C_p \mathbf{G} \operatorname{grad} T + \operatorname{div}(EC_p \operatorname{grad} T) + \operatorname{div}(k_p \operatorname{grad} T)$$
$$+ (\operatorname{grad} T) \cdot (E \operatorname{grad} \mathbf{y})\mathbf{s}^{-1}\mathbf{S}_R\mathbf{C}_p{}^T + \mathbf{R}^T\mathbf{Q} = 0 \quad (3\text{-}15)$$

In writing this equation, **Q**, the matrix of heats of reaction, has been substituted for the product $-\mathbf{SH}^T$, that is, for the negative of the isothermal enthalpy change for the reaction. It will be noticed that the third term does not appear in a so-called heat balance, which is formulated to account for the heat accumulated and produced in a region, using EC_p for the eddy thermal conductivity. Such a formulation fails because heat is not conserved in the flow.

It may be worth while to review the different kinds of multiplicity involved in the symbols appearing in Eqs. (3-6) and (3-15). Equation (3-6) is merely a shorthand way of writing the material balance for each of the key components, each term being a row matrix having as many elements as there are independent reactions. The equation asserts that when these matrices are combined as indicated, each element in the resulting matrix will be zero. The elements in the first two terms are obtained by vector differential operation, but the elements are scalars. Equation (3-15), on the other hand, is a scalar equation, from the point of view of both vector analysis and matrix algebra, although some of its terms involve vector operations and matrix products. No account need be taken of the interrelation of the vectors and matrices in these equations, but the order of vector differential operators and their operands as well as of all matrix products must be observed.

d. Equations in Terms of Virtual Rates and Conversions

An alternative formulation of these equations is in terms of the virtual conversions of selected key reactions. From Eqs. (2-8) and (2-12) it may be seen that

$$\text{grad } \mathbf{y} = \text{grad } \mathbf{x}^T \mathbf{s} \qquad (3\text{-}16)$$

and from Eq. (2-15) that

$$\mathbf{R}^T \mathbf{S}_c = \mathfrak{R}^T \mathbf{s} \qquad (3\text{-}17)$$

With these substitutions, Eq. (3-6) becomes

$$-\mathbf{G} \cdot \text{grad } \mathbf{x}^T \mathbf{s} + \text{div } (E \text{ grad } \mathbf{x}^T)\mathbf{s} + \mathfrak{R}^T \mathbf{s} = 0 \qquad (3\text{-}18)$$

On dropping the factor **s** and transposing, the equation in **x** is obtained.

$$-\mathbf{G} \cdot \text{grad } \mathbf{x} + \text{div } (E \text{ grad } \mathbf{x}) + \mathfrak{R} = 0 \qquad (3\text{-}19)$$

In order to make the corresponding transformation of Eq. (3-15), the term $\mathbf{R}^T \mathbf{Q}$ must be expressed in terms of $\mathfrak{R}$. The reduced matrix $\mathfrak{Q}$ is introduced, representing the heats of reaction of the key reactions, being equal to $-\mathbf{S}_R \mathbf{H}^T$. The sequence of substitutions is

$$\mathbf{R}^T\mathbf{Q} = -\mathbf{R}^T\mathbf{S}\mathbf{H}^T$$
$$= -\mathbf{R}^T\mathbf{S}_c\mathbf{s}^{-1}\mathbf{S}_R\mathbf{H}^T$$
$$= \mathfrak{R}^T\mathbf{s}\mathbf{s}^{-1}\mathbf{Q}$$
$$= \mathfrak{R}^T\mathbf{Q} \qquad (3\text{-}20)$$

After making this substitution and putting $\mathbf{x}^T$ for $\mathbf{y}\mathbf{s}^{-1}$ in Eq. (3-15), the equation in $\mathbf{x}$ is

$$-C_p\bar{G} \cdot \text{grad } T + \text{div } (EC_p \text{ grad } T) + \mathbf{C}_p\mathbf{S}_R{}^T(\text{grad } T) \cdot (E \text{ grad } \mathbf{x})$$
$$+ \text{div } (k_p \text{ grad } T) + \mathbf{Q}^T\mathfrak{R} = 0 \qquad (3\text{-}21)$$

e. Expansions in Terms of Measured Quantities

In using either of Eqs. (3-15) or (3-21), the term div $(C_p E \text{ grad } T)$ must be evaluated. An expansion of the term gives the expression

$$\text{div } (C_p E \text{ grad } T) = C_p \text{ div } (E \text{ grad } T) + (\text{grad } C_p) \cdot (E \text{ grad } T) \qquad (3\text{-}22)$$

The quantity that requires attention is grad C_p. C_p can be expressed —as it was in Eq. (3-11)—as a linear combination of the heat capacities of the components. When the gradient is expanded to show its dependence on composition and temperature, the result is

$$\text{grad } C_p = (\text{grad } \mathbf{y})\mathbf{s}^{-1}\mathbf{S}_R\mathbf{C}_p{}^T$$
$$+ (\mathbf{Y}^0 - \mathbf{y}^0\mathbf{s}^{-1}\mathbf{S}_R + \mathbf{y}\mathbf{s}^{-1}\mathbf{S}_R)\frac{\partial \mathbf{C}_p{}^T}{\partial T} \text{ grad } T \qquad (3\text{-}23)$$

The evaluation of this expression is time consuming, a situation that explains the fact that the variation of the heat capacity is neglected if it is not excessively large. It may be remarked that the elements of $\mathbf{S}_R\mathbf{C}_p{}^T$ are the values of ΔC_p for the key reactions, and so that $\mathbf{S}_R(\partial \mathbf{C}_p{}^T/\partial T)$ involves the variation of ΔC_p with temperature. In many cases both of these quantities are small enough to be negligible.

In the numerical solution of these equations, the heat capacity of the fluid must be calculated at every mesh point if it is not constant, so that it may be more convenient to use a finite-difference approximation for grad C_p. The factor $(E \text{ grad } \mathbf{y})\mathbf{s}^{-1}\mathbf{S}_R\mathbf{C}_p{}^T$ appearing in Eq. (3-15) must be evaluated in any case.

4. Equations for a Tubular Reactor

The flow is practically parallel to the axis of a cylindrical reactor and the whole system is symmetrical about the axis of the cylinder. Then the fluxes E grad T and E grad y have only radial components, and the product $(\text{grad } C_p) \cdot (E \text{ grad } T)$ is just the product of the radial derivative of C_p with the magnitude of E grad T. It takes the form

$$(\text{grad } C_p) \cdot (E \text{ grad } T) = \frac{\partial C_p}{\partial r} \frac{Gd_p}{N_{\text{Pe}}r_2{}^2} \frac{\partial T}{\partial r} \qquad (3\text{-}24)$$

in which r is the dimensionless radial coordinate, and $\partial C_p/\partial r$ is evaluated by putting the operator $\partial/\partial r$ for grad in Eq. (3-23). The divergence of E grad T in this case is

$$\text{div}(E \text{ grad } T) = \frac{d_p}{N_{\text{Pe}} r_2{}^2}\left[\frac{G}{r}\frac{\partial}{\partial r}\left(r\frac{\partial T}{\partial r}\right) + \frac{\partial G}{\partial r}\frac{\partial T}{\partial r}\right] \quad (3\text{-}25)$$

When the relations given above are incorporated into Eq. (3-15), and the resulting equation is divided by $C_p G$, the working equation is obtained.

$$\frac{\partial T}{\partial z} = \frac{d_p}{N_{\text{Pe}} r_2{}^2}\left[\frac{1}{r}\frac{\partial}{\partial r}\left(r\frac{\partial T}{\partial r}\right) + \frac{1}{G}\frac{\partial G}{\partial r}\frac{\partial T}{\partial r}\right] + \frac{d_p}{C_p N_{\text{Pe}} r_2{}^2}\frac{\partial \mathbf{y}}{\partial r}\mathbf{s}^{-1}\mathbf{S}_R \mathbf{C}_p{}^{\text{T}}\frac{\partial T}{\partial r}$$

$$+ \frac{d_p}{C_p N_{\text{Pe}} r_2{}^2}(\mathbf{Y}^0 - \mathbf{y}^0\mathbf{s}^{-1}\mathbf{S}_R + \mathbf{y}\mathbf{s}^{-1}\mathbf{S}_R)\frac{\partial \mathbf{C}_p{}^{\text{T}}}{\partial T}\left(\frac{\partial T}{\partial r}\right)^2$$

$$+ \frac{1}{C_p G r_2{}^2 r}\frac{\partial}{\partial r}\left(k_p r\frac{\partial T}{\partial r}\right) + \frac{\mathbf{R}^{\text{T}}\mathbf{Q}}{C_p G} \quad (3\text{-}26)$$

The corresponding equation derived from Eq. (3-6) is

$$\frac{\partial \mathbf{y}}{\partial z} = \frac{d_p}{N_{\text{Pe}} r_2{}^2}\left[\frac{1}{r}\frac{\partial}{\partial r}\left(r\frac{\partial \mathbf{y}}{\partial r}\right) + \frac{1}{G}\frac{\partial G}{\partial r}\frac{\partial \mathbf{y}}{\partial r}\right] + \mathbf{R}^{\text{T}}\mathbf{S}_c \quad (3\text{-}27)$$

The calculation of the factor $(\partial \mathbf{y}/\partial r)\mathbf{s}^{-1}\mathbf{S}_R\mathbf{C}_p{}^{\text{T}}$ may be illustrated by the example used in Section II, the catalytic oxidation of ethylene. The two modes of oxidation of ethylene are used as the key reactions, and ethylene oxide and carbon dioxide are selected as the key components. The reactions and the corresponding stoichiometric matrices are shown in Section II. In the range of temperature between 200° and 350°C., the heat capacities of the reactants are nearly enough linear functions of the temperature, and accordingly the product $\mathbf{S}_R\mathbf{C}_p{}^{\text{T}}$, the matrix of changes of heat capacity for the key reactions, will be a linear function of the temperature. The matrix is found to be

$$\mathbf{S}_R\mathbf{C}_p{}^{\text{T}} = \begin{pmatrix} -3.57 + 5.44 \times 10^{-3}T \\ 3.56 - 5.71 \times 10^{-3}T \end{pmatrix} \text{cal./mole deg.}$$

with T expressed in degrees K. Then [see Eq. (2-19)]

$$\mathbf{s}^{-1}\mathbf{S}_R\mathbf{C}_p{}^{\text{T}} = \begin{pmatrix} 1 & 0 \\ 0 & \tfrac{1}{2} \end{pmatrix}\begin{pmatrix} -3.57 + 5.44 \times 10^{-3}T \\ 3.56 - 5.71 \times 10^{-3}T \end{pmatrix}$$

$$= \begin{pmatrix} -3.57 + 5.44 \times 10^{-3}T \\ 1.78 - 2.86 \times 10^{-3}T \end{pmatrix} \text{cal./mole deg.}$$

Finally, this matrix is multiplied by the row matrix $\partial \mathbf{y}/\partial r$ as a prefactor, giving the result

$$\frac{\partial \mathbf{y}}{\partial r}\mathbf{s}^{-1}\mathbf{S}_R\mathbf{C}_p{}^{\text{T}} = \frac{\partial y_1}{\partial r}(-3.57 + 5.44 \times 10^{-3}T)$$

$$+ \frac{\partial y_2}{\partial r}(1.78 - 2.86 \times 10^{-3}T) \text{ cal./mole deg.}$$

If the bed of catalyst has a uniform cross section but is not a circular cylinder, another dimension enters into consideration. In cases where a fairly high symmetry is retained, the numerical solution of Eqs. (3-6) and (3-15) is certainly possible, but just as certainly not easy or fast. If the symmetry is low, or if the boundaries are complicated, as, for example, in a large bed through which cooling tubes are passed, the problem becomes very difficult. The best approach in such a case is probably to estimate the radius of an equivalent circular cylinder. A rough guide to such an estimate can be found in two-dimensional potential theory, but the fact that the rate is not a linear function of composition and temperature makes it impossible to form an accurate estimate in any general way.

5. *General Form of Eddy-Diffusion Operator*

If the reactor has a converging or diverging flow pattern, a more general expression for E is needed. In tensor notation, the expression is

$$E^{ij} = \frac{G d_p}{N_{\text{Pe}}} \left(g^{ij} - \frac{G^i G^j}{G^2} \right) \tag{3-28}$$

This expression is derived from the conditions given above for the application of the operator E.

There is some question whether the conditions in a nonadiabatic reactor that does not have high symmetry can be calculated with a reasonable effort. Even when the pressure is practically constant, the flow pattern and the reaction interact through the pressure drop in a very complicated way. The distribution of flow in an unsymmetrical packed bed is hard to calculate in the simplest case, when the temperature and molecular weight are constant.

6. *Boundary Conditions for the Case of a Tubular Reactor*

The conditions at the entrance to a reactor tube must be specified to fix one boundary condition. In general, the temperature and composition would be specified as functions of radial position, but the performance of the reactor depends very little on anything but the average conditions in the reactant fluid.

The boundary conditions at the wall, on the other hand, influence the performance of the reactor critically, and should be determined as accurately as possible. For the equations of concentration (or conversion) the condition at the wall is that the flux of material normal to the wall is zero, which requires that the directional derivative of concentration normal to the wall be zero. For the tubular reactor, with cylindrical symmetry, the condition is expressed by the equation

$$r = 1 : \frac{\partial y}{\partial r} = 0 \tag{3-29}$$

For the equation giving the temperature, the condition is that the flux of enthalpy normal to the wall as calculated from the eddy diffusivity and the thermal conductivity match the flux as calculated for transfer to the wall. The corresponding equation is

$$r = 1 : \frac{1}{r_2}\left(\frac{C_p G d_p}{N_{\text{Pe}}} + k_p\right)\frac{\partial T}{\partial r} = -h_w(T - T_w) \tag{3-30}$$

In this equation h_w is the coefficient for transfer of heat to the wall and T_w is the temperature of the wall.

C. Adiabatic Reactor

An adiabatic reactor with a regular flow pattern is much simpler to work with than one with exchange of heat. Except for resistance to exchange of material and heat between fluid and catalyst, the space velocity is the only significant space variable. The natural independent variable to use is σ, the reciprocal of the molar space velocity, or the volume traversed divided by the total molar flow. The relation between the concentration and the reciprocal space velocity is given by the equation

$$\frac{\partial y}{\partial \sigma} = \mathbf{R}^{\text{T}}\mathbf{S}_c \tag{3-31}$$

The temperature is fixed by the condition that the enthalpy is constant. Equation (3-8) provides the required relation.

D. The One-Dimensional Approximation

1. *Over-All Heat-Transfer Coefficient*

When most of the resistance to heat transfer between the axis of the tube and the control medium is localized at the wall of the tube, the variation of conditions in a cross section of the tube becomes relatively unimportant, and a useful approximation is obtained by using average conditions in a cross section to represent the whole cross section. The central problem in using such as approximation is to estimate the rate of heat transfer through the wall of the tube, given the average temperature. The heat transfer is determined by the temperature of the fluid next to the wall, which is between the temperature of the control medium and the average. The heat-transfer coefficient must be modified by the ratio of the differences of the wall and the average temperatures from the temperature of the control medium.

A simple approximation for this ratio, based on the assumption that the radial temperature profile is parabolic, was given by Beek and Singer (B4). Their result may be expressed in the form

$$\frac{T_{r=1} - T_c}{\overline{T} - T_c} = \frac{1}{1 + \frac{1}{4}N_{\mathrm{Bi}}} \tag{3-32}$$

where $N_{\mathrm{Bi}} = r_2 h_t/k_e$.

The temperature is averaged over the mass flow, and k_e is evaluated for the conditions close to the wall. Using the ratio given above, the overall heat-transfer coefficient U can be calculated. It is

$$U = h_t/(1 + \tfrac{1}{4}N_{\mathrm{Bi}}) \tag{3-33}$$

leading to the following equation for the average temperature.

$$\frac{d\overline{T}}{dz} = \frac{\overline{\mathbf{R}}^{\mathrm{T}}\mathbf{Q}}{C_p G} - \frac{2U(\overline{T} - T_c)}{r_2 C_p G} \tag{3-34}$$

The corresponding relation for the concentration is

$$\frac{d\overline{\mathbf{y}}}{dz} = \frac{\overline{\mathbf{R}}^{\mathrm{T}}\mathbf{S}_c}{G} \tag{3-35}$$

The simplest way to estimate the average rate is to use the rate at the average temperature in the cross section. The average rate calculated in this way is usually too low, because the rate usually varies more strongly than linearly with the temperature. It is possible to calculate the average rate in a way that is consistent with the assumption of a parabolic radial temperature profile, but the improvement obtained in this way is usually small compared to the error involved in the assumption itself. The best procedure is to use the rate corresponding to the average temperature, and to restrict the use of the one-dimensional approximation to cases in which the profile is sufficiently flat. The profile is increasingly flat when the Biot number is decreased.

2. Limitations

No general criterion can be given, with assurance that the error will be less than some specified value if the criterion is met. The reaction used below to illustrate the solution of the equations affords a comparison for one case. Table I shows that the error in this case is significant when $N_{\mathrm{Bi}} = 4.3$.

E. Calculation of the Pressure

In order to complete the description of the system, the pressure within the reactor must be known. In the present state of knowledge, it is a sufficient approximation to consider the pressure in a cross section of the reactor as constant, and to calculate it as a function of axial position

TABLE I

EFFECT OF BIOT NUMBER ON THE ACCURACY
OF THE ONE-DIMENSIONAL APPROXIMATION

N_{Bi}	$T_{max} - T_c$ (°C.)	
	One-dimensional approximation	Correct value
4.3	42.4	52.2
3	39.8	46.5
2	37.2	41.3
1	33.5	35.2

only, using an empirical friction factor to estimate the axial pressure gradient. The equation is

$$\frac{dP}{dz} = f(N_{Re})G^2/\rho_f d_p \qquad (3\text{-}36)$$

IV. Estimation of Transport Properties

A. INTRODUCTORY DISCUSSION

After the differential equations and boundary conditions have been formulated, there remain the problems of evaluating the various coefficients that appear, and carrying out the numerical solution of the equations. The following quantities are required.

 1. The velocity distribution.
 2. The Peclet number for eddy diffusion.
 3. The contribution of other modes of transport to the effective thermal conductivity.
 4. The heat-transfer coefficient at the wall.
 5. The friction factor.

It is immediately apparent that all these quantities vary to some extent with position in the reactor. The Peclet number presumably changes with the structure of the packing as the wall is approached, and changes in temperature and composition of the reacting fluid affect the intensive properties of the fluid that determine rates of transport. The only justification for ignoring the actual variation is that the uncertainty about the value of any one of the listed quantities is as large as the correction that would be made for a change in the properties of the fluid. A possible exception to this rule is encountered in the case of a liquid flowing with a Reynolds number in the range below 100. In this situation, a large change in temperature might, through its effect on the viscosity and in turn on the Reynolds number, have a significant effect on the Peclet number. In such a situation, some improvement in precision could be obtained by treating the Peclet number as a point function of conditions in the reactor.

B. Velocity Profile

Smith and his students (M4, S2, D4) have shown that the velocity in a packed tube varies strongly with radial position, and have given a qualitative explanation of the effect, relating it to the larger void fraction in the part of the bed near the wall. The experimental work that has been done does not, however, lead to an accurate prediction of the velocity profile in a given situation. It appears that the inherent lack of reproducibility in packing a tube has obscured the results of the most careful measurements. The kind of difficulty that is encountered is shown in Table II, which gives differences between measurements with cylindri-

TABLE II

Differences in Velocity Profiles in Beds of Cylinders and Spheres, at 4 Radial Positions

Average velocity (ft./sec.)	Diameter of particle (in.)		
	1/4	3/8	1/2
1.01	0.12	−0.03	0.05
	0.19	−0.24	0.03
	0.11	−0.10	−0.09
	−0.24	0.16	−0.17
1.62	0.12	−0.06	0.09
	0.20	−0.12	0.08
	0.07	−0.12	−0.12
	−0.20	−0.06	−0.15
2.14	0.14	0.00	0.13
	0.14	−0.22	0.06
	0.03	−0.09	0.09
	−0.24	0.06	0.15
2.64	0.15	0.02	0.10
	0.16	−0.12	0.06
	0.02	−0.13	−0.13
	−0.11	−0.03	−0.15

cal and spherical packings at a sequence of radial positions, taken from Table 3 of the paper by Schwartz and Smith (S2). The quantities given in the original table are ratios of velocities at a given radial position to average velocities, all measured at a distance of 2 in. above the top of the bed. Table II gives the entries for spheres minus the entries for cylinders having the same diameter, with 3 sizes of particles and 4 flow rates. The measurements were made in a 4-in. tube, at distances from the axis of 0.32, 0.55, 0.70, and 0.88 times the radius of the tube: each figure given represents the average of 3 runs.

With the measurements subject to fluctuations of 20 or 30%, no accurate description of the profile is possible. All that can be said is that with moderate ratios of tube to particle diameter, the maximum velocity is about twice the minimum, and that when the particles are relatively small, the profile is relatively flat near the axis. It is fairly well established that the ratio of the velocity at a given radial position to the average velocity is independent of the average velocity over a wide range. Another observation that is not so easy to understand is that the velocity reaches a maximum one or two particle diameters from the wall. Since the wall does not contribute any more than the packing to the surface per unit volume in the region within one-half particle diameter from the wall, there is no obvious reason for the velocity to drop off farther than some small fraction of a particle diameter from the wall. In any case, all the variations that affect heat transfer close to the wall can be lumped together and accounted for by an effective heat-transfer coefficient. Material transport close to the wall is not very important, because the diffusion barrier at the wall makes the radial variation of concentration small.

It is desirable to have some systematic representation of the velocity profile that has the right features, and preferably is easy to use. It is clear that no polynomial in r of reasonable degree will give the flat region near the axis that is required for small particles. A convenient empirical formula for the profile across a slab that would have the right properties has the form of a constant, the velocity that would be obtained with a very large thickness, plus a hyperbolic cosine function of the distance from the middle. The analogous formula for a cylinder has a constant plus the modified Bessel function of the first kind and order zero, written I_0. After imposing the condition that the velocity have the right average, an expression involving two adjustable parameters may be written. It is

$$\frac{G}{\bar{G}} = \frac{1 + \lambda\sqrt{2\pi\theta r_2/d_p}\ [\exp(-\theta r_2/d_p)]I_0(\theta r_2 r/d_p)}{1 + 2\lambda\sqrt{2\pi d_p/\theta r_2}\ [\exp(-\theta r_2/d_p)]I_1(\theta r_2/d_p)} \qquad (4\text{-}1)$$

in which I_1 denotes the first-order modified Bessel function and λ and θ are adjustable parameters. λ determines how much higher the velocity is near the wall than at the axis, and θ determines how flat the profile is in the middle region. This expression is derived from the assumptions that the limiting value of the ratio of velocity near the wall to the velocity at the axis as the radius increases is a fixed constant, equal to $1 + \lambda$; and that the limiting shape of the profile near the wall is invariant to particle diameter if the distance from the wall is measured in particle diameters. Figure 1 shows curves calculated from this empirical expression with $\lambda = 1.5$, $\theta = 1$, and two values of r_2/d_p: 12.8 and 4, together with corresponding sets of experimental results from the measurements of Schwartz and Smith (S2). The comparison shows that the expression of Eq. (4-1) does not represent these results at all well. At the same time, it is true that no expression giving a reasonable variation

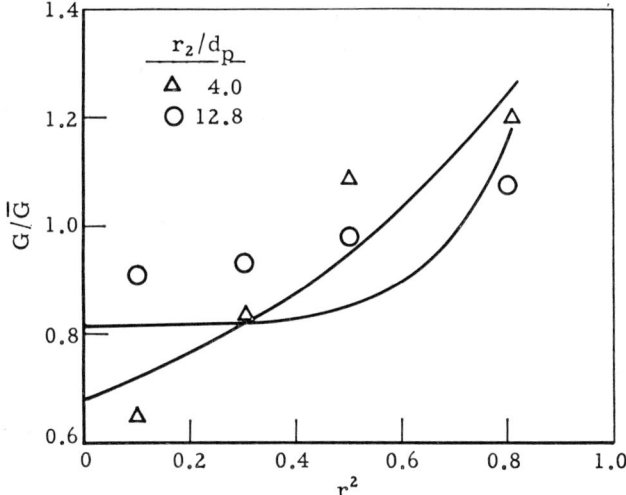

FIG. 1. Representation of velocity profile.

of the profile with conditions could give a good representation of the figures shown in Table II. What is needed is an experimental study comprehensive enough to fix the average behavior of the system on a statistical basis, with all significant systematic errors eliminated or corrected for. A qualitative representation of the trends is all that can be attained on the basis of the experiments that have been made.

C. EDDY DIFFUSIVITY

Bernard and Wilhelm (B6) showed that the radial diffusivity in a packed tube is proportional to mass velocity and particle diameter when the Reynolds number is above 200, and that the diffusivity in this range is so large that molecular diffusion could make no significant contribution to it. They inferred from these results that the diffusion in this range of Reynolds number could be related usefully to a Peclet number equal to about 10, with the diameter of the particles serving for the characteristic length. Their results for the diffusion of CO_2 in air at low Reynolds number showed the expected dependence of the Peclet number on Reynolds number, the Peclet number decreasing in a region where the effect of molecular diffusion is expected to appear. The diffusion of dye in water, however, showed an unexplained increase at very low flow rates. The experiments with water were repeated by Latinen (L1), who found that if precautions were taken to insure that no gas accumulated in the bed, the Peclet number continued to increase as the flow rate was decreased, reaching a value of about 80 at a Reynolds number of 7. Several other workers have made measurements of radial

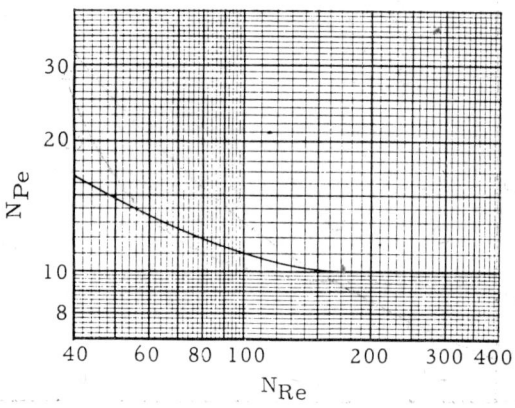

FIG. 2. Peclet number for radial diffusion.

diffusivity in packed beds, the results all being in agreement that the radial average of the Peclet number is close to 10 when the Reynolds number is above 100, and that it increases as the Reynolds number is decreased below this value. Latinen's results may be taken as representative; Fig. 4-2 is a graph of his results, as smoothed by him, but adjusted by a factor to make the Peclet number approach 10 at high Reynolds number.

It must be kept in mind that these measurements were made with water, in which the molecular diffusivity is negligibly small. When the fluid is a gas, with a Schmidt number having the order of magnitude of unity, the Peclet number for molecular diffusion, being the product of Schmidt number and Reynolds number, has the order of magnitude of the Reynolds number. Even if the effect of molecular diffusion is multiplied by the fraction void and by another factor of ⅓, to take care of the tortuosity of the diffusion path, the contribution becomes considerable when the Reynolds number is less than 20. The descriptions of heat and mass transfer assembled in this article should not be used for a Reynolds number less than 40.

D. Other Contributions to the Effective Thermal Conductivity

1. *Discussion by Argo and Smith*

A detailed discussion of the transport of heat in packed beds was given by Argo and Smith (A3). These authors recognized that no complete theory can be constructed on the basis of what is known now, but they attempted to take account of all the processes that would be expected to take part in the transport. The expression they derived was necessarily based on a much simplified description of the packed bed, but presumably has the right kind of dependence on the various quantities that appear. In the notation of this chapter, it is

$$k_e = \epsilon \left(k_f + \frac{C_p G d_p}{\epsilon N_{Pe}} + \frac{4\epsilon_r \sigma_r d_p T^3}{2 - \epsilon_r} \right) + (1 - \epsilon) \frac{h' k_s d_p}{2k_s + h' d_p} \quad (4\text{-}2)$$

The terms in order represent the effects of molecular conduction, of eddy conduction, of radiation through the void space, and of conduction through the solid. The quantity h' that appears includes all paths for the transport from one particle to the next, being composed of terms for radiation, point contacts, and convective transfer to and from the fluid. It is remarkable that this expression contains no adjustable parameters, being based completely on the theoretical analysis and on correlations established for special conditions.

Most of the measurements in this field have been made under conditions such that only the terms for eddy transport and for conduction through the solid need be considered. If it is assumed that, by analogy with mass transport, the modified Peclet number increases when the Reynolds number is decreased below 100, all the evidence (A1a, B9, C2, M1, Q1) indicates that the conduction through the solid is at least as important as the eddy transport when the Reynolds number is less than 50, even with such poorly conducting solids as glass. This feature of the experimental results is well reproduced by the Argo-Smith theory. It is in the region of very high conductivity that the theory fails.

The inclusive quantity h' is at least as large as the convective heat-transfer coefficient, so that a lower bound for the conduction term is given by calculating it using h instead of h'. A good illustration is afforded by the fact that the term calculated in this way for a bed of steel spheres with air flowing through it at a Reynolds number of 500 is 0.31 times the term for eddy transport, assuming a Peclet number of 10 for the eddy transport. This is to be compared with a calculated ratio of 0.10 for alumina, with a conductivity of 5×10^{-4} cal./cm. sec. deg. (S3). The experiments of Singer and Wilhelm (S4) and of Kwong and Smith (K3) show that the effective conductivity is not significantly changed by the substitution of steel for ceramic or glass packing in this range of Reynolds number, although there is a substantial increase in the range below 100. Chu and Storrow (C3) observed a significant increase in the over-all heat-transfer coefficient at high Reynolds numbers but this is not certainly an effect of larger effective thermal conductivity. Thus the term accounting for conduction through the solid, which is needed to explain the effective conductivity at low Reynolds numbers, leads to too great a sensitivity to high values of the solid conductivity at high Reynolds numbers.

2. Suggested Simplification

Even if it is supposed that the expression as given is a good approximation, and that the experiments somehow failed to indicate the large effect, it must be concluded that the experimental basis for the calculation of the effective conductivity is uncertain. Until some theory is developed that represents the experimental results better, or until the Argo-Smith theory is confirmed by experiment, it appears that it would be useful to have a simplified expression, containing parameters that can be adjusted to fit the experimental results. A suitable expression is

$$k_c = \frac{0.6 h d_p k_s}{2k_s + 0.7 h d_p} \quad (4\text{-}3)$$

which is exactly analogous to the last term of Eq. (4-2), but with the smaller h substituted for h', with the factor 0.6 instead of $1 - \epsilon$, and with the factor 0.7 introduced in a term in the denominator to reduce the effect of a large increase in the conductivity of the solid when the Reynolds number is high. Compared with the last term of Eq. (4-2), this empirical expression gives a somewhat poorer correspondence with most of the experiments at low Reynolds numbers but a decidedly improved correspondence for Reynolds numbers of 200 or higher. The effect of void fraction is ignored, first, because its effect has not been measured, and second, because the fraction of the volume occupied by the solid has opposing effects, by way of the area for conduction and by way of the path of conduction per particle.

3. Radiative Transport

There is no experimental basis for including the factor ϵ in the term for radiative conduction. Polack (P1) concluded from his measurements that Dämkohler's estimate of 0.5 for the coefficient (D2) is acceptable. Although the effect of radiation must ultimately increase as the fraction void is increased, the variation in the usual range encountered is presumably small and may be neglected.

4. Proposed Expression for Effective Conductivity

The effect of molecular conduction is negligible when the Reynolds number is greater than 40. These considerations lead to the following proposed working expression for the effective thermal conductivity in a packed bed.

$$k_e = \frac{C_p G d_p}{N_{Pe}} + \frac{0.6 h d_p k_s}{2k_s + 0.7 h d_p} + 2\epsilon_r \sigma_r d_p T^3 \qquad (4\text{-}4)$$

This expression is compared with the authors' correlations of their experimental results in Fig. 3, which shows the reciprocal of the apparent Peclet number.

The decision to maintain an upper limit of 10 on the apparent Peclet number for heat conduction at high Reynolds numbers is justified by

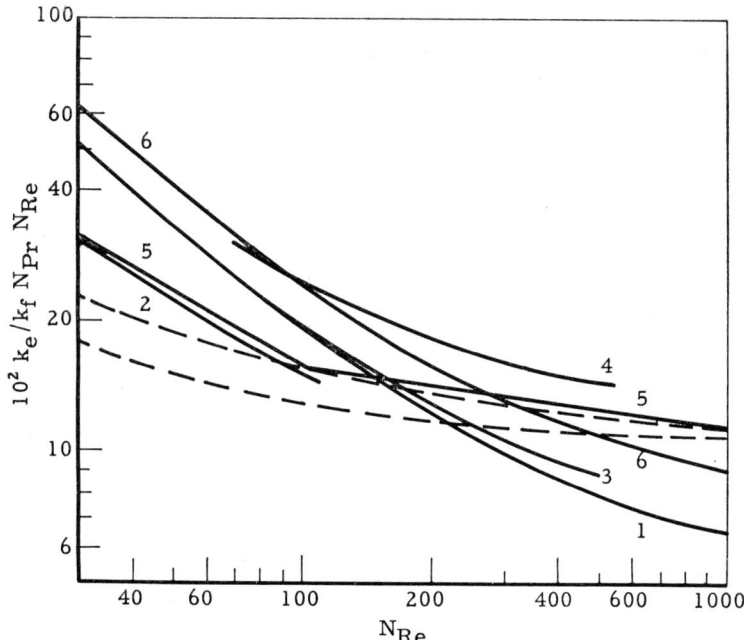

FIG. 3. Radial thermal conductivity. Broken lines are calculated for $k_s = 2 \times 10^{-3}$ and 3×10^{-4} cal./cm. sec. deg. Curves: (1) Aerov and Umnik (A1); (2) Bunnell and others (B9); (3) Campbell and Huntington (C2); (4) Coberly and Marshall (C4); (5) Maeda (M1); (6) Quinton and Storrow (Q1).

the results of the measurements on material diffusion. The effective diffusivity can be measured much more easily than the thermal conductivity, and the evidence for the eddy character of the diffusion is overwhelming. The conclusion that a Peclet number of 10 describes material diffusion at high Reynolds numbers leads directly to a limiting value of 10 for thermal conduction, since enthalpy must be transported at least as effectively as matter.

There is room for substantial improvement in our knowledge of thermal conduction in packed tubes. At present, the effect of conduction through the solid is quite uncertain, and this effect must have some importance at low Reynolds numbers. In view of the very large experimental effort that has gone into this field, it does not seem to be an inviting field to enter, but the serious questions that remain are worthy of attention.

One remark should be added to this discussion. The eddy-conductive term in k_e, containing the heat capacity of the fluid as a factor, is adequate as long as there is no reaction. When a reaction is going on, the very concept of thermal conduction loses its precision, and for an accurate description of the system the equations of Section III must be used.

E. Heat-Transfer Coefficient at the Wall

There is even more uncertainty in estimating the heat-transfer coefficient at the wall of the tube than in estimating the effective thermal conductivity in the bed of catalyst. The measurement is essentially a difficult one, depending either on an extrapolation of a temperature profile to the wall or on determining the resistance at the wall as the difference between a measured over-all resistance and a calculated resistance within the packed bed. The proper exponent to use on the flow rate to get the variation of the coefficient has been reported as 0.33 (C4), 0.47 (C2), 0.5 and 0.77 (H1), 0.75 (A2), and 1.00 (Q1).

In the face of these discrepancies, any choice of a formula for calculating h_w must be to some extent arbitrary. It seems reasonable to depend on the close similarity between the processes involved in transfer to the wall and to the packing. Thoenes and Kramers (T1) have presented a thorough discussion of mass transfer from a particle of packing to the fluid, which can be translated into a treatment of heat transfer by substituting Nusselt and Prandtl numbers for Sherwood and Schmidt numbers. These authors suggest using three terms to represent the contributions associated with conduction in laminar flow, with eddy conduction, and with stagnant conduction. The heat-transfer form of the expression they recommend for flow through a body-centered cubic lattice of spheres normal to the (100) plane is

$$N_{Nu} = 2.42 N_{Re}^{1/3} N_{Pr}^{1/3} + 0.129 N_{Re}^{0.8} N_{Pr}^{0.4} + 1.4 N_{Re}^{0.2} \quad (4\text{-}5)$$

This expression gives a rather good representation of the broken line recommended by Hougen and Watson [see (H6), p. 987] for random packing.

While it is desirable to take account of all effects in a description of the transport of heat, a complete description must await a deeper analysis

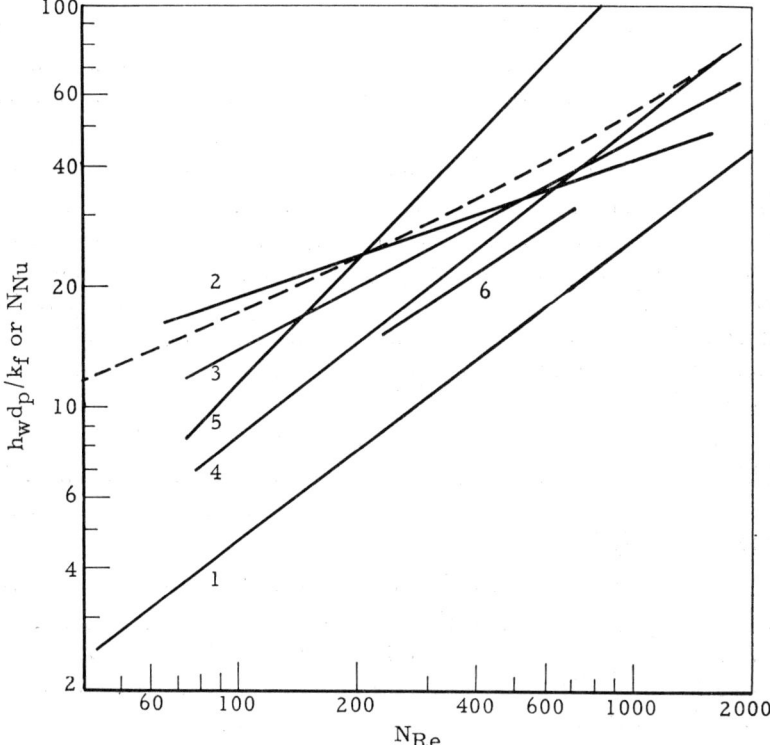

Fig. 4. Heat-transfer coefficient to wall. Broken line is N_{Nu} according to Thoenes and Kramers (T1). Curves: (1) Aerov and Umnik (A2); (2) Coberly and Marshall (C4); (3) Hanratty (cylinders) (H1); (4) Hanratty (spheres) (H1); (5) Quinton and Storrow (Q1); (6) Yagi and Wakao (Y2).

or a larger experimental effort. It is not entirely satisfactory to show the laminar, turbulent, and stagnant conductivities as being simply additive. Nevertheless, the success of the expression given by Thoenes and Kramers in representing a smooth relation between the Nusselt and Reynolds numbers shows that the form of the expression is useful, at least as an empirical formula.

Figure 4 shows some experimental measurements of h_w in comparison with the Thoenes-Kramers formula. As might be expected from the different conditions, the transfer coefficient for the packing is larger than for the wall. A reasonable approximation can be obtained by taking the coefficient for the wall as a constant multiple of the coefficient for the packing. A further approximation, involving dropping the term for stagnant conduction, is admissible if the Reynolds number is restricted to be greater than 40. The simplified expression for the Nusselt number is

$$N_{Nu} = 3.22 N_{Re}^{1/3} N_{Pr}^{1/3} + 0.117 N_{Re}^{0.8} N_{Pr}^{0.4} \qquad (4\text{-}6)$$

and, with a factor of 0.8 to account for the difference between transfer to a particle and to the wall, the heat-transfer coefficient at the wall becomes

$$h_w = (k_f/d_p)(2.58 N_{Re}^{1/3} N_{Pr}^{1/3} + 0.094 N_{Re}^{0.8} N_{Pr}^{0.4}) \qquad (4\text{-}7)$$

Hanratty (H1) has pointed out that there is some evidence that heat transfer to the wall is different with spherical and cylindrical particles. The experiments indicated that the less obstructed flow near the wall in a bed of spheres made the transport to the wall more like what it is in an empty tube, leading to an exponent of 0.77 on the Reynolds number. Hanratty recommends for air flowing among spheres the correlation

$$h_w = (k_f/d_p)(0.25 N_{Re}^{0.77}) \qquad (4\text{-}8)$$

This correlation can be well represented in the range of Reynolds numbers covered by the experiments by an expression of the Thoenes-Kramers form. It is

$$h_w = (k_f/d_p)(0.203 N_{Re}^{1/3} N_{Pr}^{1/3} + 0.220 N_{Re}^{0.8} N_{Pr}^{0.4}) \qquad (4\text{-}9)$$

Until further experimental work has given some reason to change, it would be desirable to use Eq. (4-7) for particles that pack next to the wall like cylinders, and Eq. (4-9) for particles that pack next to the wall like spheres.

F. Friction Factor

The review published by Ergun (E2) provides a definitive description of pressure drop in packed tubes when the ratio of particle diameter to tube diameter is sufficiently low. In addition, although the complicated relationship between the diameter ratio, the fraction void and the friction factor can not be accurately represented without some explicit dependence of the friction factor on the diameter ratio, Ergun showed that his correlation does work for a wide variety of experimental conditions. The friction factor is calculated from the expression

$$f = \frac{1-\epsilon}{\epsilon^3}\left(1.75 + 150\frac{1-\epsilon}{N_{Re}}\right) \qquad (4\text{-}10)$$

with the Reynolds number based on the diameter of spherical packing that gives the same superficial area per unit volume of bed as does the actual packing.

In using this friction factor for a tubular reactor, the Reynolds number is evaluated at some estimated average condition, and then the corresponding friction factor is used for the whole bed. In calculating the axial pressure profile, the average composition and temperature in a cross section are used to estimate the density of the fluid, and this density is used with the average superficial mass velocity to estimate the axial derivative of pressure.

V. Numerical Solution of Equations

A. Automatic Computer Assumed

It will be assumed in discussing the solution of the differential equations relating conditions within a tubular reactor that the calculations are to be carried out with some sort of automatic digital computer. The problem is just manageable with a desk computing machine in the one-dimensional approximation when there is only one independent reaction, but even in this simplest case, this is an expensive way to work, particularly if a fairly large number of solutions is required.

B. One-Dimensional Approximation

If the one-dimensional approximation is adequate, the problem is reduced to a routine integration of a set of ordinary differential equations. Procedures for integrating such sets of equations are given in standard works on numerical analysis (see, for example, H5, M2, and M3). The working equations for simple forward-difference equations are given in the next paragraph.

Equations (3-34), (3-35), and (3-36) may be writen in the form

$$\frac{d\phi_i}{dz} = F_i(\phi_1, \phi_2, \ldots \phi_{j+2}) \tag{5-1}$$

in which the ϕ_i are the conversions, the temperature, and the pressure, and j is the number of independent reactions. Then the F_i are

$$F_1 = fG^2/\rho_f d_p \tag{5-2}$$

$$F_2 = \frac{\Re^T \mathfrak{Q}}{C_p G} - \frac{2U}{r_2 C_p G}(\bar{T} - T_c) \tag{5-3}$$

$$F_i = R/^i G \qquad (2 < i < j + 3) \tag{5-4}$$

It must be remembered that heat capacity, heat of reaction, density, and the rates are, in general, variables in these equations, being known functions of the dependent variables.

When the cooling medium is a boiling liquid, the dependence of its temperature on the length of the tube must be calculated from the conditions in the pool of liquid. When it is a circulating liquid, flowing concurrently with the reacting fluid, its temperature can be calculated from a heat balance. This heat balance is expressed by the equation

$$W_c C_c \frac{dT_c}{dz} = U(2\pi r_2)(T - T_c) \qquad (5\text{-}5)$$

in which W_c is the mass rate of flow of the cooling medium per reactor tube and C_c is its specific heat capacity. With the definition

$$W = 2\pi r_2 U / W_c C_c \qquad (5\text{-}6)$$

the derivative of T_c can be written as

$$F_{j+3} = W(T - T_c) \qquad (5\text{-}7)$$

The working equation, based on the form of Eq. (5-1), is

$$\phi_{i,n+1} = \phi_{i,n} + k F_i(\phi_{1n}, \phi_{2n}, \ldots, \phi_{j+3,n}) \qquad (5\text{-}8)$$

where k is written for the increment in z. As the truncation error in this expression is of the order of k^2, so that the error in the solution is of the order of k, it is advantageous to calculate the solution with two different values of k, and to extrapolate linearly to a zero value for k. Methods having smaller truncation errors are given in the references cited above.

C. Sources of Error

1. Introduction

When the radial variation of temperature must be taken into account, the problem assumes an entirely different character. Each of the equations is now a partial differential equation, and both radial and axial profiles must be calculated; a mesh or network of radial and axial lines is set up, and the temperature and composition are calculated for each intersection. A great deal of work has been done on the formulation of difference equations for solving the related diffusion or heat-conduction equations; most of this has been directed towards the case in which there is only one dependent variable and in which the source is a linear function of that variable. Although the results obtained for one dependent variable are only partially applicable to the multiple-variable problem, some discussion of the way the formulation of the difference equations affects the accuracy and stability of solutions is in order here.

The analysis of a difference equation is greatly simplified if it is thought of as an approximation to the corresponding differential equation at a specified point, which will usually be a mesh point. From this point of view, the approximation to each term arises naturally from considerations of symmetry or convenience, and just as naturally the error in the approximation can be related to the Taylor expansion of the term in question about the reference point.

2. Truncation Error

The accuracy of a solution is affected by two properties of the difference equations and by the rounding error involved in the calculation. The first property of the difference equations is the truncation error, that is, the discrepancy between the differential equation written for some point and the difference equation that is supposed to correspond to it. The second property is the order of the difference equation with respect to the axial step. These effects can be illustrated by the way they appear in the solution of a first order ordinary differential equation.

Suppose the equation

$$y' + by = 0 \qquad (5\text{-}9)$$

with b positive, is to be solved. Let k be the step length. Then a forward-difference approximation to the derivative at $z = nk$ is

$$y_n' \cong (y_{n+1} - y_n)/k \qquad (5\text{-}10)$$

which leads to the difference equation

$$y_{n+1} - y_n + bky_n = 0 \qquad (5\text{-}11)$$

This has the solution

$$y_n = y_0(1 - bk)^n \qquad (5\text{-}12)$$

which may be expanded in the form

$$y_n = y_0 e^{-bkn}(1 - \tfrac{1}{2}b^2k^2n + \cdots) \qquad (5\text{-}13)$$

As the solution of the differential equation is

$$y = y_0 e^{-bz} \qquad (5\text{-}14)$$

and kn is the same as z, the term $-\tfrac{1}{2}b^2k^2n$ gives a measure of the fractional error in the approximate solution, when there is no rounding error. This error is proportional to k when z is fixed. At the same time, it may be seen that there is no tendency for a systematic growth of the rounding error.

3. Satellite Solutions

An alternative approximation to the derivative is the central-difference form

$$y_n' \cong (y_{n+1} - y_{n-1})/2k \tag{5-15}$$

which gives the second-order difference equation

$$y_{n+1} + 2bky_n - y_{n-1} = 0 \tag{5-16}$$

Assuming that an estimate of y_1 is somehow available, the solution of this difference equation is [see (H4), p. 238].

$$y_n = B(-bk + \sqrt{1 + b^2k^2})^n + C(-bk - \sqrt{1 + b^2k^2})^n \tag{5-17}$$

where

$$B = \frac{1}{2}y_0 + \frac{bky_0 + y_1}{2\sqrt{1 + b^2k^2}} \tag{5-18}$$

and

$$C = \frac{1}{2}y_0 - \frac{bky_0 + y_1}{2\sqrt{1 + b^2k^2}} \tag{5-19}$$

The first thing to be observed about Eq. (5-17) is that the second quantity in parentheses is both negative and greater in magnitude than unity. As a result of this situation, the successive powers of this quantity will oscillate increasingly and without limit as n is increased. It is clear, then, that the approximate solution will not even resemble the solution of the differential equation unless the coefficient C is made sufficiently small by a proper choice of y_1. It does not do any good to make the step small, because the quantity multiplied by C is $(-1)^n e^{bz}$ in the first approximation, and so nearly independent of k.

The spurious or satellite term in the solution is introduced by using a second-order difference equation to approximate a first-order differential equation. An extra condition is needed to fix the solution of the second-order equation, and this condition must be that the coefficient of the spurious part of the solution is zero. In the general case of a nonlinear difference equation, no method is available for meeting this condition exactly.

If the coefficient C is exactly zero, B becomes y_0. With this value of B, the solution of Eq. (5-17) may be expanded to the form

$$y_n = y_0 e^{-bkn}(1 + \tfrac{1}{6}b^3k^3n + \cdots) \tag{5-20}$$

In this result, the fractional error at a given value of z, that is, of kn, is proportional in the first approximation to k^2, instead of to k, as in Eq. (5-13). This means that the approximation to the derivative given by Eq. (5-15) is in this sense better than that of Eq. (5-10). Before discussing the difference between these forms in terms of the truncation error, the effect of rounding errors on the numerical solution of Eq. (5-16) must be considered.

It was seen that if y_1 is not chosen to make C practically zero, the solution of Eq. (5-16) oscillates widely. The same requirements apply to the subsequent steps: thus, if a significant error is introduced at any step by rounding, that error will be magnified as the solution proceeds. For this reason, a high accuracy in the individual numerical operations is required when a difference equation that suffers from this sort of instability is used. In using this central-difference approximation with nonlinear equations, however, the problem of getting started with a proper value of y_1 is usually more serious than the problem of controlling roundoff errors.

4. Estimation of Order of Error

The approximations of both Eqs. (5-10) and (5-15) are for the derivative of y at $z = nk$. The error involved in the approximations may be estimated by comparing the value of y_{n+1} calculated from the corresponding difference equation with the exact value. The forward-difference formula gives

$$y_{n+1} \cong y_n + k y_n' \qquad (5\text{-}21)$$

which is to be compared with the Taylor expansion

$$y_{n+1} = y_n + k y_n' + \tfrac{1}{2} k^2 y_n'' + O(k^3) \qquad (5\text{-}22)$$

In this case, then, the truncation error in the expression for y_{n+1} is dominated by a term proportional to k^2. The central-difference formula gives

$$y_{n+1} \cong y_{n-1} + 2 k y_n' \qquad (5\text{-}23)$$

instead of the complete expression

$$y_{n+1} = y_{n-1} + 2 k y_n' + \tfrac{1}{3} k^3 y_n''' + O(k^5) \qquad (5\text{-}24)$$

from which it may be seen that the truncation error is of the order of k^3.

These truncation errors in the stepwise calculation lead plausibly to the magnitude of the errors at the nth step that are indicated in Eqs. (5-13) and (5-20). The accumulation at each step of an error of a certain magnitude gives after n steps an error of n times the magnitude. Since the number of steps to reach a given length is inversely proportional to the size of the step, it may be expected that the order of the truncation error in the solution will be one less than the order of the truncation error for one step. Accordingly, the second-order truncation error in the forward-difference expression gives an error of the order of k in the solution, while the third-order truncation error in the central-difference formula gives an error of the order of k^2 in the solution.

D. Stability in Partial Difference Equations

Each of the kinds of error exhibited in the numerical solution of the ordinary differential equation can appear in the solution of the partial difference equations used to solve a partial differential equation. In addition, a different kind of instability can appear, arising from a different kind of spurious or satellite solution. A good discussion of this kind of instability is given in a paper by O'Brien et al. (O1); a more rigorous treatment is given by Lowan (L2). The essential point is that the step from one profile to the next is made by calculating each value on the new profile as a linear combination of the values on old profiles, together with any source term involved. Equation (5-35) given below is an example of the nature of the step. An equivalent representation of a step, then, is to say that the set of values on the old profile is multiplied by a square matrix to get the set of values on the new profile. If the set of values on a profile is thought of as a linear combination of characteristic vectors of the matrix, it may be seen that one contribution to the solution after a large number of steps is the same linear combination of characteristic vectors, each now multiplied by a large power of the corresponding characteristic value. From this observation it may be inferred that if any characteristic value of the matrix of coefficients in the partial difference equation is greater in magnitude than unity, the solution cannot be stable.

The classical example of the instability that results is the difference formulation given by Richardson (R2) for the heat-conduction equation. He proposed using a central-difference formula [as in Eq. (5-15)] for the derivative with respect to time, together with the usual central-difference formula for the space derivative. The resulting equation has a truncation error of the third order in both time and space steps, but the solution is unstable for any length of step. Thus, this natural and accurate formulation is not available for use if more than a few steps are to be taken.

E. Introduction of r^2 as Radial Variable

In working with a tubular reactor, or with some other form that can be well enough represented by an equivalent tubular reactor, a significant advantage in accuracy is gained by taking account of the symmetry about the axis. Since all the properties in the system must be even functions of radial position, the square of the radial coordinate is preferable to the first power as one of the independent variables. If the substitution $u = r^2$ is made, the operators $(1/r)(\partial/\partial r)r$ and $(\partial/\partial r)$ change to $2(\partial/\partial u)u^{1/2}$ and $2u^{1/2}(\partial/\partial u)$. This means that in Eqs. (3-26) and (3-27), quantities like $(1/r)(\partial/\partial r)[r(\partial T/\partial r)]$ become $4(\partial/\partial u)[u(\partial T/\partial u)]$, and quantities like $(\partial G/\partial r)(\partial T/\partial r)$ become $4u(\partial G/\partial u)(\partial T/\partial u)$.

F. Explicit Partial Difference Equations

1. General Discussion

There are two practical approaches in formulating the working difference equations for the packed tubular reactor. The simple one is to use a forward-difference equation for the axial derivative and central-difference formulas for the radial derivatives. The leading terms in the truncation error are then proportional to k^2 and to kh^2, where h and k are written for the radial and axial steps. This means that, in order to take advantage of the accuracy of the approximations for the radial derivatives, k must have the same order of magnitude as h^2, so that k^2 and kh^2 will be comparable. This is a serious limitation on the length of the axial step that can be used.

2. Condition for Stability

There is another limitation that has nearly the same effect. If the equation is written

$$\frac{\partial \phi}{\partial z} = \omega u \frac{\partial^2 \phi}{\partial u^2} + \text{(other terms)} \qquad (5\text{-}25)$$

so that the corresponding difference equation has the form

$$\frac{\phi_{m,n+1} - \phi_{m,n}}{k} = \omega u \frac{\phi_{m-1,n} - 2\phi_{m,n} + \phi_{m+1,n}}{h^2} + \text{(other terms)} \qquad (5\text{-}26)$$

the solution may be unstable if

$$k > h^2/2\omega \qquad (5\text{-}27)$$

In Eq. (5-26), the first index is used for the radial position, the second for the axial position. The limitation of the inequality (5-27) does not restrict the order of magnitude of k, but it does put a quantitative limit on its value.

When more than one dependent variable appears, with the variables linked through source terms, that is, through the rates, the limit on the axial step may be greater or less. The conditions imposed at the boundary can also have some effect on the stability. All that can be said in the more general case is that the inequality (5-27) provides a rough rule, but that somewhat shorter steps may be required.

Two considerations regarding truncation error that enter into the derivation of the partial difference equations should be pointed out. In some published formulations of these equations, the first radial derivative has been approximated by a forward-difference expression (K1, S5, W1). This unsymmetrical formula has no advantage over a symmetrical or central-difference expression, but has a greater—lower order—truncation error. The central-difference approximation

$$\left.\frac{\partial \phi}{\partial u}\right|_{mn} \cong (\phi_{m+1,n} - \phi_{m-1,n})/2h \tag{5-28}$$

is used in the equations given below.

When equal increments in r are used, the symmetry about the axis can be depended on to derive an accurate expression for the required radial derivative at the axis. When, as in the present formulation, equal increments in r^2 are used, the first increment in r is so large that the analogous formula gives too great a truncation error. If r^2 or u is considered as the radial variable, no useful additional property of symmetry about the axis is available. In order to keep the truncation error small, a three-point forward-difference approximation is used for the radial derivative at the axis. This accounts for the appearance of $\phi_{2,n}$ in the formula for $\phi_{0,n+1}$.

3. Boundary Condition at the Wall

A convenient way to take care of the boundary condition at the wall is to introduce a so-called image point, a virtual axial profile one increment outside the wall. The value of the variable on this profile is found from the boundary condition, and then used just as though it were at a regular mesh point in calculating the value at the wall for the next step.

4. Formulation of Equations

For a reason that will become apparent when the other type of difference equation is given, the terms that are not linear in the dependent variables are separated out for special treatment. With this distinction in mind, Eqs. (3-26) and (3-27) are written schematically in the form

$$\frac{\partial \phi}{\partial z} = \omega\left(u\frac{\partial^2 \phi}{\partial u^2} + \frac{\partial \phi}{\partial u}\right) + \eta u \frac{\partial \phi}{\partial u} + S \tag{5-29}$$

with the definitions

$$\omega = 4d_p/N_{Pe}r_2^2 \tag{5-30}$$

and
$$\eta = \frac{4d_p}{N_{Pe}r_2{}^2 G}\frac{\partial G}{\partial u} \qquad (5\text{-}31)$$

ω and η as thus defined are independent of z. When the variable is conversion, the term S is just the virtual rate divided by the molar mass velocity. When the variable is temperature, S includes the terms associated with the variation of heat capacity. The relations are

$\phi = \mathbf{x}: \quad S = \mathfrak{R}/G \qquad (5\text{-}32)$

$$\phi = T: \quad S = \frac{\mathfrak{R}^T \mathfrak{Q}}{C_p G} + \frac{4u d_p}{C_p N_{Pe} r_2{}^2}(\mathbf{Y}^0 + \mathbf{x}^T \mathbf{S}_R)\frac{\partial \mathbf{C}_p{}^T}{\partial T}\left(\frac{\partial T}{\partial u}\right)^2$$
$$+ \frac{8u d_p}{C_p N_{Pe} r_2{}^2}\frac{\partial \mathbf{x}^T}{\partial u}\mathbf{S}_R \mathbf{C}_p{}^T \frac{\partial T}{\partial u} + \frac{4}{C_p G r_2{}^2}\frac{\partial}{\partial u}\left(u k_p \frac{\partial T}{\partial u}\right) \qquad (5\text{-}33)$$

With these definitions, the difference equations are the following:

$$\phi_{0,n+1} = \left(1 - \frac{3k\omega_0}{2h}\right)\phi_{0,n} + \frac{2k\omega_0}{h}\phi_{1,n} - \frac{k\omega_0}{2h}\phi_{2,n} + kS_{0,n} \qquad (5\text{-}34)$$

$$\phi_{m,n+1} = \frac{k}{2h}[\omega_m(2m-1) - mh\eta_m]\phi_{m-1,n} + \left(1 - \frac{2km\omega_m}{h}\right)\phi_{m,n}$$
$$+ \frac{k}{2h}[\omega_m(2m+1) + mh\eta_m]\phi_{m+1,n} + kS_{m,n} \qquad (0 < m < M) \qquad (5\text{-}35)$$

$$\phi_{M,n+1} = 2kM^2\omega_M\phi_{M-1,n}$$
$$+ \left(1 - 2kM^2\omega_M - \frac{kN_{Bi}}{2}[\omega_M(2M+1) + \eta_M]\right)\phi_{M,n}$$
$$+ \frac{kN_{Bi}}{2}[\omega_M(2M+1) + \eta_M]T_{C,n} + kS_{M,n} \qquad (5\text{-}36)$$

In these equations, the index M is used for the points at the wall. Since u, the square of the dimensionless radius, varies from zero to unity, M is just the reciprocal of h.

In practice, this set of equations is written with ϕ representing each conversion in turn, and finally the temperature. When ϕ represents a conversion, the value of N_{Bi} is taken to be zero; when it is temperature, N_{Bi} takes the value appropriate to the conditions next to the wall.

In conjunction with these equations, the temperature of the cooling medium and the pressure may be followed with the simple forward-difference equations, neglecting any radial variation of pressure. The equations are

$$T_{C,n+1} = T_{C,n} + kW(T_{M,n} - T_{C,n}) \qquad (5\text{-}37)$$

and

$$P_{n+1} = P_n + kf\overline{G}_n{}^2 \overline{M}_n{}^2/\rho_n d_p \qquad (5\text{-}38)$$

In Eq. (5-37), W is defined in terms of the heat-transfer coefficient across the wall instead of an over-all coefficient to the average temperature in the cross section of the tube. The definition is

$$W = 2\pi r_2 h_t / W_c C_c \tag{5-39}$$

The derivatives appearing in S can be approximated by finite differences on the nth profile. The quantity k_p can be treated as a constant, since the uncertainty in its determination is so great that any refinement would be wasted.

G. Implicit Partial Difference Equations

1. General Discussion

An alternative approach to the formulation of the difference equations is to sacrifice some convenience in solving the equations for the advantage of being able to take longer steps. It has been found that a central-difference approximation to the axial derivative can be used if the radial derivatives are approximated by suitable averages of differences calculated on the new profile and on one or more old profiles. To fix the thoughts, suppose that the $(n + 1)$th profile is to be calculated. Then a suitable average to use for the radial derivative would have equal weights on the $(n + 1)$th and the $(n - 1)$th profiles, and the rest on the nth profile. This arrangement is symmetrical about the nth profile, where the differential equation is being approximated, so that the truncation error is small. If the weights on the $(n + 1)$th and $(n - 1)$th profiles are larger than $\frac{1}{4}$, that is, if the weight on the nth profile is less than $\frac{1}{2}$, the solution of the difference equation is stable for any length of step (O1).

A convenient choice of weight is $\frac{1}{2}$ for the new profile, giving a weight of zero on the nth profile. In the general case, the approximation to the second radial derivative at the point m,n is

$$\frac{\partial^2 \phi}{\partial u^2} \cong \frac{1}{h^2} [w(\phi_{m-1,n-1} - 2\phi_{m,n-1} + \phi_{m+1,n-1}$$
$$+ (1 - 2w)(\phi_{m-1,n} - 2\phi_{m,n} + \phi_{m+1,n})$$
$$+ w(\phi_{m-1,n+1} - 2\phi_{m,n+1} + \phi_{m+1,n+1})] \tag{5-40}$$

and the approximation to the first derivative takes a similar form. With expressions like these for the radial derivatives, and with a central-difference approximation to the axial derivative, the difference equation corresponding to Eq. (5-29) when $w = 0.5$ is

$$\frac{\phi_{m,n+1} - \phi_{m,n-1}}{2k}$$

$$= \frac{m\omega_m}{2h}(\phi_{m-1,n-1} - 2\phi_{m,n-1} + \phi_{m+1,n-1} + \phi_{m-1,n+1} - 2\phi_{m,n+1} + \phi_{m+1,n+1})$$

$$+ (\omega_m + mh\eta_m)\frac{1}{4h}(\phi_{m+1,n-1} - \phi_{m-1,n-1} + \phi_{m+1,n+1} - \phi_{m-1,n+1}) + S_{m,n}$$

(5-41)

At this point it may be seen why some terms in addition to those involving reaction rates are grouped in S. The coefficients ω and η may depend on the radial position, but they can be treated as independent of temperature and composition, and accordingly they can be assigned values at each mesh point. The terms appearing in S, on the other hand, have coefficients that are affected by the dependent variables; since the unknown values of the dependent variables at three mesh points in the new profile appear in the difference equation, it is important for them to have known coefficients, so that the set of simultaneous equations can be solved in a straightforward way.

2. Boundary Conditions

The boundary conditions are used in setting up equations for the values at the axis and the wall much as for the forward-difference equations. The principal change is that, in order to control the truncation error, the first difference equation represents the differential equation localized on the nth profile a distance $(\frac{1}{4})h$ from the axis, instead of on the axis. The difference between the temperature of the reacting fluid and of the cooling medium is taken as the average at the $(n-1)$th and $(n+1)$th profile, and similarly the density and molecular weight used to calculate the change in pressure are obtained as averages for the profiles.

3. Formulation of Equations

The working equations are

$$\left[\frac{3}{4} + \frac{2kw}{h}\left(\omega_{1/4} + \frac{1}{4}h\eta_{1/4}\right)\right]\phi_{0,n+1} + \left[\frac{1}{4} - \frac{2kw}{h}\left(\omega_{1/4} + \frac{1}{4}h\eta_{1/4}\right)\right]\phi_{1,n+1}$$

$$= \frac{2k}{h}(1-2w)\left(\omega_{1/4} + \frac{1}{4}h\eta_{1/4}\right)\phi_{0,n} + \frac{2k}{h}(1-2w)\left(\omega_{1/4} + \frac{1}{4}h\eta_{1/4}\right)\phi_{1,n}$$

$$+ \left[\frac{3}{4} - \frac{2kw}{h}\left(\omega_{1/4} + \frac{1}{4}h\eta_{1/4}\right)\right]\phi_{0,n-1} + \left[\frac{1}{4} + \frac{2kw}{h}\left(\omega_{1/4} + \frac{1}{4}h\eta_{1/4}\right)\right]\phi_{1,n-1}$$

$$+ \frac{3}{2}kS_{0,n} + \frac{1}{2}kS_{1,n}$$

(5-42)

$$-\frac{wk}{h}[\omega_m(2m-1) - mh\eta_m]\phi_{m-1,n+1} + \left(1 + \frac{4wkm\omega_m}{h}\right)\phi_{m,n+1}$$

$$-\frac{wk}{h}[\omega_m(2m+1) + mh\eta_m]\phi_{m+1,n+1}$$

$$= \frac{k(1-2w)}{h}[\omega_m(2m-1) - mh\eta_m]\phi_{m-1,n}$$

$$-\frac{4km\omega_m(1-2w)}{h}\phi_{m,n} + \frac{k(1-2w)}{h}[\omega_m(2m+1) + mh\eta_m]\phi_{m+1,n}$$

$$+\frac{wk}{h}[\omega_m(2m-1) - mh\eta_m]\phi_{m-1,n-1} + \left(1 - \frac{4wkm\omega_m}{h}\right)\phi_{m,n-1}$$

$$+\frac{wk}{h}[\omega_m(2m+1) + mh\eta_m]\phi_{m+1,n-1} + 2kS_{m,n} \qquad (5\text{-}43)$$

$$-4wkM^2\omega_M\phi_{M-1,n+1}$$

$$+\left\{1 + 4wkM^2\omega_M + \frac{wkN_{\text{Bi}}}{1+kW}[\omega_M(2M+1) + \eta_M]\right\}\phi_{M,n+1}$$

$$= 4kM^2\omega_M(1-2w)\phi_{M-1,n}$$

$$- k(1-2w)\{4M^2\omega_M + N_{\text{Bi}}[\omega_M(2M+1) + \eta_M]\}\phi_{M,n} + 4wkM^2\omega_M\phi_{M-1,n-1}$$

$$+\left\{1 - 4wkM^2\omega_M - \frac{wkN_{\text{Bi}}}{1+kW}[\omega_M(2M+1) + \eta_M]\right\}\phi_{M,n-1}$$

$$+ kN_{\text{Bi}}[\omega_M(2M+1) + \eta_M]\left[(1-2w)T_{C,n} + \frac{2w}{1+kW}T_{C,n-1}\right]$$

$$+ 2kS_{M,n} \qquad (5\text{-}44)$$

$$T_{C,n+1} = \frac{1-kW}{1+kW}T_{C,n-1} + \frac{kW}{1+kW}(T_{M,n-1} + T_{M,n+1}) \qquad (5\text{-}45)$$

$$p_{n+1}^2 = p_{n-1}^2 + 2kf\left(\frac{\overline{G}_{n-1}^2\overline{M}_{n-1}^2 P_{n-1}}{\bar{\rho}_{n-1}d_p} + \frac{\overline{G}_{n+1}^2\overline{M}_{n+1}^2 P_{n+1}}{\bar{\rho}_{n+1}d_p}\right) \qquad (5\text{-}46)$$

4. *Pressure*

Equation (5-46) is written for the case of a gaseous reactant, in which P/ρ is rather insensitive to P. This form of the equation for pressure drop is used, rather than one analogous to Eq. (5-38), because the long steps that can be taken with this set of difference equations demand an equation with a small truncation error. At the same time, the requirement for stability makes it impossible to use the conditions on the nth profile to fix the difference in pressure, so that an average on the $(n-1)$th and $(n+1)$th profiles must be used. Equation (5-46) is an explicit equation for P_{n+1}^2 if P_{n+1}/ρ_{n+1} can be estimated without using an accurate value of P_{n+1}. If the reactant is a liquid, the appropriate equation is

$$P_{n+1} = P_{n-1} + kf\left(\frac{\overline{G}_{n-1}{}^2\overline{M}_{n-1}{}^2}{\rho_{n-1}d_p} + \frac{\overline{G}_{n+1}{}^2\overline{M}_{n+1}{}^2}{\rho_{n+1}d_p}\right) \quad (5\text{-}47)$$

5. Stability

Although the solution of these equations is stable in the technical sense that the fluctuations do not grow without limit, it must be observed that there is a serious practical limitation on their use when the source term is small, and that they are useless when the source term is extremely small. The limitation is associated with the appearance of fluctuations of limited magnitude, which may be large enough to obscure the results. The common situation in a heat-exchanger type of reactor is that the rate of heat production and of heat transfer are fairly well balanced, in which case the solutions are smooth. In the extreme case of the heat-conduction equation without source, however, the fluctuations quickly develop to the same magnitude as the desired solution. Jim Douglas (D5) has devised a cure for this difficulty. His scheme is to calculate the radial derivatives with weights $\frac{1}{6}$, $\frac{1}{6}$, $\frac{1}{6}$, and $\frac{1}{2}$ on the $(n-2)$th, $(n-1)$th, nth, and $(n+1)$th profiles, respectively. In spite of the fact that more profiles must be stored and used in the calculation, this formulation is better than the forward-difference or explicit scheme when the storage space is available.

6. Technique for Solving the Simultaneous Equations

Equations (5-42) to (5-44) constitute a set of simultaneous linear equations in the unknown values of the quantity on the new profile, each of the sets for temperature and the conversions being independent. The matrix of coefficients is the same for each conversion, and differs for the temperature only in one element, the one containing the heat-transfer coefficient. An important property of each matrix of coefficients is that it is independent of the axial position, so that for the purpose of the calculation it is a constant matrix. Another important property is that only three diagonals of the matrix contain elements that are not zero. These properties make it possible to throw the calculation of the unknown quantities into a very simple form. The essential feature of the calculation is that coefficients in two two-term recursion formulas are constructed from the matrix elements, and then these coefficients are used to calculate, first, a set of ancillary quantities, and then the desired quantities. The procedure is exactly what would be used in eliminating the unknown quantities successively from the equations and then substituting the quantities that have been calculated back into the equations in order to calculate the others.

Equations (5-42) to (5-44) are transformed by dividing each one by the magnitude of the first coefficient appearing on the left-hand side. After this transformation, they can be written schematically in the form

$$\phi_0 - b_0\phi_1 = C_0 \tag{5-48}$$

$$-\phi_{m-1} + a_m\phi_m - b_m\phi_{m+1} = C_m \tag{5-49}$$

$$-\phi_{M-1} + a_M\phi_M = C_M \tag{5-50}$$

In these equations, the second subscript, $n + 1$, has been dropped; all the terms on the right-hand side have been collected into the C_i, and the a_i and b_i are constants for a given mesh size, defined by the relation of these equations to the set (5-42) to (5-44). The next operation is to construct a set of coefficients from the a_i and b_i, by the following procedure.

$$p_0 = b_0 \tag{5-51}$$

$$p_m = b_m/(a_m - p_{m-1}) \quad (m = 1, 2, 3, \ldots, M - 1) \tag{5-52}$$

$$q_m = 1/(a_m - p_{m-1}) \quad (m = 1, 2, 3, \ldots, M) \tag{5-53}$$

These coefficients are computed at the start of the calculation and used in each succeeding step that is based on the same mesh size.

The unknown quantities are calculated by using two recursion formulas in succession. They are

$$D_0 = C_0 \tag{5-54}$$

$$D_m = q_m(D_{m-1} + C_m) \quad (m = 1, 2, 3, \ldots M) \tag{5-55}$$

$$\phi_M = D_M \tag{5-56}$$

$$\phi_m = D_m + p_m\phi_{m+1} \quad (m = 0, 1, 2, \ldots, M - 1) \tag{5-57}$$

With this method of calculation, the unknown quantities are found from the C_m with only two multiplications per mesh point, and the rounding errors are kept small.

7. Technique for Starting

There is still one problem in connection with using Eqs. (5-42) to (5-46) that has not been touched upon, namely, the selection of a procedure for getting started. The difference equations show how to get the solution for a profile when the values on the two preceding profiles are known, but do not tell how to take a step when only the initial conditions are given. Equations (5-34) to (5-38) cannot be used, because they give too large a truncation error with a large axial step. One simple way to proceed is to start with short steps, using the explicit formulas, and then to double the length of the step repeatedly until the desired length of step is reached. The fastest and most convenient procedure would be to double the length of the step at every step after duplicating the first short step. Unfortunately, this procedure leads to irregular behavior of the solution, apparently because the initial profile represents the $(n-1)$th profile for all the steps taken until the final step length is reached. In order to get a good approximation to the solution of the differential equation, it is necessary to take two or three steps of each size during the starting period.

8. *Special Treatment of Radiative Heat Transfer*

A troublesome situation arises when radiation makes a large contribution to the transport of heat. The corresponding term in the difference equation is one of those that were grouped with the source term, to be evaluated on the nth or intermediate profile, because it has a variable coefficient. If this term becomes as large as the term accounting for eddy transport, the solution of the difference equation becomes unstable. The difficulty can be surmounted by representing the coefficient of $(\partial T/\partial u)$ as the sum of a constant value somewhere near the middle of its range, and the difference from this constant value. Then the part of the term having the constant coefficient can be calculated as the average of the approximations on the $(n-1)$th and $(n+1)$th profiles, leaving the remainder to be evaluated on the nth profile. The variable factor in the coefficient is then broken into two parts, as follows:

$$\frac{T^3}{C_p} = \left(\frac{T^3}{C_p}\right)_{T_1} + \left[\frac{T^3}{C_p} - \left(\frac{T^3}{C_p}\right)_{T_1}\right] \tag{5-58}$$

By choosing T_1, the reference temperature, so that the value of T^3/C_p at T_1 is somewhat higher than one-half of its maximum value, the variable part is made smaller than the constant part. The variation of heat capacity with conversion is not likely to be large enough to make trouble, but in any case a value of heat capacity corresponding to an intermediate conversion can be used.

H. Adiabatic Reactor

1. *General Case*

By using the reciprocal space velocity as an independent variable, the adiabatic reactor can be treated as a one-dimensional reactor. Equations (3-31) and (3-8) provide the basis for the calculation. Put in terms of the virtual conversions, these equations are

$$\frac{d\mathbf{x}}{d\sigma} = \mathcal{R} \tag{5-59}$$

and
$$H = Y^0H^T + x^TS_RH^T \tag{5-60}$$
$$= \text{constant}$$

Equation (5-60) can be represented adequately in nearly all cases with terms of not more than the second degree in temperature, so that the temperature required to determine the rate can be calculated. If the heat capacity does not vary too much, Eq. (5-60) can be used as a linear relation in temperature.

2. *Only One Independent Reaction*

When there is only one independent reaction, Eq. (5-60) establishes a unique relation between conversion and temperature for a given initial condition. In this case, it may be advantageous to calculate the reciprocal space velocity from the conversion by a quadrature, instead of integrating (Eq. 5-59) as written. The relation is

$$\sigma(x_1) = \int_0^{x_1} \frac{dx}{R} \tag{5-61}$$

Some quadrature formula, such as Simpson's rule, gives the solution faster than does numerical integration of the differential equation by a general method.

I. Illustrations

As an illustration of the use of these methods, a simple example of a schematic catalytic reaction may be treated. Two reactions are supposed to occur in a constant ratio. The reactions are

and
$$A + B = C + D \quad (\Delta H = -84.7 \text{ kcal.})$$
$$A + 2B = 2E \quad (\Delta H = -79.3 \text{ kcal.})$$

The combined reaction is

$$A + 1.068B = 0.932C + 0.932D + 0.136E \quad (\Delta H = -84.3 \text{ kcal.})$$

The rate of the reaction has been found to be

$$R = \frac{k_R P N_A}{1 + \beta P N_c} \exp\left(\frac{A}{600} - \frac{A}{T}\right)$$

with the parameters

$$k_R = 6.69 \times 10^{-7} \text{ mole./cm.}^3 \text{ sec. atm.}$$
$$\beta = 0.69 \text{ atm.}^{-1}$$
$$A = 1.158 \times 10^{-4} \text{ deg.}$$

Although the rate of the side reaction is specified only approximately as being proportional to the rate of the main reaction, the maximum temperature is limited to 380°C.

The feed to the reactor is a mixture of A and B with an inert gas, with the composition

$$N_A = 0.095$$
$$N_B = 0.887$$

The following properties are estimated for this mixture at 600°K.

$$\overline{M} = 41.4 \text{ g./mole}$$
$$C_p = 28.7 \text{ cal./mole deg.}$$
$$\mu = 1.54 \times 10^{-4} \text{ g./cm. sec.}$$
$$k_f = 1.24 \times 10^{-4} \text{ cal./cm. sec. deg.}$$

The following dimensions and conditions for the reactor are to be studied. The catalyst pellets are in the form of short cylinders.

$r_2 = 1.87$ cm.
$Z = 10^3$ cm.
$d_p = 0.318$ cm.
$\epsilon = 0.39$
$k_s = 1.6 \times 10^{-3}$ cal./cm. sec. deg.
$T_c = 320°C. = 593°K.$
$P = 14.62$ atm.
$G = 1.942 \times 10^{-3}$ (initial mol. wt.)/cm.² sec.

The tubes are cooled by a boiling liquid, with an effective heat-transfer coefficient of 2.85×10^{-2} cal./cm.² sec. deg.

From these conditions and properties, the transport properties are calculated.

$$N_{\text{Pr}} = \frac{C_p \mu}{\overline{M} k_f} = 0.86$$

$$N_{\text{Re}} = \frac{\overline{M} G d_p}{\mu} = 166$$

$N_{\text{Pe}} = 10.1$ \hfill (Fig. 2)

$N_{\text{Nu}} = 23.4$ \hfill [Eq. (4-6)]

$h = 9.1 \times 10^{-3}$ cal./cm.² sec. deg.

$h_w = (0.8)h = 7.3 \times 10^{-3}$ cal./cm.² sec. deg. \hfill [Eq. (4-7)]

$$\frac{1}{h_t} = \frac{1}{h_w} + \frac{1}{2.85 \times 10^{-2}}, \text{ or}$$

$$h_t = 5.8 \times 10^{-3} \text{ cal./cm.}^2 \text{ sec. deg.}$$

$$k_c = \frac{0.6 h d_p k_s}{2k_s + 0.7 h d_p} = 5.3 \times 10^{-4} \text{ cal./cm. sec. deg.} \quad \text{[Eq. (4-3)]}$$

The conduction by radiation is based on an average temperature of 630°K and an emissivity of unity for broken-in catalyst. It is

$$k_r = (2)(1.37 \times 10^{-9})(630)^3(0.318) \quad \text{[Eq. (4-4)]}$$
$$= 2.2 \times 10^{-4} \text{ cal./cm. sec. deg.}$$

Then the effective thermal conductivity is

$$k_e = \frac{C_p G d_p}{N_{Pe}} + k_c + k_e$$
$$= 1.75 \times 10^{-3} + 0.53 \times 10^{-3} + 0.22 \times 10^{-3}$$
$$= 2.50 \times 10^{-3} \text{ cal./cm. sec. deg.}$$

$$N_{Bi} = \frac{r_2 h_t}{k_e} = 4.32$$

The first questions to be answered are whether the pressure drop and the difference in temperature between the fluid and the catalyst are large enough to be important. Again using 630°K. as a representative temperature, the density of the mixture is found to be 1.17×10^{-2} g./cm.³ The friction factor is

$$f = \frac{1 - \epsilon}{\epsilon^3}\left(1.75 + 150 \frac{1 - \epsilon}{N_{Re}}\right) \quad \text{[Eq. (4-10)]}$$
$$= 23.8$$

Then the pressure drop is

$$\Delta P = (23.8) \frac{[(41.4)(1.942 \times 10^{-3})]^2}{(1.17 \times 10^{-2})(0.318)}$$
$$= 2.46 \times 10^4 \text{ g./cm. sec.}^2$$
$$= 2.4 \times 10^{-2} \text{ atm.}$$

which is a small enough fraction of the pressure to be negligible.

A representative rate of reaction is for the conditions $N_A = 0.06$, $T = 630$. The rate is then about 7×10^{-7} mole/cm.³ sec., or, referred to the outside surface area of the catalyst particles, $7 \times 10^{-7}/13$, or 6×10^{-8} mole/cm.² sec. The corresponding flux of heat is 5×10^{-3} cal./cm.² sec., which, with a heat-transfer coefficient of 9×10^{-3} cal./cm.² sec. deg., gives a temperature difference of about 0.6°C.

DESIGN OF PACKED CATALYTIC REACTORS

The example will be carried through without taking account of the variation of heat capacity with conditions or of the variation of mass velocity with radial position, because the computing program that was available for this calculation does not accommodate these effects. As pointed out above, the extra terms arising from these effects are treated as contributing to the source term in the equation for the temperature.

The essential details of the calculation with two radial increments are shown. First the difference equations are written in literal form [see Eqs. (5-42) to 5-44)].

$(0.75 + k\omega/h)\phi_{0,n+1} + (0.25 - k\omega/h)\phi_{1,n+1}$
$= (0.75 - k\omega/h)\phi_{0,n-1} + (0.25 + k\omega/h)\phi_{1,n-1} + 1.5kS_{0,n} + 0.5kS_{1,n}$
$-0.5(k\omega/h)\phi_{0,n+1} + (1 + 2k\omega/h)\phi_{1,n+1} - 1.5(k\omega/h)\phi_{2,n+1}$
$= 0.5(k\omega/h)\phi_{0,n-1} + (1 - 2k\omega/h)\phi_{1,n-1} + 1.5(k\omega/h)\phi_{2,n-1} + 2kS_{1,n}$
$-2kM^2\omega\phi_{1,n+1} + [1 + 2kM^2\omega + k\omega N_{\text{Bi}}(M + 0.5)]\phi_{2,n+1} = 2kM^2\omega\phi_{1,n-1}$
$+ [1 - 2kM^2\omega - k\omega N_{\text{Bi}}(M + 0.5)]\phi_{2,n-1} + 2kS_{2,n} + 2k\omega N_{\text{Bi}}(M + 0.5)T_c$

The table below shows what the symbols represent in the separate equations for conversion and temperature.

ϕ	x	T
ω	$4d_p/r_2^2 N_{\text{Pe}}$	$4k_e/r_2^2 C_p G$
N_{Bi}	0	4.32
S	R/G	$QR/C_p G$

The numerical values for h and M were not used in these equations because the literal form indicates better how they enter. The values 0.5 and 2 are inserted for h and M, and k is taken to be 5 cm. Each of the resulting equations is divided through by the absolute value of, the coefficient of the first term on the left, giving the set of working equations. The working equations for the temperature are written in terms of the variable $t = T - T_c$, a transformation that has the effect of making T_c the origin for the scale of temperature.

$x_{0,n+1} - 0.09951 x_{1,n+1}$
$= 0.35074 x_{0,n-1} + 0.54975 x_{1,n-1} + 3.4777 \times 10^3 R_{0,n} + 1.1592 \times 10^3 R_{1,n}$
$-x_{0,n+1} + 9.5478 x_{1,n+1} - 3x_{2,n+1}$
$= x_{0,n-1} + 1.5478 x_{1,n-1} + 3x_{2,n-1} + 2.8567 \times 10^4 R_{1,n}$
$-x_{1,n+1} + 1.6935 x_{2,n+1}$
$= x_{1,n-1} - 0.30653 x_{2,n-1} + 3.5709 \times 10^3 R_{2,n}$

$t_{0,n+1} - 0.20944 t_{1,n+1}$

$= 0.18585 t_{0,n-1} + 0.60472 t_{1,n-1} + 8.9742 \times 10^6 R_{0,n} + 2.9914 \times 10^6 R_{1,n}$

$-t_{0,n+1} + 7.8841 t_{1,n+1} - 3 t_{2,n+1}$

$= t_{0,n-1} - 0.11589 t_{1,n-1} + 3 t_{2,n-1} + 5.8788 \times 10^7 R_{1,n}$

$-t_{1,n+1} + 2.8350 t_{2,n+1} = t_{1,n-1} - 1.8639 t_{2,n-1} + 7.3485 \times 10^6 R_{2,n}$

From this point on, the calculation is best shown in tabular form. The profiles at $z = 90$ cm. and $z = 95$ cm. are used to calculate the profiles at $z = 100$ cm.

	$z = 90$ cm.		$z = 95$ cm.		
m	$10^2 x$	$t\ (C)$	$10^2 x$	$t\ (C)$	$10^7 R$ mole/cm.³ sec.
0	4.7618	42.607	4.8869	39.615	8.1925
1	4.5518	25.158	4.7060	23.489	5.3394
2	4.5038	11.469	4.6645	10.720	3.6302

These values are used to calculate the terms on the right-hand side of the difference equations. The quantities a, b, and C appear in Eqs. (5-48) to (5-50), p and q in (5-51) to (5-53), and D in (5-54) to (5-57).

Calculation of conversion profile at $z = 100$ cm.:

m	a_m	b_m	$10^2 C_m$	p_m	$1/q_m$	$10^2 D_m$	$10^2 x_m$
0	—	0.099510	4.5193	0.099510	—	4.5193	5.0016
1	9.5478	3.0000	26.844	0.31752	9.4482	3.3194	4.8472
2	1.6935	—	3.3008	—	1.3760	4.8114	4.8114

Calculation of temperature profile at $z = 100$ cm.:

m	a_m	b_m	C_m	p_m	$1/q_m$	D_m	t_m
0	—	0.20944	32.081	0.20944	—	32.081	36.652
1	7.8841	3.0000	105.484	0.39090	7.6747	17.925	21.823
2	2.8350	—	6.4505	—	2.4441	9.9732	9.9732

The following table shows how the temperature at the axis of the tube and the average conversion vary along the reactor.

z (cm.)	Axial temperature (°C.)	Average conversion ($\times 10^2$)
0	320	0.00
10	333	0.44
20	345	0.97
30	356	1.57
40	365	2.18
50	370	2.75
60	372	3.27
70	371	3.72
80	367	4.09
90	362	4.41
100	356	4.65
150	337	5.53
200	330	6.14
300	326	6.97
400	324	7.56
500	323	7.99
600	322	8.31
700	322	8.56
800	321	8.76
900	321	8.91
1000	321	9.02

The conversion at 1000 cm., 9.02×10^{-2}, is to be compared with the initial mole fraction of substance A, 9.5×10^{-2}. The fractional conversion of 0.95 is in a suitable range. The maximum temperature, 372°C., is below the limit imposed, but not enough below to leave very much room for increasing the production rate.

Some effects of changing the size of the steps used in the calculation on the accuracy of the results are shown below, using the axial temperature at a length of 60 cm. as illustration.

	k (cm.)				
h ($= 1/M$)	10	5	2.5	1.25	0.625
0.5	376.42	373.73	373.05	—	—
0.25	—	373.07	372.47	372.30	—
0.125	—	—	372.38	372.24	372.20

An obvious method for increasing the production rate is to raise the temperature of the cooling medium, and at the same time to improve the heat transfer by decreasing the radius of the reactor tubes and increasing the flow rate. The performance of a reactor with the radius decreased to 1.27 cm. and with the mass velocity doubled has been calculated. With the larger mass velocity the pressure drop is multiplied by 4, but it is still not excessive.

The following new parameters are obtained.

$r_2 = 1.27$ cm.

$G = 3.884 \times 10^{-3}$ (initial mol. wt.)/cm.² sec.

$N_{Re} = 332$

$N_{Pe} = 10.0$

$N_{Nu} = 32.7$

$h_t = 1.02 \times 10^{-2}$ cal./cm.² sec. deg.

$k_e = 4.41 \times 10^{-3}$ cal./cm. sec. deg.

$N_{Bi} = 2.94$

With these parameters and with the cooling medium at 350°C., the maximum temperature is 385°C., above the limit. With T_c at 345°C., the maximum temperature is 371°C., which is safe enough, and the conversion at a length of 1000 cm. is 0.0902. This choice of conditions then seems to be acceptable, and has the advantage that the production rate is twice as high as in the previous set of conditions, while the number of tubes required is only about 10 percent larger.

By exploring the effects of changing mass velocity, radius of the tube, and temperature of the control medium, an optimum set of conditions can be determined. In addition to these variables, the diameter of catalyst pellets and the pressure are sometimes disposable parameters, and can be adjusted to improve the performance of the reactor.

There is a simple method for speeding the calculation of conversion that should be pointed out. After the conditions in the upstream end of a reactor have been calculated, including a region well beyond the point where the temperature is a maximum, the conditions in the rest of the reactor can be calculated accurately enough with only 2 or 3 radial increments, or, if the Biot number is not too large, with the one-dimensional approximation. In this region, the temperature has no intrinsic interest, and it is only necessary to estimate the average conversion.

In the case of the first example given above, the accuracy of this kind of continuation was tested, starting in each case with substantially correct profiles. The intervals over which the calculation was carried and the results of the calculation of the average conversion are shown in the following table.

Interval (cm.)	$M = 2$	One-dimensional	Correct
1–80	0.04206	0.03986	0.04179
80–300	0.07094	0.07037	0.07087
100–300	0.07090	0.07061	0.07087
300–1000	—	0.09024	0.09023

The temperature reached a maximum in this case at a length close to 60 cm.

VI. Stability of Packed Tubular Reactors

A. STATEMENT OF THE PROBLEM

No complete picture of the stability of a reactor can be worked out without specifying the kind of control that is operating, and then, using this specification together with the properties of the reactor, studying the transient behavior of the system, at least in the locally linearized form. The method of Deans and Lapidus (D3) for finding the transient conditions in a packed tubular reactor can be adapted to this kind of study, but at great expense in computing time. It must be recognized, however, that there is no simple solution to the problem, and that the best procedure for checking a combination of reactor and control system may be to use a single tube of a multitube reactor as an element in an analog computer. Even this scheme is not always adequate, as may be seen by considering as an example the dynamic response of a multitube reactor that is cooled by a boiling liquid, coupled with the condenser that removes the heat and returns the liquid. In this case, the resistance to two-phase flow and mixing offered by the bank of tubes cannot be simulated accurately with a single tube.

In critical cases it may well be worthwhile to make a complete analysis of stability. In many cases, however, enough can be learned by studying what Bilous and Amundson (B7) called parametric sensitivity. These authors derived formulas for calculating the amplification or attenuation of disturbances imposed on an unpacked tubular reactor originally in a steady state, with the idea that if the disturbances grow unduly the performance of the reactor is too sensitive to the conditions imposed on it, that is, to the parameters of the system. The effect of feedback from a control system was not considered. As pointed out by the authors, it would be a much more complicated task to apply their procedure to a packed reactor, but it still would entail far less computation than a study of the transient response.

A simple criterion for easy controllability was proposed by Wilson (W4). He pointed out that if

$$E_a(T_m - T_c)/R_g T_m^2 < 1 \qquad (6\text{-}1)$$

where T_m is the maximum temperature in the reactor, there can be no problem about control. This is an excessively conservative requirement, as was shown by Barkelew. It can be thought of as a sufficient but not a necessary condition for controllability.

B. Description of Barkelew's Criterion

1. *Assumptions Involved*

Wilson's criterion is based on the one-dimensional reactor. A more useful criterion of parametric sensitivity based on this model was given by Barkelew (B2). Barkelew proceeded by assuming that only one reaction is important, that its rate may be expressed by the form

$$R = kcg\left(\frac{c_0 - c}{c_0}\right) e^{\gamma T} \qquad (6\text{-}2)$$

and that the temperature of the cooling medium is constant. He then defined a new set of variables and parameters:

$$X_f = 1 - y/y_0 \qquad (6\text{-}3)$$

$$\tau = \gamma(T - T_c) \qquad (6\text{-}4)$$

$$N = 2U/r_2 C_p k e^{\gamma T_c} \qquad (6\text{-}5)$$

$$S = Q y_0 \gamma/C_p \qquad (6\text{-}6)$$

With these variables, the relation between dimensionless temperature τ and the fractional conversion X_f is

$$\frac{d\tau}{dX_f} = S - \frac{N\tau e^{-\tau}}{(1 - X_f)g(X_f)} \qquad (6\text{-}7)$$

2. *Nature of the Criterion*

The dependence of τ_m, the maximum dimensionless temperature, on the parameters N and S was studied by carrying out numerical integrations of Eq. (6-7), using either $1 + \alpha X$ or $(1 + \beta X)^{-1}$ for $g(X)$. The key to the criterion developed by Barkelew is in the observation that the family of curves showing τ_m/S as a function of N/S for constant values of S is bounded below by an envelope, and that there is for each curve a rather sharp change from a small slope to a large one in the neighborhood of the contact with the envelope. A large slope for this curve means, of course, a maximum temperature that is sensitive to the parameters.

This sharp curvature is particularly pronounced when S is large, which is to say when the product of activation energy and heat of reaction is large. Barkelew proposed as a condition for sensitivity that the combination of N and S be such that the value of N/S is above the point of contact with the envelope. Given this criterion, other families of curves were constructed, showing the critical relationship between N/S and S for various combinations of the parameters α, β, and τ_0, the initial dimensionless temperature. Several such families of curves are shown in the original paper.

The introduction of the factor $g(X_f)$ adds a considerable generality to the treatment. Examples given by Barkelew of the application of these forms of $g(X_f)$ are:

Second-order reaction, $-1 \leq \alpha < 0$, $\beta = 0$.
Product-inhibited reaction, $\alpha = 0$, $\beta > 0$.
Reactant-inhibited reaction, $\alpha = 0$, $-1 < \beta < 0$.
Autocatalyzed reaction, $\alpha > 0$, $\beta = 0$.

VII. Scale Models of Packed Tubular Reactors

A. GENERAL DISCUSSION

1. *Requirements for a Useful Scale Model*

The concept of a scale model of a reactor is useful because in some situations of practical importance it permits us to predict rather accurately the effects of changing certain design parameters of a reactor even though the kinetics of the reactions are not known, provided other parameters are changed in a corresponding way. A familiar problem is that of increasing the diameter of a reactor without changing the course of the reactions going on, and without any detailed information about how the rates depend on local conditions. In spite of the fact that no exact scale model of a heat-exchanger type of reactor can be made, the requirements for a good approximation can be met in many cases when only a moderate change in scale is required.

The rules given by Damköhler (D1) for changing the scale of catalytic reactors without changing the course of the reaction were derived primarily by dimensional analysis. A better idea of the requirements for scaling up can be obtained by a detailed examination of the coefficients in the differential equations and boundary conditions describing the reactor, with the independent variables in the equations transformed to a modified reciprocal space velocity and a dimensionless radial variable. In an exact scale model, these coefficients are all the same as they are in the prototype, so that the temperature, pressure, and composition depend in just the same way on the new variables, and no essential feature of the reacting system is changed. The construction of a useful scale model depends on the possibility of choosing the design parameters in such a way that the more important coefficients are approximated rather well, without changing any coefficient enough to change the desired relationship seriously. It will be seen that no useful model can be made when either the pressure drop through the reactor or the temperature difference between the catalyst and the fluid has a significant effect on the reactions, and that the construction of the model is much simplified when the radial variation of conditions can be ignored.

2. *Role of the Activity of the Catalyst*

In the treatment of the scaling problem given by Johnstone and Thring (J1), the activity of the catalyst is considered as a disposable parameter. Since the rate of reaction appears in the dimensionless quantities chosen by Damköhler (D1) to characterize a reactor, the activity of the catalyst enters implicitly. The situations in which the activity of a catalyst can be changed without affecting the selectivity or apparent activation energy are practically restricted to those in which the catalyst particles are impermeable, and the activity can be adjusted by changing the size of the particles. This means, in effect, that the activity is at our disposal only to the extent that it can be changed by making the reciprocal change in the particle diameter. In principle, the catalyst could be diluted with an inert packing, but this scheme is not useful in designing new equipment. In order to show how it enters the relations, an activity factor a, which would normally be unity, is carried through the calculations.

3. *Parameters to be Chosen*

The parameters nominally at our disposal are the diameter of the reactor, the mass velocity, and the particle diameter. The problem of changing the scale will be discussed from the point of view of an engineer who wants to match the performance of two reactors of different diameters, and who must then choose a corresponding mass velocity and an accessible—and acceptable—combination of particle diameter and activity of catalyst.

In all that follows, it will be assumed that neither radiation nor molecular conduction will affect the performance of a scale model. If either is so important that its variation cannot be ignored, no useful scale model can be designed.

B. Derivation of Requirements for the General Case

1. Transformation of Axial Variable

The axial variable in Eqs. (3-6), (3-15), (3-28), and (3-30) is transformed by the substitution

$$z = (\overline{G}/a)\zeta \qquad (7\text{-}1)$$

This substitution is made, and the terms involving the rates are multiplied by the activity factor a. After some rearranging, the equations take the form

$$\frac{\partial \mathbf{Y}}{\partial \zeta} = \frac{\overline{G}d_p}{ar_2^2}\left[\frac{1}{N_{\text{Pe}}r}\frac{\partial}{\partial r}\left(r\frac{\partial \mathbf{Y}}{\partial r}\right) + \frac{1}{N_{\text{Pe}}G}\frac{\partial G}{\partial r}\frac{\partial \mathbf{Y}}{\partial r}\right] + \frac{\overline{G}}{G}\mathbf{R}^{\mathrm{T}}S \qquad (7\text{-}2)$$

$$\frac{\partial T}{\partial \zeta} = \frac{\overline{G}d_p}{ar_2^2}\left[\left(\frac{1}{N_{\text{Pe}}} + \frac{k_p}{C_p G d_p}\right)\frac{1}{r}\frac{\partial}{\partial r}\left(r\frac{\partial T}{\partial r}\right) + \frac{1}{N_{\text{Pe}}G}\frac{\partial G}{\partial r}\frac{\partial T}{\partial r}\right.$$
$$\left. + \frac{1}{C_p G d_p}\frac{\partial k_p}{\partial r}\frac{\partial T}{\partial r} + \frac{1}{N_{\text{Pe}}C_p}\frac{\partial T}{\partial r}\left(\frac{\partial C_p}{\partial r} + \frac{\partial \mathbf{Y}}{\partial r}\mathbf{C}_p^{\mathrm{T}}\right)\right] + \frac{\overline{G}}{C_p G}\mathbf{R}^{\mathrm{T}}Q \qquad (7\text{-}3)$$

$$\rho_f \frac{d\overline{P}}{d\zeta} = f(N_{\text{Re}})\frac{\overline{G}^3}{ad_p} \qquad (7\text{-}4)$$

$$\left.\frac{\partial T}{\partial r}\right|_{r=1} + [T(1) - T_c]r_2 h_t \Big/ \left[\frac{C_p d_p G(1)}{N_{\text{Pe}}} + k_p(1)\right] = 0 \qquad (7\text{-}5)$$

2. Necessary Approximations

As the model is to be constructed so that the intensive properties of the reacting fluid are to be invariant to the change of scale, such quantities as the heat capacity and the rate and heat of reaction are also invariant. In his treatment of packed catalytic reactors, Bosworth [see (B8), p. 318] assumes that the diffusivities and the thermal conductivity remain constant when the scale is changed. Since these quantities are approximately proportional to the mass velocity and the particle diameter, the resulting rules for scaling can not be correct. The presence of such factors as $(\partial G/\partial \rho)/G$ and $\dfrac{\overline{G}}{G}$ in the first two equations requires that the velocity profile must retain its shape when the scale is changed, that is, that G must change in the same ratio everywhere.

There are two quantities appearing in Eq. (7-3) that cannot be kept invariant. These are k_p/Gd_p and $(\partial k_p/\partial r)/Gd_p$. Owing to the fact that k_p and G vary with r in different ways, there is no way to keep the ratios fixed everywhere when G is changed. The best that can be done is to keep their average values unchanged.

The difference in temperature between the surface of a catalyst pellet and the fluid surrounding the pellet can have an important effect on the rate, and can be changed significantly by changing the particle diameter or the flow. The only quantity that is in question here is the heat-transfer coefficient which, for a given composition and temperature of fluid, can be written in the form

$$h = g(N_{Re})/d_p \qquad (7\text{-}6)$$

where the function of Reynolds number written as $g(N_{Re})$ can be expressed as proportional to some power of the Reynolds number, with the exponent between 0.5 and 0.6 in the range of general interest.

3. *List of Invariant Quantities*

The following quantities are then to remain unchanged in any exact scale model:

$$k_1 = \overline{G}d_p/ar_2^2 N_{Pe} \qquad (7\text{-}7)$$

$$k_2 = \overline{r_2 h_t N_{Pe}/\overline{G}d_p} \qquad (7\text{-}8)$$

$$k_3 = \overline{k_p/\overline{G}d_p} \qquad (7\text{-}9)$$

$$k_4 = \overline{(\partial k_p/\partial r)/\overline{G}d_p} \qquad (7\text{-}10)$$

$$k_5 = k_p(1)/G(1)d_p \qquad (7\text{-}11)$$

$$k_6 = f(N_{Re})\overline{G}^3/ad_p \qquad (7\text{-}12)$$

$$k_7 = g(N_{Re})/d_p \qquad (7\text{-}13)$$

It is obvious that no scale model can be constructed when all these quantities are important. The question that must be answered is under what conditions useful models can be designed. In the discussion, it will be convenient to replace the second quantity by the product of the first and second, that is, by

$$k_2' = h_t/ar_2 \qquad (7\text{-}14)$$

If it is assumed that h_t/h is independent of Reynolds number, k_2' may be replaced by another ratio,

$$k_2'' = g(N_{Re})/ar_2 d_p \qquad (7\text{-}15)$$

C. Selection of Important Quantities

When the resistance to heat transfer is appreciable both within the bed and at the wall, all the ratios except k_6 and k_7 are important, and some approximation must be made. Some progress may be made by observing that the performance of such a reactor is much more critically determined by heat transfer than by mass transfer, and that eddy transport, which provides mass transfer, makes the major contribution to heat transfer. The problem can be attacked, then, by giving attention to the principal terms arising from the effective thermal conductivity, and using an average value of mass velocity as the governing quantity. The quantity that replaces k_1, k_3, k_4, and k_5 in this approximation is

$$k_1' = k_1\left(1 + \frac{N_{Pe}k_p}{C_p G d_p}\right) \tag{7-16}$$

This quantity is obtained by dividing the effective thermal conductivity by ar_2^2, and then by C_P. As the problem is only approximately solved in this way, k_1' may be calculated with average values of C_P and G.

There are just two cases in which anything simple can be said about the way the activity of a catalyst changes when the particle diameter is changed. One is when the particles are porous and small enough to make the activity independent of size, and the other is when the particles are impermeable, in which case the product of activity and diameter is constant. These will be considered in turn.

1. The activity is constant. The k's are fixed by the prototype reactor. The new G and d_p are chosen to give the fixed values of k_1' and k_2'' with the new r_2. In order to illustrate how the required changes in G and d_p affect the requirements on k_6 and k_7, an approximate calculation can be carried through, neglecting conduction through the solid, and assuming that $g(N_{Re}) = $ (const) N_{Re}^n, where n is between 0.5 and 0.6. The equations for G and d_p are then

$$\left(\frac{G}{G^0}\right)\left(\frac{d_p}{d_p^0}\right) = \left(\frac{r_2}{r_2^0}\right)^2 \tag{7-17}$$

and

$$\left(\frac{G}{G^0}\right)^n\left(\frac{d_p}{d_p^0}\right)^{n-1} = \left(\frac{r_2}{r_2^0}\right) \tag{7-18}$$

These yield the relations

$$\left(\frac{G}{G^0}\right) = \left(\frac{r_2}{r_2^0}\right)^{3-2n} \tag{7-19}$$

and

$$\left(\frac{d_p}{d_p^0}\right) = \left(\frac{r_2}{r_2^0}\right)^{2n-1} \tag{7-20}$$

It may be seen that G varies nearly as the square of r_2, but that d_p does not change very much. This is favorable for the assumption that the activity is constant. It is immediately obvious that, on the other hand, neither k_6 nor k_7 will be very nearly constant, k_6 changing like at least the cube of r_2 and k_7 being proportional to r_2. In this case, then, neither pressure drop nor heat transfer to or from the catalyst can have any influence on the reaction on either scale. If these effects are negligible, a good scale model can be designed.

2. The activity of the catalyst is inversely proportional to the diameter of the particles. Using the same assumptions as for case 1, the relations become

$$\left(\frac{G}{G^0}\right)\left(\frac{d_p}{d_p^{\,0}}\right)^2 = \left(\frac{r_2}{r_2^{\,0}}\right)^2 \tag{7-21}$$

and

$$\left(\frac{G}{G_0}\right)^n\left(\frac{d_p}{d_p^{\,0}}\right)^n = \left(\frac{r_2}{r_2^{\,0}}\right) \tag{7-22}$$

These give the results

$$\left(\frac{G}{G^0}\right) = \left(\frac{r_2}{r_2^{\,0}}\right)^{\frac{2}{n}(1-n)} \tag{7-23}$$

and

$$\left(\frac{d_p}{d_p^{\,0}}\right) = \left(\frac{r_2}{r_2^{\,0}}\right)^{\frac{2n-1}{n}} \tag{7-24}$$

Here again, G varies as something like the square of r_2, and d_p varies but little, leading to substantially the same conclusions for this case as for the other.

The problem of establishing the proper axial temperature profile and heat transfer coefficient in the heating or cooling medium has not been touched upon. If the temperature and transfer coefficient are determined by flow conditions, the total flow rate of the medium is kept in a constant ratio to the total flow rate of reactants, and the size of the channels in which the medium flows is then chosen to match the original value of the ratio k_2'. If heat is transferred to a boiling liquid, there is either the obvious solution of changing nothing or there is no solution, depending on whether the boiling point at the top and bottom of the pool of liquid is or is not practically the same. In the second case, the vertical temperature profile might be matched approximately by using a liquid with a different normal boiling point, and at a different pressure.

When conduction through the solid is taken into account, the variations of both G and d_p are somewhat increased, but the variation of d_p is still gratifyingly small. In both cases considered, it is advantageous to have the required increase in particle diameter small, in the first case because it makes the assumption of constant activity more secure, and the second because the accompanying decrease in activity is small. In some situations a multitubular reactor with tubes of larger diameter may demand less metal in the tubes even though it contains a somewhat larger volume of catalyst.

D. Limitations

The results reached for the packed reactor in which both conduction through the bed and heat transfer at the wall are important may now be summarized. When the diameter of the reactor is changed, the mass velocity and particle diameter are adjusted to keep k_1' and k_2'', defined by Eqs. (7-16) and (7-15), at their values in the prototype reactor, any variation of the activity with particle diameter being taken into account. The effectiveness of this procedure depends on the assumption that several conditions that actually change in going from one scale to the other change little enough to have a negligible effect. These conditions are: the pressure drop in the reactor, the shape of the velocity profile, the ratio of the diffusivity for heat to the diffusivity for material, and the temperature difference between the reacting fluid and the catalyst. The first and fourth of these must be negligible on both scales. The condition on the pressure drop will usually make it impossible to change the scale by any large factor, because of the strong dependence of pressure drop on mass velocity.

E. The One-Dimensional Approximation

The requirements for approximating the radial heat transfer in a packed reactor with a constant over-all heat-transfer coefficient are discussed in Section III, above. When these requirements are met, Eqs. (3-32), (3-33), and (3-34) are used to describe conditions in the reactor. With the substitution of Eq. (7-1), these equations take the form

$$\frac{d\overline{T}}{d\zeta} = \frac{\mathbf{R}^{\mathrm{T}}\mathbf{Q}}{C_p} - \frac{2U}{ar_2 C_p}(\overline{T} - T_c) \qquad (7\text{-}25)$$

$$\frac{d\mathbf{y}}{d\zeta} = \mathbf{R}^{\mathrm{T}}\mathbf{S}_c \qquad (7\text{-}26)$$

$$\rho_f \frac{dP}{d\zeta} = f(N_{\mathrm{Re}})\frac{\overline{G}^3}{ad_p} \qquad (7\text{-}27)$$

The only quantities that must be kept constant in this case are

$$k_8 = U/ar_2 \qquad (7\text{-}28)$$

and the k_6 of Eq. (7-12) to take care of the pressure drop. In addition, there is the requirement that the catalyst and the fluid surrounding it be at practically the same temperature, or that, alternatively, the k_7 of Eq. (7-13) remain unchanged.

With only one principal quantity k_8 to be kept invariant, something may be done about one or the other of the quantities k_6 and k_7. Because of the complicated dependence of U on $\overline{G}$ and d_p in the range of practical concern, a simple approximate relation cannot be constructed. Some idea of the possibilities may be indicated with two illustrations.

First, consider the case when k_6 is important but k_7 is not. Nothing can be done if the product ad_p is constant and the Reynolds number is high, because then the friction factor is insensitive to Reynolds number, and $\overline{G}$ must be nearly constant. If the activity is constant and the Reynolds number is high, d_p varies about as the cube of $\overline{G}$. With the same assumptions about the effective thermal conductivity and the heat transfer coefficient that were used for the three-dimensional reactor, with the additional assumption that in the prototype reactor $h_w r_2/k_e = 2$, the relation between $\overline{G}$ and r_2 obtained by eliminating d_p is

$$(\overline{G}/G^0)^{-0.6}/[1 + 0.5(r_2/r_2^0)(\overline{G}/\overline{G}^0)^{-4.6}] = 0.67(r_2/r_2^0) \qquad (7\text{-}29)$$

From this relation it is found that if r_2 is to be multiplied by the factor 1.1, $\overline{G}$ is multiplied by 1.3 and d_p by 2.2. This result shows that even in the case of the one-dimensional reactor, no useful increase in scale is possible if pressure drop in the reactor has a significant effect on the reaction.

If the extra degree of freedom is needed to keep the coefficient for heat transfer to the catalyst constant, the situation is not quite so bad. In order to fix k_7, d_p must vary as something between the first and the three-halves power of $\overline{G}$. Continuing with the same assumptions as in the case above, short tables have been calculated to show the relations when the activity is constant and when it is inversely proportional to the diameter of the particle.

TABLE III

PARAMETERS IN SCALE MODELS OF A ONE-DIMENSIONAL REACTOR

$a = a^0$			$a = a^0 d_p^0/d_p$		
r_2/r_2^0	G/G^0	d_p/d_p^0	r_2/r_2^0	G/G^0	d_p/d_p
1.1	1.16	1.25	1.18	1.1	1.15
1.2	1.42	1.69	1.37	1.2	1.31
1.3	1.78	2.39	1.58	1.3	1.48
—	—	—	1.79	1.4	1.66

It may be seen that there are disadvantages in making any considerable increase of scale in either case. If the activity is constant, the particle size increases so much that great stress is put on the approximations used in calculating the parameters. If the activity is inversely proportional to the diameter of the particles, the production rate per unit volume of reactor is seriously decreased.

Kjaer (K1, p. 79) has given the appropriate modification for over-all heat-transfer coefficients to the average temperature and to the temperature of the thermowell when a reactor tube contains a coaxial thermowell.

F. Example

The use of the foregoing rules for constructing scale models may be illustrated by changing the scale of the reactor used as the principal example in Section V. Consider the problem of making a scale model in which the mass velocity is 0.6 times what it is in the prototype, with the same activity of catalyst, so that the length of the reactor is changed from 10 meters to 6 meters. The corresponding radius of tube and diameter of catalyst particle are to be calculated.

The quantities to be kept constant are k_1', and k_2', defined as

$$\frac{Gd_p}{ar_2^2 N_{Pe}}\left(1 + \frac{N_{Pe}k_p}{C_p G d_p}\right) \text{ and } h_t/ar_2$$

Because of the fact that the Nusselt number and the Peclet number are not simply expressed in terms of the particle diameter it is convenient to use them as intermediate variables in the calculation. Since the mass velocity is fixed, the Reynolds number and the Nusselt number are determined by the particle diameter. The calculation proceeds by finding the values of r_2 from selected values of d_p, as determined from k_1', and k_2' separately. By inverse interpolation, the value of d_p that makes the two values of r_2 equal is found.

The quantities that are calculated in turn from a tentative value of d_p are N_{Re}, N_{Pe}, N_{Nu}, k_p, $r_2(k_1')$; and for k_2', h_t and then $r_2(k_2')$. The calculation gives

$d_p = 0.326$ cm.
$r_2 = 1.53$ cm.
$N_{Pe} = 11.0$
$G = 1.165 \times 10^{-3}$ (initial mol. wt.)/cm.^{2} sec.
$k_e = 1.680 \times 10^{-3}$ cal./cm. sec. deg.

The calculated profiles given below show how the scale model compares with the prototype in this case. The distance from the entrance is 60 cm. for the prototype and 36 cm. for the model.

This illustration does not pretend to represent the correspondence to be expected between a real scale-model reactor and its prototype, but to show that if the transport properties can be accurately estimated, a useful scale model can be constructed in simple cases. In the real situation, there may be large uncertainties in the best obtainable estimates of the transport properties, and there may be in addition significant effects of

Dimensionless radius	Conversion		$T - T_c$ (°C.)	
	Model	Prototype	Model	Prototype
0.00	0.03593	0.03556	52.19	52.47
0.50	0.03366	0.03357	40.80	40.94
0.71	0.03222	0.03231	30.81	30.88
0.87	0.03144	0.03164	21.93	21.97
1.00	0.03121	0.03144	13.99	14.01

the variation of heat capacity with temperature and composition, and of the radial variation of mass velocity.

Nomenclature

Note: definitions which are not followed by parenthetical dimensions refer to dimensionless entities.

- a activity factor
- a_m coefficients in implicit difference equations
- A rate of production of an entity [(amount) $L^{-3}t^{-1}$]
- b_m coefficients in implicit difference equations
- C concentration of an entity [(amount) M^{-1}]
- C_c heat capacity of control medium ($L^2t^{-2}T^{-1}$)
- C_m known terms in implicit difference equations (T or none)
- C_p heat capacity of reacting fluid ($L^2t^{-2}T^{-1}$)
- $\mathbf{C}_p$ matrix of molar heat capacities of components ($ML^2t^{-2}T^{-1}$ mole^{-1})
- D_m auxiliary quantity in solution of difference equations (T or none)
- d_p diameter of catalyst pellet (L)
- E modified eddy-diffusion coefficient ($ML^{-1}t^{-1}$)
- E_a energy (enthalpy) of activation (ML^2t^{-2} mole^{-1})
- f friction factor
- F_i general expression for a derivative (various)
- g functions defined in text, Sections VI and VII (various)
- G superficial mass velocity ($ML^{-2}t^{-1}$)
- h heat-transfer coefficient, Section IV; radial increment, Section V ($Mt^{-3}T^{-1}$: none)
- h_t heat-transfer coefficient between reacting fluid and control medium ($Mt^{-3}T^{-1}$)
- h_W heat-transfer coefficient between reacting fluid and wall ($Mt^{-3}T^{-1}$)
- H enthalpy of reacting fluid (L^2t^{-2})
- $\mathbf{H}$ matrix of molar enthalpies of components (ML^2t^{-2} mole^{-1})
- J diffusive flux of an entity [(amount) $L^{-2}t^{-1}$]
- k axial step, Section V; rate constant, Section VI (L: t^{-1})
- k_e effective radial thermal conductivity in packed tube ($MLt^{-3}T^{-1}$)
- k_f thermal conductivity of reacting fluid ($MLt^{-3}T^{-1}$)
- k_i invariants in Section VII (various)
- k_p radial thermal conductivity in excess of eddy conductivity ($MLt^{-3}T^{-1}$)
- m index of radial step
- M number of radial increments
- $\overline{M}$ average molecular weight of mixture (M mole^{-1})
- n index of axial step, Section V; numerical constant, Section VII
- N parameter used in Barkelew's criterion

N_i	mole fraction of the ith component	W	coefficient in axial derivative of T_C (L^{-1})
N_{Bi}	Biot number, $h_t r_2/k_e$	W_C	mass rate of flow of control medium (Mt^{-1})
N_{Nu}	Nusselt number, hd_p/k_f		
N_{Pe}	modified Peclet number, Gd_p/E	X	conversion, defined in Section II (mole M^{-1})
N_{Re}	Reynolds number, Gd_p/μ		
P	pressure ($ML^{-1}t^{-2}$)	X_f	fractional conversion
p_m, q_m	coefficients in recursion formulas	x	matrix of virtual conversions for key reactions (mole M^{-1})
Q	heat of reaction ($-\Delta H$) (ML^2t^{-2} mole^{-1})	$\mathbf{X}$	matrix of conversions (mole M^{-1})
$\mathbf{Q}$	matrix of heats of reaction (ML^2t^{-2} mole^{-1})	Y	concentration (mole M^{-1})
		y	matrix of key concentrations (mole M^{-1})
$\mathfrak{Q}$	matrix of heats of key reactions (ML^2t^{-2} mole^{-1})	$\mathbf{Y}$	matrix of concentrations (mole M^{-1})
r	dimensionless radial coordinate		
r_2	radius of tubular reactor (L)	z	coordinate in direction of flow (L)
R	rate of a reaction (mole $L^{-3}t^{-1}$)		
R_g	molar gas constant ($ML^2t^{-2}T^{-1}$ mole^{-1})	Z	length of reactor (L)
		α	parameter in rate expression
$\mathbf{R}$	matrix of rates of reactions (mole $L^{-3}t^{-1}$)	β	parameter in rate expression
		γ	parameter in rate expression (T^{-1})
$\mathfrak{R}$	matrix of virtual rates of key reactions (mole $L^{-3}t^{-1}$)	ζ	modified reciprocal space velocity ($M^{-1}L^3t$)
S	generalized source term, Section V; parameter in Barkelew's criterion, Section VI (T or none: none)	η	coefficient in schematic differential equation (L^{-1})
		θ	constant in expression for velocity profile
s	reduced stoichiometric matrix	λ	constant in expression for velocity profile
$\mathbf{S}$	matrix of stoichiometric coefficients		
$\mathbf{S}_c$	matrix of stoichiometric coefficients for key components	μ	viscosity ($ML^{-1}t^{-1}$)
		ν	change in number of moles in a reaction
$\mathbf{S}_R$	matrix of stoichiometric coefficients for key reactions	$\boldsymbol{\nu}$	matrix of changes in number of moles for key reactions
T	absolute temperature (T)		
T_C	temperature of control medium (T)	ρ_f	density of fluid (ML^{-3})
		σ	reciprocal space velocity ($M^{-1}L^3t$)
T_W	temperature of wall of a tube (T)	τ	dimensionless temperature
U	over-all heat-transfer coefficient ($Mt^{-3}T^{-1}$)	ϕ	general dependent variable (various)
w	weighting coefficient on radial profile	ω	coefficient in schematic differential equation (L^{-1})

References

A1. Acrivos, A., *Ind. Eng. Chem.* **47**, 1553 (1955).
A1a. Aerov, M. E., and Umnik, N. N., *J. Tech. Phys. (U.S.S.R.)* **21**, 1351 (1951).
A2. Aerov, M. E., and Umnik, N. N., *J. Tech. Phys. (U.S.S.R.)* **21**, 1364 (1951).
A3. Argo, W. B., and Smith, J. M., *Chem. Eng. Progr.* **49**, 443 (1953).
A4. Aris, R., and Amundson, N. R., *A.I.Ch.E. Journal* **3**, 280 (1957).
B1. Bakhurov, V. G., and Boreskov, G. K., *J. Appl. Chem. (U.S.S.R.)* **20**, 721 (1947).

B2. Barkelew, C. H., *Chem. Eng. Progr. Symposium Ser.* **55**, No. 25 (Reaction Kinetics and Unit Operations), 37 (1959).
B3. Baron, T., *Chem. Eng. Progr.* **48**, 118 (1952).
B4. Beek, J., Jr., and Miller, R. S., *Chem. Eng. Progr. Symposium Ser.* **55**, No. 25 (Reaction Kinetics and Unit Operations), 23 (1959).
B5. Beek, J., Jr., and Singer, E., *Chem. Eng. Progr.* **47**, 534 (1951).
B6. Bernard, R. A., and Wilhelm, R. H., *Chem. Eng. Progr.* **46**, 233 (1950).
B7. Bilous, O., and Amundson, N. R., *A.I.Ch.E. Journal* **2**, 117 (1956).
B8. Bosworth, R. C. L., "Transport Processes in Applied Chemistry." Wiley, New York, 1956.
B8a. Brötz, W., "Grundriss der chemischen Reaktionstechnik," pp. 35–44. Verlag Chemie, Weinheim, Germany, 1958.
B9. Bunnell, D. G., Irvin, H. G., Olson, R. W., and Smith, J. M., *Ind. Eng. Chem.* **41**, 1977 (1948).
C1. Cairns, E. J., and Prausnitz, J. M., *Chem. Eng. Sci.* **12**, 20 (1960).
C2. Campbell, J. M., and Huntington, R. L., *Petroleum Refiner* **31**(2), 123 (1952).
C3. Chu, Y. C., and Storrow, J. A., *Chem. Eng. Sci.* **1**, 230 (1952).
C4. Coberly, C. A., and Marshall, W. R., Jr., *Chem. Eng. Progr.* **47**, 141 (1951).
C5. Converse, A. O., *A.I.Ch.E. Journal* **6**, 344 (1960).
D1. Damköhler, G., *Z. Elektrochem.* **42**, 846 (1936).
D2. Damköhler, G., in "Der Chemie-Ingenieur," Vol. 3, p. 421. Akademische Verlagsgesellschaft, Leipzig, 1937.
D3. Deans, H. A., and Lapidus, L., *A.I.Ch.E. Journal* **6**, 656, 663 (1960).
D4. Dorweiler, V. P., and Fahien, R. W., *A.I.Ch.E. Journal* **5**, 139 (1959).
D5. Douglas, J., personal communication.
E1. Ebach, E. A., and White, R. R., *A.I.Ch.E. Journal* **4**, 161 (1958).
E2. Ergun, S., *Chem. Eng. Progr.* **48**, 89 (1952).
F1. Fahien, R. W., and Smith, J. M., *A.I.Ch.E. Journal* **1**, 28 (1955).
G1. Gamson, B. W., Thodos, G., and Hougen, O. A., *Trans. A.I.Ch.E.* **39**, 1 (1943).
H1. Hanratty, T. J., *Chem. Eng. Sci.* **3**, 209 (1954).
H2. Hartman, M. E., Wevers, C. J. H., and Kramers, H., *Chem. Eng. Sci.* **9**, 80 (1958).
H3. Hatta, S., and Maeda, S., *Chem. Eng. (Tokyo)* **13**, 79 (1949).
H4. Hildebrand, F. B., "Methods of Applied Mathematics." Prentice-Hall, New York, 1952.
H5. Hildebrand, F. B., "Introduction to Numerical Analysis." McGraw-Hill, New York, 1956.
H6. Hougen, O. A., and Watson, K. M., "Chemical Process Principles." Wiley, New York, 1947.
J1. Johnstone, R. E., and Thring, M. W., "Pilot Plants, Models, and Scale-up Methods in Chemical Engineering." McGraw-Hill, New York, 1957.
K1. Kjaer, J., "Measurement and Calculation of Temperature and Conversion in Fixed-bed Catalytic Reactors." Gjellerups, Copenhagen, 1958.
K2. Kramers, H., and Alberda, G., *Chem. Eng. Sci.* **2**, 173 (1953).
K3. Kwong, S. S., and Smith, J. M., *Ind. Eng. Chem.* **49**, 894 (1957).
L1. Latinen, G. A., Dissertation, Princeton University, Princeton, New Jersey, December, 1951.
L2. Lowan, A. N., "The Operator Approach to Problems of Stability and Convergence," Scripta Mathematica, New York, 1957.
M1. Maeda, S., *Tech. Repts. Tohoku Univ.* **16**, 1 (1952).

M2. Mickley, H. S., Sherwood, T. K., and Reed, C. E., "Applied Mathematics in Chemical Engineering." McGraw-Hill, New York, 1957.
M3. Milne, W. E., "Numerical Calculus." Princeton Univ. Press, Princeton, New Jersey, 1949.
M4. Morales, M., Spinn, C. W., and Smith, J. M., *Ind. Eng. Chem.* **43,** 225 (1951).
O1. O'Brien, G. G., Hyman, M. A., and Kaplan, S., *J. Math. Phys.* **29,** 223 (1951).
P1. Polack, J. A., Dissertation, Massachusetts Inst. Technol., Cambridge, 1948.
Q1. Quinton, J. H., and Storrow, J. A., *Chem. Eng. Sci.* **5,** 245 (1956).
R1. Ranz, W. E., *Chem. Eng. Progr.* **48,** 247 (1952).
R2. Richardson, L. F., *Phil. Trans. Roy. Soc. London, Ser. A210,* 307 (1910).
S1. Schuler, R. W., Stallings, V. P., and Smith, J. M., *Chem. Eng. Progr. Symposium Ser. 48,* No. 4 (Reaction Kinetics and Transfer Processes), 19 (1952).
S2. Schwartz, C. E., and Smith, J. M., *Ind. Eng. Chem.* **45,** 1209 (1953).
S3. Sehr, R. A., *Chem. Eng. Sci.* **9,** 145 (1958).
S4. Singer, E., and Wilhelm, R. H., *Chem. Eng. Progr.* **46,** 343 (1950).
S5. Smith, J. M., "Chemical Engineering Kinetics." McGraw-Hill, New York, 1956.
S6. Strang, D. A., and Geankoplis, C. J., *Ind. Eng. Chem.* **50,** 1305 (1958).
T1. Thoenes, D., Jr., and Kramers, H., *Chem. Eng. Sci.* **8,** 271 (1958).
W1. Walas, S. M., "Reaction Kinetics for Chemical Engineers." McGraw-Hill, New York, 1959.
W2. Wehner, J. F., and Wilhelm, R. H., *Chem. Eng. Sci.* **6,** 89 (1956).
W3. Wilke, C. R., and Hougen, O. A., *Trans. A.I.Ch.E.* **41,** 445 (1945).
W4. Wilson, K. B., *Trans. Inst. Chem. Engrs.* **24,** 77 (1946).
Y1. Yagi, S., and Kunii, D., *A.I.Ch.E. Journal* **6,** 97 (1960).
Y2. Yagi, S., and Wakao, *A.I.Ch.E. Journal* **5,** 79 (1960).
Y3. Yoon, C. Y., Ph.D. Thesis, Massachusetts Inst. Technol., Cambridge, September, 1959.

OPTIMIZATION METHODS

Douglass J. Wilde

Department of Chemical Engineering
The University of Texas, Austin 12, Texas

I. Introduction	273
A. The Best	273
B. Operations Analysis and Chemical Engineering	274
C. Search, Interaction, and Feasibility	275
D. Presentation	276
II. Search Problems	277
A. General Remarks	277
B. One-Variable Search Methods	279
C. Multivariable Search Methods	286
III. Interaction Problems	292
A. General Remarks	292
B. Simple Optimization	293
C. Series Optimization	295
D. Dynamic Programming	297
IV. Feasibility Problems	314
A. General Remarks	314
B. Finding a Feasible Combination	316
C. Linear Programming	320
D. Quadratic Programming	323
E. Feasibility in Interacting Systems	328
Acknowledgment	331
References	331

I. Introduction

A. THE BEST

The primary goal of chemical engineering is to understand the mechanisms underlying chemical manufacturing processes. Such basic knowledge is indispensable to a chemical engineer; for, whether designing a new plant or operating an old one, he must know how the process works. But even this is not always enough—he often wants to know how the process works *best*.

This article reviews the growing body of theory on optimization—powerful new ways of finding the combination of adjustable operating variables to give the best possible value of some selected criterion of process performance. The meaning of "best" depends upon what sort of criterion is chosen. If the criterion is considered desirable, process yield for example, then the greatest value is best. On the other hand, one would wish to minimize criteria measuring loss or cost. The desired extreme value (either maximum or minimum) of the criterion is known as the *optimum,* and the values of the adjustable operating variables that produce this optimum are called the *optimum conditions.*

One cannot begin to solve an optimization problem without first choosing a single, quantitative, criterion of optimality. An equally important prerequisite is a means of finding the value of the criterion for any combination of the operating variables. Such a means, which may be a set of graphs, tables, mathematical functions, or computer codes, is often called a *model* of the system to be optimized.

At first glance, optimization theory all by itself may not sound very much like conventional chemical engineering. This is because it is precisely during construction of the model that an engineer uses the physical, chemical, and mathematical knowledge he learned as a student. By divorcing the process of optimization from the system being optimized, we bypass most of the problems usually associated with chemical engineering. But any experienced engineer will recognize that the mathematical model of a process is more often a means to an end than an end in itself; most technical studies are made to help someone decide something. Since the ultimate decision is usually based on what is "best," optimization theory provides the finishing touch to any engineering project.

B. Operations Analysis and Chemical Engineering

Optimization theory has been developed mainly by mathematicians investigating problems in *operations analysis* (also called operations research), a relatively new applied science dealing with the rational attainment of decisions. Since chemical engineers also are charged with arriving at decisions, a few words about operations analysis and its relation to chemical engineering may be in order here.

The operations analyst, just like the chemical engineer, must do three things to reach a decision: he must build a mathematical model, choose a criterion of effectiveness, and then optimize the criterion.

In the first step, construction of the mathematical model, operations analysis quite naturally tends to concern itself with decisions not already highly rationalized by other professions. Thus the most successful applications of operations analysis have been in such unusual domains as military tactics (M3) and toll bridge management (E1) where quantitative models had never before been constructed. In the chemical industry, the models describing production processes already have been developed by chemical engineers to the point where there would seem to be little room for improvement by an operations analyst not already trained as a chemical engineer. There are, however, a few important chemical plant operations which have not been well enough described by the engineers, where operations analysts with a good understanding of probability theory may be of service. These are the operations involving such uncertain factors as supply, demand, random upsets, breakdowns, accidents, prediction error, and experimental imprecision.

The choice of a measure of effectiveness, the second of the three steps in decision-making, is often difficult for a chemical engineer, who may not have the information necessary to see how his plant fits into the company, industry, or the economy as a whole. If the proper measure of effectiveness is not already obvious to the engineer concerned, often no one else will be able to find it either. In practice, operations analysts seem to be the professional group most concerned with selecting practical measures of effectiveness.

In optimization, the final stage of decision preparation, major contributions have been made by mathematicians associated with operations analysis. Most engineers know only one mathematical method—the differential calculus—for finding an optimum. This technique involves expressing the criterion of effectiveness as a function of the operating variables, differentiating the function with respect to each variable, setting each derivative equal to zero, and solving the resulting simultaneous equations. Only rarely can all of these steps be carried out in an industrial problem; even when they can, the "optimum" conditions solved for are often unattainable because of physical or economic restrictions on the ranges of the operating variables. Under these conditions, engineers have been limited to picking a few promising combinations of the variables at random, calculating the criterion of effectiveness for each case, and choosing the one with the most optimal value. In contrast to this, the goal of optimization theory is to develop methods which not only will find the optimum quickly and efficiently, but also will indicate when the optimum has been found and it is time to stop looking.

C. Search, Interaction, and Feasibility

One can distinguish several branches of optimization theory according to the mathematical form of the model being optimized, or, from another point of view, according to which obstacle in the model prevents the use of differential calculus. Three kinds of optimization problem will be discussed here: search problems, interaction problems, and feasibility problems.

In a search problem, almost nothing is known in advance about how the criterion of effectiveness depends upon the operating variables, the only way to learn being to perform experiments. Here the obstacle to using the calculus is the complete lack of a function that can be differentiated. The objective of the search is to get as close as possible to the optimum after only a limited number of experiments. Box and Wilson, with their paper "The Experimental Attainment of Optimum Conditions" published in 1951, were the first to interest engineers in search problems (B4).

In interaction problems the criterion of effectiveness depends upon so many factors that it is impractical, if not impossible, to find the optimum by conventional methods. Successful techniques for solving such problems involve "decomposing" a big system into several smaller ones. This is advantageous because the number of computations tends to increase with the cube of the number of operating variables. Whether or not a system can be decomposed depends upon how the subsystems interact. Bellman has shown how a series of decisions, each depending upon the one preceding it, can be successfully partitioned by his method of *dynamic programming* (B2); the middle third of our article will be devoted to this powerful technique. Another decomposition method, recently developed by Dantzig and Wolfe (D1), will be described at the very end of this study.

When the operating variables have limited ranges of variation, as they often do in industrial situations, it may not be physically possible to operate so that all the first derivatives of the criterion vanish. Restrictions on the variables thus give rise to *feasibility* problems; for only operating conditions which are feasible (that is, which lie within all the prescribed limits) can be considered. Operations analysts generally use the term *mathematical programming* to describe such constrained-optimization problems. In 1951 Dantzig published his *simplex method,* the first practical technique for solving feasibility problems (D2). We shall show how the simplex method can be used to optimize either linear or quadratic criteria of effectiveness, when the constraints are linear inequalities. The applications of Dantzig's work are so widespread that over one thousand technical articles stemming from it have now been published.

D. Presentation

Professional operations analysts may be surprised by our treatment of advanced topics, such as quadratic programming, which we show not to require the formidable mathematical apparatus that is customarily employed. We treat such topics by using the techniques of partial differentiation, with which most engineers are already familiar, instead of the matrices and Lagrange multipliers favored by operations analysts. Another help to engineers, including undergraduates to whom optimization theory should be made available, is our explanation of dynamic programming by flow diagrams instead of by functional equations. (To help the reader understand the existing literature on dynamic programming, a parallel development using functional equations is also included.)

In the treatment of search theory we will entirely neglect the question of experimental error. The no-error case, because of its design applications, is nevertheless of some interest to the engineer. The treatment including errors would require us to introduce probability theory.

Part of our search-theory section deals with steep-ascent methods. Contrary to the beliefs of most engineers, the "method of steep ascent" is not always the best way to search for an optimum. By clarifying this point we hope to stimulate further investigation of search methods, a subject erroneously considered by many research workers to be a closed book.

A fourth branch of optimization theory, not discussed here, deals with *logical* problems. Such problems arise when operations must be performed in a predetermined order (A1, G2, J1, M3, S4); however, the methods presently known are not powerful enough to solve practical scheduling problems. This review of optimization theory is thus by no means complete, but we will cover almost everything well enough developed now to be useful to the engineer. The reader will gradually come to realize that no engineering study is complete without some form of optimization.

II. Search Problems

A. General Remarks

We begin by considering situations in which nothing is known about how the criterion of effectiveness depends on the operating variables. This means that to find the value of the criterion at any combination of operating variables we must take an experimental measurement. In this case, optimization involves using information from past experiments to direct the search for successively better values of the criterion. For this reason such situations are called *search* problems.

There are several reasons why functions to be optimized, such as yield, profit, throughput, or cost, are often unknown in advance. Physical theory describing a process is usually imperfect, containing many approximations and uncertainties. System parameters often change as time passes. And even if the theory were perfect and the parameters constant, systematic or random errors in the process instrumentation would distort the engineer's picture of the system. Thus, although physical theory can give a rough idea of which operating conditions are optimum, the true optimum can only be found by direct experimentation on the full-scale operation.

If experiments at full scale cost nothing, finding the optimum would simply be a matter of trying out all possible combinations of the operating variables. In practice, however, it is usually expensive to change operating conditions. If this changeover cost were the only expense, it would never be worthwhile to search for an optimum. But there is also a counterbalancing loss, in the integrated difference between the unknown optimal value and the actual values of the correct criterion, which justifies a search for the optimum by means of methods that will reach near-optimum conditions as quickly as possible.

Search problems can be divided into two groups, depending on whether or not random experimental error is associated with each measurement. There are, indeed, significant problems that have no experimental error; as when the function in question is given as an exact mathematical expression, but one too complicated to be optimized directly by calculus or by known methods of mathematical programming. Design problems are often of this latter nature. We shall discuss mainly the no-error case, since its principles are simple and can be used even in the presence of experimental error.

The simplest search problem arises, of course, when there is only one operating variable to be adjusted. In this one-dimensional case the optimal search policy has been established under certain conditions by Keifer (K1) and Johnson (J2) independently, and we shall discuss their results here. Although the search problem has not been solved for cases with more independent variables, Box and Wilson (B4, B3) have suggested methods for choosing directions in which to search in the multidimensional case. We shall discuss their methods, and also propose some new ones. As Shapiro *et al.* (S2) have remarked, once a search direction is chosen, finding the point in that direction where the unknown function is nearest the optimum is a one-dimensional search problem. It therefore seems natural to combine direction-finding methods with the one-dimensional search technique, when exploring a function of several operating variables; in this way, the multidimensional problem is reduced to a series of one-dimensional searches.

B. One-Variable Search Methods

1. *Unimodality*

Consider a function y, of a single independent variable x restricted to the interval (a, b); that is, $a \leqslant x \leqslant b$. We shall consider the greatest value of y that can occur in this interval to be the *optimal* value y^*. The value of x for which y is optimal will be denoted by x^*; that is,

$$y^* = y(x^*)$$

Suppose that the function y is not known in advance and that the only way to obtain information about y is to measure it at different

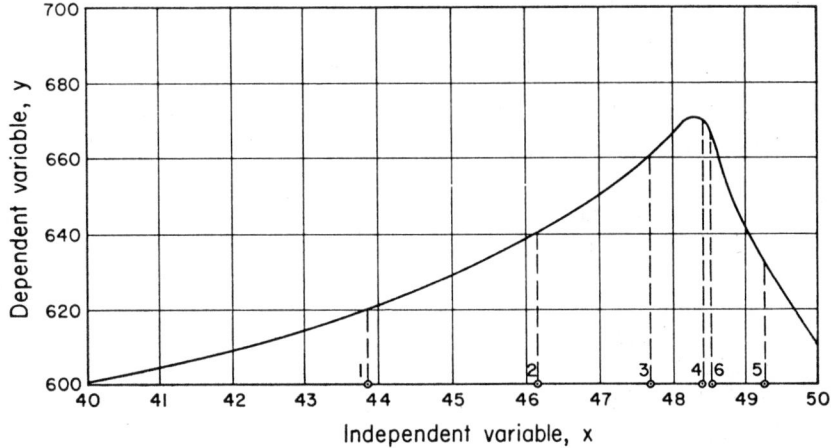

Fig. 1

values of the independent variable x. Assume that y *strictly decreases* as x moves away from x^* in either direction. Such a function is said to be *unimodal*, since under this hypothesis a graph of y versus x would only have one hump in the interval (see Fig. 1). This is a strong assumption, but in many practical situations it is justified. Notice, however, that y is not assumed to be smooth or continuous. This unimodality property is described mathematically as follows:

(1) If $x_1 < x_2 < x^*$, then $y(x_1) < y(x_2) < y(x^*)$, and
(2) if $x^* < x_1 < x_2$, then $y(x^*) < y(x_1) < y(x_2)$.

Suppose that we can make n measurements of y with negligible experimental error, anywhere in the interval. At the end of this sequence of measurements, we shall be able to say that x^* is in some subinterval of (a, b). To see why this is so, consider any two measurements taken at points x_1 and x_2, where

$$a \leqslant x_1 < x_2 \leqslant b$$

and suppose that $y(x_1) < y(x_2)$. Then since y is unimodal, the optimum x^* could not be less than x_1 without contradicting property (2) given in the definition of unimodality. That is,

(3)   if $y(x_1) \leqslant y(x_2)$,
              then $x_1 \leqslant x^* \leqslant b$.

By similar reasoning we see that

(4)   if $y(x_1) \geqslant y(x_2)$,
              then $a \leqslant x^* \leqslant x_2$.

Thus any time more than one measurement is taken $(n > 1)$, the unimodality of y permits us to reduce the interval in which we are sure to find the optimum value x^*. Let I_n be the length of this interval after n measurements, and let I_0 be $(b - a)$, the length of the original interval. Then the ratio I_0/I_n measures the effectiveness of any search procedure. Since a single measurement yields no information about x^*,

$$I_0/I_1 = 1$$

In order to have a basis for comparing proposed search methods, let us start with the worst possible technique. This would be to take all n measurements at the same value of x^*. Naturally, in this case $I_0/I_n = 1$ for all n, and we see that we should always compare I_0/I_n with unity no matter how many measurements are taken.

2. Uniform Search

Probably the first search method occurring to an engineer would be to space the n measurements uniformly in the interval (a, b), i.e., at

$$x_k = a + kI_0/(n+1) \quad \text{for} \quad k = 1, 2, \ldots, n$$

For this *uniform search* x^* is narrowed down to be inside two adjacent intervals each $I_0/(n+1)$ units long. Hence

$$I_0/I_n = (n+1)/2$$

We can naturally visualize search schemes where the measurements are spaced irregularly and the size of the final interval obtained would depend on the experimenter's luck. In such cases it is difficult to decide *a priori* which interval should be used to measure the effectiveness of the scheme. We shall take as our final interval the *largest* interval that we

3. Uniform Dichotomous Search

It is easy to find the minimax experimental plan for only two experiments on the interval (a, b). We wish to know the locations x_1 and x_2 for which the greater of the two intervals (a, x_2) and (x_1, b) is as small as possible. This condition is satisfied when both x_1 and x_2 are as close as possible to the midpoint of the segment, with x_1 remaining less than x_2. That is, the minimax plan for $n = 2$ is

$$x_1 = [(a+b)/2] - \epsilon \quad \text{and} \quad x_2 = [(a+b)/2] + \epsilon,$$

where ϵ is a small positive number. For this *dichotomous search* policy, $I_2 = 1/2 + \epsilon \approx 1/2$ and $I_0/I_2 = 2$. This is a clear improvement over the uniform policy, for which $I_0/I_2 = 3/2$.

Suppose now that more than two experiments are to be made. If the entire experimental plan must be laid out in advance, the minimax technique is to pair off the experiments, locating each pair as close as possible and distributing the pairs uniformly over the interval. That is, if $[k/2]$ represents the largest integer less than or equal to $k/2$, then

$$x_k = a + [k/2]I_0/([n/2] + 1) + (-1)^k \epsilon$$

For this *uniform dichotomous search*, the final interval I_n is $1/([n/2] + 1)$ and

$$I_0/I_n = \left[\frac{n}{2}\right] + 1$$

When n is even, this technique is slightly more effective than an ordinary uniform search, the advantage diminishing as n becomes large.

4. Sequential Dichotomous Search

Suppose that it is not necessary to plan the entire set of experiments in advance. If we are allowed to look at the results of past experiments before planning the next one, we can take advantage of some very effective search plans. This is because the interval to be searched shrinks after each experiment when we can profit by past information. Such plans are said to be *sequential*.

Consider the sequential extension of the dichotomous search. After the first pair of experiments the new interval to be searched is only half as long as the original one. This new interval can be halved again by another pair of experiments at *its* midpoint, and so on. The final interval will be $(1/2)^{[n/2]}$, for which the effectiveness ratio is

$$I_0/I_n = 2^{[n/2]}$$

Thus the *sequential dichotomous search* is considerably more powerful than the preplanned *uniform dichotomous search*, the advantage growing exponentially with the number of experiments. To reduce the interval of uncertainty to 1% of the original would require 200 preplanned dichotomous experiments. The same job could be done with only 14 sequential dichotomous experiments.

5. Fibonacci Search

Keifer (K1) and Johnson (J2) have shown that there is another sequential method that is even better. It involves using *Fibonacci numbers* F_n, which are defined inductively as follows:

$$F_0 \equiv F_1 \equiv 1$$
$$F_n = F_{n-1} + F_{n-2} \quad \text{for} \quad n > 0$$

Thus each Fibonacci number is the sum of the two preceding it. The Fibonacci numbers associated with the first eleven integers are given in the following table.

THE FIBONACCI SEQUENCE

n	0	1	2	3	4	5	6	7	8	9	10	11
F_n	1	1	2	3	5	8	13	21	34	55	89	144

The sequence having been named after Leonardo Fibonacci (1180–1225), a pioneer in the study of infinite series, the technique about to be described is called the *Fibonacci search method*.

Consider the placement x_1 and x_2 of the first two experiments in the original interval I_0. I_0 may also be written as I_1 because a single experiment cannot reduce the interval of uncertainty. Let the two ends of the interval be denoted by a_1 and b_1. Define the distance Δ_2 by

$$\Delta_2 \equiv I_1 F_{n-2}/F_n$$

Locate the two experiments a distance Δ_2 from each end of the interval. That is, set

$$x_1 = a_1 + \Delta_2$$

and

$$x_2 = b_1 - \Delta_2$$

Since Δ_2 is never more than half the length of the original interval, either $(a_1, a_1 + \Delta_2)$ or $(b_1 - \Delta_2, b_1)$ will be eliminated by examining the results $y(x_1)$ and $y(x_2)$. Hence I_2, the interval of uncertainty after two trials, will be shorter than I_1 by the amount Δ_2. Thus

$$I_2 = I_1 - \Delta_2 = I_1(1 - F_{n-2}/F_n) = I_1 F_{n-1}/F_n$$

since Fibonacci numbers are defined to be such that

$$F_n - F_{n-2} = F_{n-1}$$

Let a_2 and b_2 be the end points of I_2 and define the new distance Δ_3 by

$$\Delta_3 \equiv I_2 F_{n-3}/F_{n-1}$$

The term Δ_3 is, in fact, the distance between the first two experiments x_1 and x_2, for

$$x_2 - x_1 \equiv b_1 - \Delta_2 - (a_1 + \Delta_2) = I_1 - 2\Delta_2$$
$$= I_1((F_n - F_{n-2}) - F_{n-2})/F_n = I_1(F_{n-1} - F_{n-2})/F_n$$
$$= I_1 F_{n-3}/F_n = I_2 F_{n-3}/F_{n-1} \equiv \Delta_3$$

One of the preceding experiments will be at the end of the new interval I_2, the other being in position for another Fibonaccian search cycle. It only remains to place the third experiment a distance Δ_3 from the *other* end of I_2. That is, x_3 is located as follows:

(1) If $y(x_1) \geqslant y(x_2)$, then $a_2 = a_1, b_2 = x_2$, and $x_3 = a_2 + \Delta_3$.
(2) If $y(x_1) \leqslant y(x_2)$, then $a_2 = x_1, b_2 = b_1$, and $x_3 = b_2 - \Delta_3$.

As before, the interval of uncertainty remaining after the third experiment is

$$I_3 = I_2 F_{n-2}/F_{n-1} = I_1 F_{n-2}/F_n$$

Repeating this process for the remaining experiments gives finally

$$I_n = I_1 F_0/F_n$$

from which we see that the effectiveness ratio for a Fibonaccian search with n experiments is equal to the nth Fibonacci number:

$$I_0/I_n = F_n$$

The last cycle of the method will be a dichotomous search, which we know is minimax for two experiments. Keifer (K1) and Johnson (J2) show that the Fibonacci search is in fact minimax among all sequential techniques. In order to reduce the interval of uncertainty to less than 1% it only takes 11 Fibonacci experiments, three less than for a sequential dichotomous search. The advantage increases with the number of experiments.

6. Search with a Variable Number of Experiments

So far we have only discussed methods where the number of experiments, or equivalently, the size of the final interval desired, is known in advance. It often happens, however, that the experimenter does not decide in advance how many trials to perform. He just keeps trying until satisfied. If a dichotomous search is used, the sequence does not depend on the ultimate number of experiments, since at each stage the remaining interval is bisected. On the other hand, the exact location of the first two Fibonacci trials in principle depends on n, the total number of experiments to be performed. Fortunately, the ratio F_{n-2}/F_n is very near its limit 0.38 for n greater than 4. A very nearly optimal Fibonaccian method would be to take

$$\Delta_k = 0.38 I_{k-1}$$

during the early stages of the search. When the experimenter decides not to use more than five additional trials to complete the search, he can switch to the true Fibonacci sequence, which will reduce the remaining interval by almost an order of magnitude ($F_5 = 8$).

It should be remembered that all of these methods are very conservative, since they are all based on the assumption that nothing is known about the function y except that it is unimodal. If, as is often the case with physical systems, the function is known to be smooth and continuous, the engineer may wish to fit a curve to his points and estimate the maximum by ordinary differentiation. When doing this, however, it is worthwhile to locate the points according to the Fibonacci sequence so as to be able to shift to a Fibonacci search if the function does not behave according to preliminary estimates.

7. Example: Finding the Hot Spot in a Tubular Reactor

a. Statement of Problem. In some chemical processes carried out in tubular reactors, the decision to change the catalyst is based on the location of the highest temperature in the tube. When the catalyst is new, the "hot spot," or position of maximum temperature, is near the entrance of the tube. As the catalyst ages, the hot spot migrates toward the exit. The temperature profile can be assumed unimodal. In a given process, theory predicts that the hot spot is between 40 and 50 inches from the entrance. Find the hot spot in seven sequential experiments.

b. Solution. If a sequential search is permitted, our strategy will be to reduce the interval of uncertainty as much as possible with the first six experiments, make an estimate of where the hot spot is in this remaining interval, and then perform our last trial at that estimated position. For a six-point sequential dichotomous experiment, the final interval of uncertainty is

$$I_6 = I_0/2^{[n/2]} = 10/2^3 = 1.25 \text{ inches}$$

half as long as that for a uniform search with seven experiments. For a six-point Fibonaccian experiment, the final interval is

$$I_6 = I_0/F_6 = 10/13 = 0.77 \text{ inches}$$

39% shorter than for a dichotomous search. We shall therefore use a six-point Fibonacci search, fit a quadratic approximation to the three points in the final interval, estimate the hot-spot location by differentiating the empirical function, and place the last experiment at the point estimated. Since the sixth, fifth, and fourth Fibonacci numbers are 13, 8, and 5, the first two experiments x_1 and x_2 are located 5/13 of the distance from each end of the ten-inch interval, that is,

$$x_1 = 40 + (5)(10)/13 = 43.85$$

and

$$x_2 = 50 - (5)(10)/13 = 46.15$$

The corresponding temperatures are $T(x_1) = 620°F.$ and $T(x_2) = 641°F.$ Hence the maximum cannot be less than 43.85 inches from the entrance. The third experiment should be placed a distance Δ_3 from the right end of the new interval, where

$$\Delta_3 = x_2 - x_1 = 2.30$$

Hence

$$x_3 = 50 - 2.30 = 47.70$$

The rest of the calculations and experimental results are given in the following table. x'_k represents the location at which y is maximum in the first k experiments.

FIBONACCIAN SEARCH

Experiment Number k	1	2	3	4	5	6
I_k, Interval of uncertainty	40.00 50.00	40.00 50.00	43.85 50.00	46.15 50.00	47.70 50.00	47.70 49.25
$\Delta_k = \|x_{k-1} - x_{k-2}\|$	—	3.85	2.30	1.55	0.75	0.80
x_k, Location	43.85	46.15	47.70	48.45	49.25	48.50
$T(x_k)$, Temperature, °F.	620	641	662	669	632	668
F_{n-k}	8	5	3	2	1	1
$I_k/I_0 = F_{n-k}/F_n$	8/13	5/13	3/13	2/13	1/13	1/13
x'_k	x_1	x_2	x_3	x_4	x_4	x_4

Thus, after six trials, we know that the hot spot is between 47.70 and 48.50 inches from the entrance and that its temperature is at least 669°F.

To locate the last trial we fit a quadratic function through experiments 3, 4, and 6, which are in the final interval. That is, we assume that

$$T(x) = a_0 + a_1 x + a_2 x^2 \quad \text{for} \quad 47.70 \leqslant x \leqslant 48.50$$

with the coefficients a_0, a_1, and a_2 to be determined from the three experiments. The desired quadratic function is

$$T = 662 + 52.1(x - 47.70) - 111.5(x - 47.70)^2$$

for which the maximum is at $x = 48.17$. The estimated hot-spot temperature is 674°F. Taking

$$x_7 = 48.17$$

we find that

$$T(x_7) = 670°F.$$

Hence our best Fibonaccian estimate is that the hot spot occurs 48.17 inches from the entrance and has a temperature of at least 670°F. The actual profile used to generate this problem is given in Fig. 1.

C. Multivariable Search Methods

1. Response Surfaces

The methods just described do not work when there is more than one independent variable. There is certainly a need for techniques which can be extended to problems with many operating variables, most industrial systems being quite complicated. We shall now consider methods which reduce an optimization problem involving many variables to a series of one-dimensional searches. For simplicity we shall discuss optimization of an unknown function y of only two independent variables x_1 and x_2, indicating later how to extend the techniques to more general problems where possible.

Let us begin with some definitions. Any combination of numerical values of x_1 and x_2 will be called an *experiment* or a *point*. If x_2 is plotted against x_1, the experimental points will lie in a bounded region of the $x_1 - x_2$ plane, called the *experimental region*. Each possible experiment can be represented by a vector **x** whose components are x_1 and x_2. That is,

$$\mathbf{x} \equiv (x_1, x_2)$$

Imagine the experimental region to be horizontal with the dependent variable y plotted in the vertical direction. Since y is assumed to take on only one value for each experiment **x**, this construction would give a two-dimensional sheet called a *response surface* floating above the experimental region. The goal of the search is to find the maximum value of y, or in geometric terms, to find the highest point on the response surface. As y is unknown in advance, the only way we can obtain information about it is to make experiments. As usual, we assume experimental error to be negligible, so as to avoid statistical questions and simplify the discussion.

2. *Lines of Search*

Consider any two different experiments **a** and **b**, with

$$\mathbf{a} \equiv (a_1, a_2)$$

and

$$\mathbf{b} \equiv (b_1, b_2).$$

These two points determine a straight line in the experimental region. Any experiment on this line fits the relation,

$$\begin{aligned}\mathbf{p} &= \mathbf{a} + (\mathbf{b} - \mathbf{a})\lambda \\ &= \mathbf{a}(1 - \lambda) + \mathbf{b}\lambda\end{aligned}$$

As the parameter λ varies, all points on the line are generated. Such a line will be called the *line of search from* **a** *to* **b**. Points for which $\lambda > 1$ will be said to be "beyond **b**"; points for which $\lambda < 0$ will be "beyond **a**"; while those for $0 < \lambda < 1$ will be "between **a** and **b**." Since the experimental region is itself bounded, λ is finite.

3. *Unimodality*

Since **p** is a function of the single parameter λ along any given line of search, the dependent variable y may be plotted against λ. Geometrically speaking, this graph would be the intersection of the response surface with a vertical plane passing through **a** and **b**. We shall say that a response surface is *unimodal* if the function y is unimodal along every possible line of search. In order to use the search procedures already developed for one-variable problems, we shall only consider unimodal response surfaces. This is certainly a strong assumption, but it seems to hold in many industrial situations. If there is reason to believe that the unknown function has more than one bump, the search methods to be described can be used to explore each peak separately.

The maximum-seeking methods developed here will involve using a sequence of lines of search. To start the exploration, a line of search is chosen and the maximum for the line found by the one-variable methods already described. A new line of search starting from the maximum is chosen and a new high point found. This procedure is continued until either all the experiments are used up or the experimenter is satisfied with his results. The important problem which remains to be discussed is how to choose each new line of search, since only one point on it is given from the preceding exploration.

4. Contours

Suppose, then, that we have a point **a** and the value of the unknown function y at **a**. In order to define a line of search through **a** we must choose some other point **b** in the experimental region. With the limited information given there is certainly no basis for this choice, and we would be forced to choose **b** at random. However, if we can assume that the unknown function and its first partial derivatives are continuous in the neighborhood of **a**, we can eliminate roughly half of the experimental region by finding the trend of the *contour* through **a**. This *contour* is the set of experimental points **c** for which the dependent variable has the same value as at **a**, i.e., where

$$y(\mathbf{c}) = y(\mathbf{a})$$

Geometrically speaking, the contour would be the curve obtained by passing the horizontal plane on which $y = y(\mathbf{a})$ through the response surface and projecting the intersection on to the experimental region.

Consider a line of search from **a** to some other point **c** on the same contour. Since y is unimodal, every point **p** between **a** and **c** must have

$$y(\mathbf{p}) > y(\mathbf{a}),$$

while every point **p′** either beyond **a** or beyond **c** must have

$$y(\mathbf{p'}) < y(\mathbf{a})$$

Hence no line of search can cut a contour in more than two places, and all of the contours are *ovals*, as in Fig. 2. We are of course interested in the points having higher values of y than at **a**. All such points will be *inside* the oval contour passing through **a**.

5. The Contour Tangent

Consider now the *tangent* to the contour at **a**. This tangent divides the experimental region into two subregions, and the oval contour at **a** will lie entirely inside one of them, which will be called the *eligible region*. Any experiment having a higher value of y must be inside the contour, which is itself contained in the eligible region. There is clearly no use in searching the other subregion, which we shall call the *noneligible region*. It is therefore of interest to locate the contour tangent at **a**.

Let m_1 and m_2 respectively be the first partial derivatives of y with respect to x_1 and x_2 at the point **a**. That is,

$$m_1 \equiv \left(\frac{\partial y}{\partial x_1}\right)_{\mathbf{x}=\mathbf{a}}$$

and

$$m_2 \equiv \left(\frac{\partial y}{\partial x_2}\right)_{x=a}$$

A linear approximation to the function y for points **x** in the neighborhood of **a** is given by the first-order Taylor series expansion

$$y - y(\mathbf{a}) = m_1(x_1 - a_1) + m_2(x_2 - a_2)$$

This is the equation of the plane tangent to the response surface at **a**. The intersection of this tangent plane with the horizontal plane $y = y(\mathbf{a})$

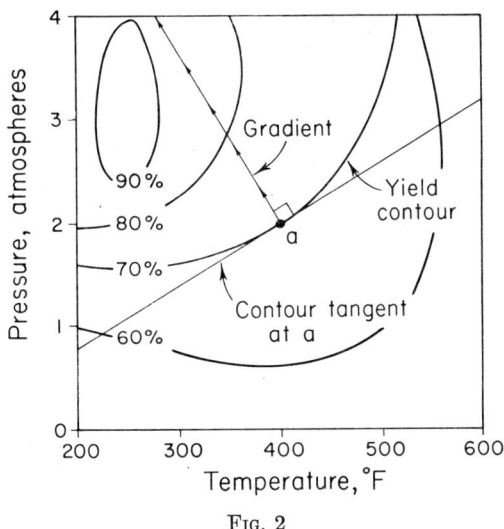

Fig. 2

gives a straight line which when projected onto the experimental region becomes the tangent to the contour at **a**. This tangent line has the equation

$$m_1(x_1 - a_1) + m_2(x_2 - a_2) = 0,$$

and the eligible region is that for which

$$m_1(x_1 - a_1) + m_2(x_2 - a_2) > 0$$

The partial derivatives m_1 and m_2 may be estimated by measuring y at two points (p_1, p_2) and (q_1, q_2) as close as possible to **a**. The derivatives m_1 and m_2 will be the solutions to the two simultaneous equations

$$y(\mathbf{p}) - y(\mathbf{a}) = (p_1 - a_1)m_1 + (p_2 - a_2)m_2$$
$$y(\mathbf{q}) - y(\mathbf{a}) = (q_1 - a_1)m_1 + (q_2 - a_2)m_2$$

To have these equations linearly independent, the three points **a**, **p**, and **q** must not lie on a straight line.

If **a** is itself the highest point on a previous line of search, there is already an experiment p very near **a**; for the last stage of any one-dimensional search always involves two points as close together as possible. Hence in this case only one additional experiment **q** would be needed to determine the contour tangent.

The region of search has now been cut down as far as possible with the information available near the single experiment **a**. In the eligible region remaining, there is no reason to consider any particular line of search better than any other, unless more assumptions are made about the system.

6. *Direction of Steepest Ascent*

Consider, however, the rather special case where the problem is to find the summit of a hill on the surface of the earth. In this situation x_1 and x_2 would represent respectively distances in the westerly and northerly directions, while y would be the elevation above sea level. The special feature of such a system is that distance is well defined in *any* horizontal direction, not just along the x_1 and x_2 axes. Hence it is meaningful in this case to draw a circle of radius r around any point **a**. There will be one particular point **b** on such a circle where y is the highest. A line of search from **a** to **b** will give the greatest increase in y for the distance r. As r becomes very small, the line of search approaches the *gradient direction* or *direction of steepest ascent*. This direction is perpendicular to the contour tangent at **a**. Any point **p** on this *gradient line*, as we shall call it, will have its coordinates p_1 and p_2 given by

$$(p_2 - a_2)/(p_1 - a_1) = m_2/m_1$$

where m_1 and m_2 are the slopes to the north and to the west respectively (P1). In this geographical example it would certainly be reasonable to conduct the new exploration along the gradient line.

7. *Scale Effects*

No sane cartographer, in drawing a contour map of a hill, would ever choose a scale in the north-south direction different from that in the east west direction. The concept of *length*, well-defined in the science of geometry, forces him to make the scale independent of the direction. Consequently any cartographer is within his rights to draw small circles and to speak of a direction of steepest ascent.

Unfortunately, in most systems of interest to chemical engineers, the idea of distance is meaningless and the choice of scales therefore arbitrary.

Suppose, for example, an engineer wants to plot process yield contours as a function of temperature and pressure. He can hardly say that, for instance, the experiment (200°F., 3 atm.) is "farther" from (400°F., 1 atm.) than from (500°F., 4 atm.). For this reason he will probably select temperature and pressure scales to fit whatever graph paper happens to be in his desk.

Figure 2 shows a yield contour for a particular choice of scales for x_1 and x_2. The same contour is plotted in Fig. 3 for which the horizontal scale has been doubled. The contour tangent and gradient line at the same point **a** are given for each choice of horizontal scale. The two gradient lines obviously do not contain the same points. Thus the "direction of steepest ascent" depends entirely on the relative scales of the

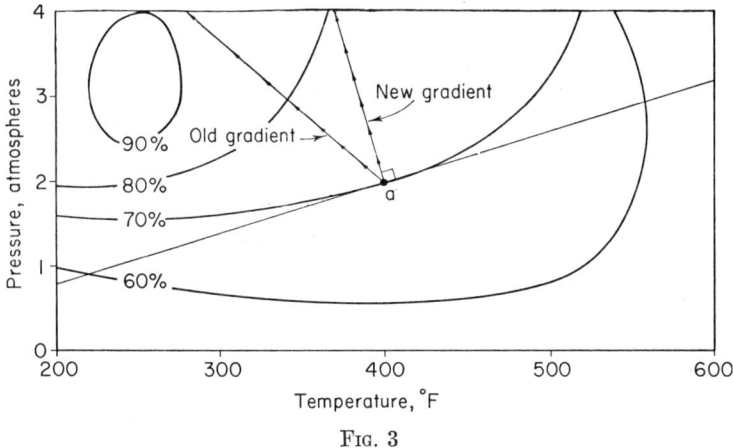

Fig. 3

independent variables. In other words, *any* line of search in the eligible region can be made a "direction of steepest ascent" by a proper choice of scales. For this reason the gradient line is usually no better and no worse than any other eligible line of search. Notice however that no eligible line of search will ever become ineligible (or vice versa) because of scale changes.

The reader may wonder why so much time has been spent in discussing what may seem obvious. Our reason is that we believe, from discussions with many engineers, that the effect of scale on the gradient is widely misunderstood. The use of the gradient as a line of search was first proposed for nongeographical problems ten years ago in a widely read and discussed paper by Box and Wilson (B4). Their method of overcoming the scale problem is to choose scales such that at the point **a** the slopes

m_1 and m_2 along the two axes are *numerically* equal. They used *numerical* values because in physical problems the equation of the gradient line is not always dimensionally homogeneous. For instance, if x_1 represents pressure in p.s.i.a. and x_2 temperature in °F., the left side $(p_2 - a_2)/(p_1 - a_1)$ of the gradient equation has units °F./p.s.i.a. while the right side m_2/m_1 has units (yield/°F.)/(yield/p.s.i.a) = p.s.i.a./°F. While we personally feel that this procedure is arbitrary, others have applied it to physical problems (S2, S1).

8. *Two-Variable Dichotomous Search*

When there are only two independent variables, the basic idea of the dichotomous search can be extended, giving a technique we shall call the *two-dimensional dichotomous search*. Notice that if the high point **a** of a line of search is not on the boundary of the experimental region, the line will be precisely tangent to the contour at **a**. Hence any such line of search itself is a contour tangent. When there are only two independent variables, every line of search divides the experimental region into two areas, one of which is ineligible for further search. It would be reasonable then to direct any new line of search so as to divide the remaining eligible region into two equal areas. After n such searches, the remaining eligible area would only be 2^{-n} times as large as the original experimental region. This is an obvious extension of the one-dimensional dichotomous search technique discussed previously. This idea cannot be applied to problems with more than two independent variables because a line cannot be a boundary between two volumes as it can between two areas.

III. Interaction Problems

A. General Remarks

In our study of search problems we have seen that single variable systems can be optimized with ease; two-variable systems, with some effort; and multivariable systems, only with extreme difficulty if at all. As more variables enter a search problem, the number of experiments needed grows rapidly, and the unimodality assumption becomes less and less plausible. Thus our investigation of search problems leads directly to interaction problems, where the criterion of effectiveness depends on so many factors that it is impractical, or even impossible, to find the optimum by conventional methods. Successful techniques for solving interaction problems involve decomposing a big system into several smaller ones, as we have already done with our lines of search.

Section III deals with one very effective method for partitioning a particular kind of interaction problem arising often in engineering and economics. This technique, called *dynamic programming* by its inventor Richard Bellman (B2), has been applied to such diverse chemical engineering problems as scheduling batch reactors (A2), control of a distillation column (S2), and replacing catalyst in a reformer (R1). A construction and maintenance scheduling application suggested by Dreyfus (D4) will be used as a numerical example.

In developing the theory of dynamic programming here we will use diagrams similar to the flow sheets and control-system block diagrams familiar to chemical engineers. This approach is quite unlike the functional-equations method used in the current technical literature. Although the diagrammatic approach could stand by itself, we include a parallel explanation using functional equations, primarily to help the reader understand other articles on dynamic programming. Our functional notation is more complicated than Bellman's because we feel that the existing discussions of dynamic programming are too abbreviated. Readers who are repelled by superscripts and subscripts are advised to direct their initial efforts toward understanding the diagrams and the example, reserving study of the equations for their first encounter with the mathematical literature.

Before discussing dynamic programming, we will describe the sort of problem lending itself to this kind of decomposition. Since the results of any particular decision are influenced by all previous decisions, operations analysts call such problems *multistage decision problems*. We prefer the name *series-optimization problems,* the word "multistage" being associated in the minds of chemical engineers with such unit operations as distillation, absorption, and extraction. In defining such problems we shall define precisely certain characteristics of simple optimization which were developed only intuitively in Section II.

Another interesting interaction problem arises when decisions are to be made at decentralized locations in a system, as at local plants in a large company. Such *decentralized decision problems* can be solved by the Dantzig-Wolfe *decomposition principle* (D1). Although this problem certainly involves interaction, it involves feasibility as well, and so we defer its discussion to the very end of the monograph.

B. SIMPLE OPTIMIZATION

The simplest sort of optimization problem arises when the criterion of effectiveness y depends only on the decision d to be made. This situation can be described by the equation

$$y = y(d)$$

or by the diagram

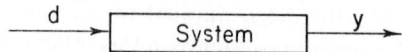

where the block represents the function giving the yield y for any decision d. The problem is to find the optimal decision D—that giving the optimal yield Y. This can be symbolized by the definition

$$Y \equiv \operatorname*{opt}_{d} \{y(d)\}$$

where "opt" stands for the mathematical operator "optimize with respect to d." This operator is considered to operate on y, varying d over its entire range, measuring the resulting values of y, and choosing the optimal one. The optimal decision D is the one associated with the optimal yield Y. D is defined mathematically by

$$y(D) \equiv Y$$

How this optimization is accomplished is not pertinent to the present development. The operator may work by Fibonacci search, differential calculus, or even brute-force evaluation of all possible decisions; all that is important is the resultant finding of the optimal yield Y and the associated decision D.

Taking a mechanistic rather than a mathematical point of view, we can represent optimization by adding an *optimizer* to the system as shown below.

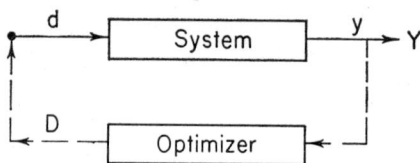

Just like the operator "optimize with respect to d," the optimizer measures the output y and varies the input decision d until y is optimal. The final output of the optimizer is D, the optimal decision. The optimal yield Y is the output of the system for the optimal decision D. It is important to notice that while d and y are variables, D and Y are constants. The optimization operator selects particular values D and Y from among all the possible values d and y.

Suppose now that the yield y depends not only on the decision d but on the *state* s of the system as well. In this case the function relating y to d is different for every possible state. For example, the yield y from a catalytic reactor is a function not only of the reactor temperature, the decision variable d under our control, but also of the catalyst activity, which is the state s of the system, at the moment the decision is taken. In mathematical terms,

$$y = y(d, s)$$

There are thus two inputs to the system—the state s and the decision d.

If the state of the system is itself changed by the decision, then the new state s' is an output as well as the yield y. This happens in a catalytic reactor whenever its catalyst activity is affected by its temperature history. This new state is expressed by a second equation

$$s' = s'(d, s)$$

The schematic relation between initial and final states is shown in the diagram below.

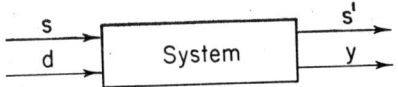

C. Series Optimization

Consider now a set of simple systems of the foregoing type in series, the initial state of each system being determined by the system immediately preceding it in the chain. The individual systems will be called "stages" and numbered consecutively from 1 to M, where M is the total number of stages. The state of stage i will be denoted by s^i. Similarly d^i and y^i represent the decision at stage i and its resulting yield. A schematic diagram of this series optimization is

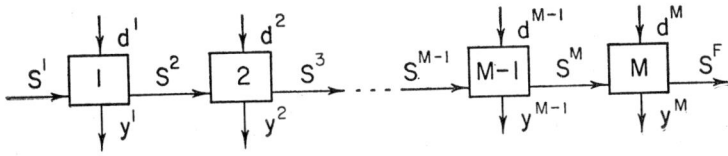

s^F stands for the final state after the last stage. In terms of functional equations, for every stage i

$$y^i = y^i(d^i, s^i)$$

and

$$s^{i+1} = s^{i+1}(d^i, s^i)$$

The stages may be different systems considered all at the same time, as for a string of reactors in which the product from one reactor is the feed for the next (A2). On the other hand, the stages may be the same system considered at consecutive time intervals, as for a single reactor whose state (catalyst activity) depends on its temperature history (past decisions) (R1). In the latter case we consider the decisions to be made at discrete time intervals.

It is usual in series-optimization processes to seek, for a given initial state s^1, the set of decisions $d^1, d^2, \ldots, d^M$ which give the highest total yield $y^1 + y^2 + \ldots + y^M$. To show that the initial state s^1 is fixed, we shall capitalize it: S^1. Let y be the total yield, Y its optimal value, and let the optimal decisions be denoted by capital letters $D^1, D^2, \ldots, D^M$. The problem is illustrated schematically in the following diagram.

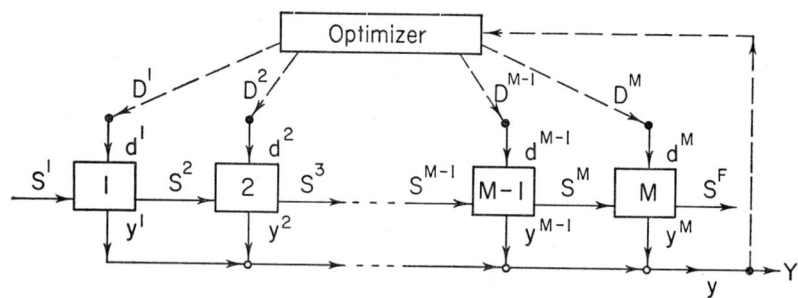

In principle, the optimum sequence of decisions $D^1, D^2, \ldots, D^{M-1}, D^M$ could be found by following the instructions in this diagram literally, that is, by manipulating all M decisions *simultaneously*. This direct approach, which could involve the study of all possible decision sequences, is naturally rather impractical. An engineer facing such a problem would be forced to limit his investigation to a few selected cases. With good judgment based on sufficient past experience, he might find a set of decisions that is at least nearly optimal. But he cannot be sure how good his guess is without evaluating all possible cases.

D. Dynamic Programming

We shall now describe a much more efficient way of analyzing the series-optimization problem. This method was conceived by Bellman, who called it *dynamic programming* (B2).

1. The Last Stage

Consider the optimal sequence of decisions $D_1, D_2, \ldots D^{M-1}, D^M$, all as yet unknown. The first $(M - 1)$ of these optimal decisions, together with the given initial state S^1, not only would determine an optimal state S^M for the last stage, but also yield $Y^1, Y^2, \ldots, Y^{M-1}$ associated with the optimal decisions for the first $(M - 1)$ stages. The situation at the last stage would be as shown below.

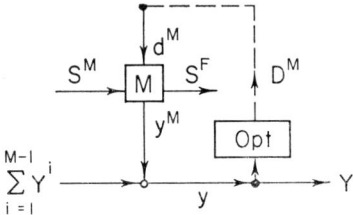

That is, the optimal final decision D^M could be found by manipulating d^M until the sum of the yields y reached its optimum value Y.

However, since the optimal sum

$$\sum_{i=1}^{M-1} Y^i$$

for the first $(M - 1)$ stages is constant, optimizing the *total* yield y is equivalent to optimizing y^M, the yield from the last stage alone. We see that it is not at all necessary to consider the yields from the first $M - 1$ stages to find the last optimal decision D^M. *All* of the information needed is contained in the optimal state S^M.

Of course there is no way of knowing in advance which of the possible states s^M happens to be the optimal one S^M. For this reason we must find an optimal decision $D_1{}^M$ and an optimal yield $Y_1{}^M$ for every possible state s^M. The subscript 1 indicates that this is the *first* stage analyzed, the optimizer carrying the same number.

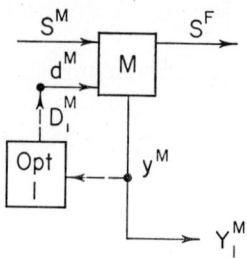

In mathematical terms,

$$Y_1^M(s^M) \equiv \operatorname*{opt}_{d^M} \{y^M(d^M, s^M)\}$$

and

$$y^M(D_1^M, s^M) \equiv Y_1^M(s^M)$$

The result of this analysis of the last stage is a decision and a yield for every possible state s^M.

To appreciate the effectiveness of this method, consider the case where the decision d^i at each stage must be chosen from among k possibilities. Then there are in all k^M different sequences of decisions possible for the M stage problem. For every possible state s^M at the last stage, all but one of the k possibilities is eliminated by analysis of the last stage. Thus the number of sequences to be considered is reduced by a factor of $1/k$. In fact, the number of nonoptimal sequences eliminated is $k^M - k^M/k = (k-1)k^{M-1}$. When k is 10, for example, the analysis of the last stage of a five-stage decision process eliminates 90,000 sequences out of the 100,000 possible.

2. *The Last Two Stages*

The number of possible sequences remaining after the analysis of the last stage can be reduced further by a similar analysis of the last *two*

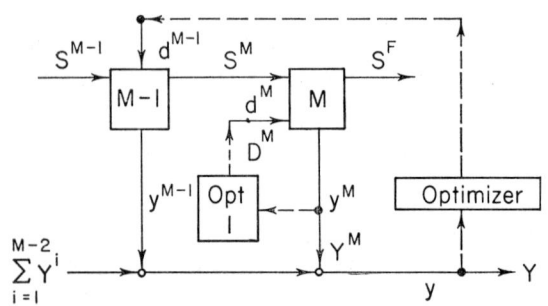

stages. If the decisions for the first $(M-2)$ stages were optimal, then the optimal decision D^{M-1} could be found by the scheme shown below.

Notice that the decision d^{M-1} affects not only stage $M-1$, but stage M as well. This is because the decision d^{M-1}, together with the optimal state S^{M-1}, determines the state s^M for the last stage. This state s^M in turn fixes the yield Y_1^M from the last stage because of the optimization 1 already carried out on the last stage. It follows that the total yield y is a function *only* of the next to last decision d^{M-1}.

Hence the optimal decision D^{M-1} can be found by directly optimizing the total yield y.

Let

$$f^{M-1} \equiv y^{M-1} + Y^M$$

f^{M-1} is the yield from the last two stages *when the last decision is optimal*. The total yield y is given by

$$y = \sum_{i=1}^{M-2} Y^i + f^{M-1}$$

Since the summation term is constant, optimizing y is equivalent to optimizing f^{M-1} alone. Hence another way of finding D^{M-1} is as shown below:

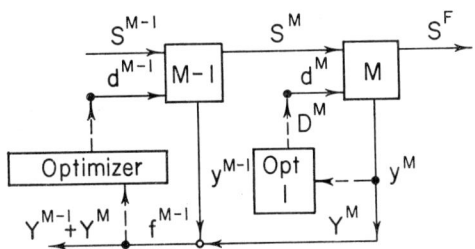

As before, we see that it is not necessary to know the yields from the first $M-2$ stages to find the next optimal decision D^{M-1}. The optimal state S^{M-1} contains all the necessary information. Notice that the optimal decision D^{M-1} puts stage M in its optimal state S^M, allowing optimizer 1 to generate the last optimal decision D^M.

Since the optimal state S^{M-1} is unknown in advance, the state of stage $M-1$ must be treated as a variable s^{M-1}. The new optimizer will be given the number 2. The optimization scheme is given below.

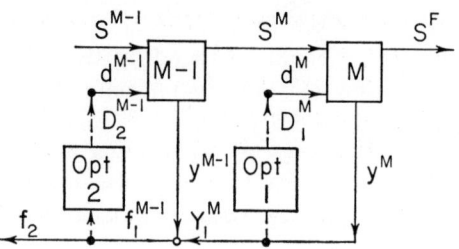

The variable to be optimized is f_1^{M-1}, defined by

$$f_1^{M-1} \equiv y^{M-1} + Y_1^M$$

The subscript 1 indicates that the quantity depends on the output of the first optimizer but not the second. Its optimum values, which are functions only of the states s^{M-1}, will be written f_2 to conform with Bellman's notation. In precise terms,

$$f_2(s^{M-1}) \equiv \underset{d^{M-1}}{\text{opt}} \{f_1^{M-1}(d^{M-1}, s^{M-1})\}$$

and

$$f_1^{M-1}(D_2^{M-1}, s^{M-1}) \equiv f_2(s^{M-1})$$

The two optimizers together have reduced the number of possible decision sequences by a factor of k^2, where k is the number of decisions possible at each stage. For a five-stage process having ten possible decisions at each stage, 99,000 nonoptimal sequences are eliminated by the two optimizers—99% of the 100,000 possible.

3. Generalization

This method of analysis can be applied to each stage in turn, working backwards from the last stage to the first as shown below schematically. The order of optimization is from right to left, following the optimization numbers of the optimizers, not the stages.

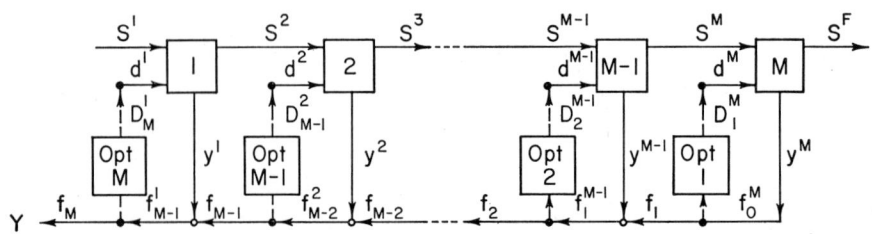

This diagram shows that the only input to any stage i is the state s^i. Dynamic programming takes advantage of this structure by treating the states rather than the decisions as variables. In this way the problem, originally involving optimizing a function of M decision variables, is decomposed into M optimization problems each involving only a single state variable.

To make the notation symmetric, we have introduced the functions f_1^M and f_1 at the last stage, setting

$$f_1^M \equiv y^M$$

and

$$f_1 \equiv Y_1^M$$

In general, the functions f_i can be defined recursively as follows:

$$f_1 \equiv \underset{d^M}{\text{opt}}\ \{y^M\}$$

and

$$f_{i+1} \equiv \underset{d^{M-i}}{\text{opt}}\ \{y^{M-i} + f_i\}$$

for all i with $1 < i < M$.

The last of these functions, f_M, depends only on the initial state S^1. Since this initial state is given in advance, f_M is a constant.

$$f_M \equiv \underset{d^M}{\text{opt}}\ \{y^1(d^M, S^1) + f_2(d^M, S^1)\}$$

This constant is in fact Y, the optimum sum of yields.

$$f_M = Y \equiv \sum_{i=1}^{M} Y^i$$

The associated decision $D_M{}^1$ is also unique, being the first optimal decision D^1 of the optimal sequence $D^1, D^2, \ldots, D^{M-1}, D^M$.

$$f_M^1(D_M^1, S^1) = f_M^1(D^1, S^1) = Y$$

4. Finding the Optimal Decisions

D^1 can be considered as the output of optimizer M when the input is the initial state S^1. The optimal state S^2 for the second stage is found by applying S^1 and D^1 to stage 1.

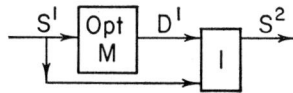

The optimum second-stage decision D^2 is the output of optimizer $M-1$ for this optimal state. D^2 and S^2 together give optimal state S^3. The entire optimal sequence of states and decisions is found in this way, as shown below.

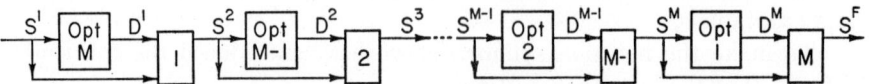

5. Example: A Long-Range Construction and Repair Program

a. Statement of problem. Every two years a chemical manufacturer must choose between three alternatives. He may (1) *continue* operating his old plant at a reduced profit, (2) *repair* the old plant to bring it back to its original efficiency, or (3) *build* a new plant which, because of technological improvements, will be more profitable than an overhauled plant. The manufacturer has a plant built in 1960, and he is to decide which of the three alternatives to take for the two-year period beginning in 1962. He has all the cost and profit predictions he needs up through the period 1970–71. For the sake of simplicity it is assumed that the process will be worthless after 1971, so that only the years 1962–1971 need be considered. The manufacturer wishes to maximize his total profit during this ten-year interval.

The primary problem is to decide what to do in 1962. But we shall see that to make this decision the manufacturer must lay down a policy giving the decisions not only for 1962, but for 1964, 1966, 1968, and 1970 as well. And in order to be able to evaluate possible alternative plans, he must know what the profit from this optimal policy will be. Hence the problems are: (1) to find the optimal sequence of construction and maintenance decisions, and (2) to calculate the profit associated with this optimal policy.

In order to describe the data given, let us introduce the following nomenclature. Let t be the year beginning a two-year period under consideration, let t_b be the year when the plant was built, and let t_r be the year when the plant was last repaired. If the plant has never been overhauled, we take t_r equal to t_b, the year of construction. In Table I is given the cost c_r of repairing a plant. This cost is a function of the time $(t - t_r)$ since the last overhaul. As might be expected, the repair cost c_r increases as the plant ages after an overhaul.

On the other hand, the net cost c_b of building a new plant is a function of the age $(t - t_b)$ of the old plant. This net construction cost, given in Table II, represents the difference between the actual price of a new

TABLE I
Repair Cost

$(t - t_r)$, years since last repair	2	4	6	8	10
c_r, repair cost	65	75	85	90	95

plant and the salvage value of the old. In this case the price of a new plant is relatively constant, and the salvage value naturally decreases as the old plant ages. Hence the net construction cost c_b increases with the age of the old plant.

TABLE II
Net Construction Cost

$(t - t_b)$, age of old plant in years	2	4	6	8	10
c_b, net cost of building new plant	150	160	180	200	250

Table III gives the net profit p_0 in the two-year period beginning in year t for a plant built in 1960. This profit p_0 is a function of $(t - t_r)$, the time elapsed since the plant was last repaired. The profit naturally decreases as the plant ages, but since we have assumed that an overhaul restores the plant to its original condition, p_0 depends only on $(t - t_r)$. The net profit is here taken as the return from sales for the two-year period less the operating costs.

TABLE III
Net Profit for a Plant Built in 1960

$(t - t_r)$, years since last repair	0	2	4	6	8	10
p_0, net profit for a plant built in 1960	100	60	40	30	20	15

Because of technological advances, plants built later will give higher profits. In general, we would have a table similar to Table III for every possible year of construction. To simplify the problem, we are assuming that the effect of technological improvement can be taken into account adequately by adding to the net profit p_0 a sort of technological bonus p_b which depends on the year of construction t_b. Thus the actual profit p can be represented by a sum of p_0, the net profit for a 1960 plant, and p_b, the technological improvement associated with a plant built $(t_b - 1960)$ years later. The technological improvement terms are given in Table IV. Notice that this improvement increases rapidly with time, as is often the case with a new process.

TABLE IV
Technological Improvement

t_b, construction date	1960	'62	'64	'66	'68	'70
p_b, technological improvement	0	20	50	80	120	180

The actual profit $p(t, t_b, t_r)$ for the period beginning in year t, produced by a plant built in year t_b and last repaired in year t_r, is therefore given by

$$p(t, t_b, t_r) = p_0(t - t_r) + p_b(t_b)$$

For example, the actual profit in the period 1968–69 for a plant built in 1962 and overhauled in 1964 would be

$$p(\text{'68}, \text{'62}, \text{'64}) = p_0(4) + p_b(\text{'62})$$
$$= 40 + 20 = 60$$

This would be the return in 1968–69 from choosing alternative 1, that is, neither to repair the old plant nor to build a new one.

Consider next alternative 2, that of repairing the old plant in year t. The profit will be $p(t, t_b, t)$, that for a plant built in year t_b and overhauled in year t. The return $r(t, t_b, t_r)$ will be this profit less the repair cost $c_r(t - t_r)$. That is,

$$r(t, t_b, t_r) = p(t, t_b, t) - c_r(t - t_r)$$
$$= p_0(t - t) + p_b(t_b) - c_r(t - t_r)$$
$$= 100 + p_b(t_b) - c_r(t - t_r)$$

For example, when $t = \text{'68}$, $t_b = \text{'62}$, and $t_r = \text{'64}$, this return is

$$r(\text{'68}, \text{'62}, \text{'64}) = 100 + p_b(\text{'62}) - c_r(\text{'68} - \text{'64})$$
$$= 100 + 20 - 75 = 45$$

Finally consider the third alternative, that of building a new plant in year t. The actual profit will be $p(t, t, t)$. The return $b(t, t_b)$, which does not depend on when the old plant was overhauled, will be this profit less the construction cost $c_b(t - t_b)$. That is,

$$b(t, t_b) = p(t, t, t) - c_b(t - t_b)$$
$$= p_0(0) + p_b(t) - c_b(t - t_b)$$
$$= 100 + p_b(t) - c_b(t - t_b)$$

For example, when $t = \text{'68}$ and $t_b = \text{'62}$, this return is

$$b(\text{'68}, \text{'62}) = 100 + p_b(\text{'68}) - c_b(\text{'68} - \text{'62})$$
$$= 100 + 120 - 180 = 40$$

b. Solution. The problem here is not so much to *find* the solution as to find it *efficiently*, for one could always as a last resort evaluate all possible policies and pick out the best. This would involve the study of $3^5 = 243$ cases, which at ten minutes per case could be completed by one man in about one working week.

We shall solve this problem by dynamic programming. Using this method, we are able to find the optimal policy and the maximum return after about four hours of manual calculation, about 10% of the time required to search all possible cases. A ten-stage problem would take roughly four times as long: 16 hours. This is less than 0.2% of the 10,000 hours which would be required to search all policies. Thus as the complexity of a problem increases, the advantages of dynamic programming become more and more evident.

The key idea of dynamic programming is to consider the *last* time period *first*. Thus, in this problem, we start by analyzing the possibilities facing us in 1970. At this point of the analysis we know nothing about the plant that will be available at the beginning of 1970; neither when it will have been built, nor when it will have been last repaired. In this case there will be five possible overhaul dates for a 1960 model, four for a 1962 model, etc., giving $(5 + 4 + 3 + 2 + 1) = 15$ combinations in all. One of these combinations will be the result of an optimal sequence of earlier construction and repair decisions. But since we don't know at this point which of the fifteen combinations is the best, we must study all of them separately.

To illustrate the method of analysis, let us study one combination in detail, a plant built in 1962 and last overhauled in 1968 for instance. There are the usual three alternatives available: continue operating, repair the old plant, or build a new one. If we continue operating, the profit earned in 1970 and 1971 will be

$$p('70, '62, '68) = p_o(2) + p_b('62)$$
$$= 60 + 20 = 80$$

If we overhaul the plant, the return for 1970–71 will be

$$r('70, '62, '68) = 100 + p_b('62) - c_r(2)$$
$$= 100 + 20 - 65 = 55$$

If we build a new plant, the return for 1970–71 will be 80.

Thus, if in 1970 we have a 1962 model plant overhauled in 1968, it doesn't matter whether we continue operating or build a new plant, for in both events we will realize the maximum return of 80. However, it is clearly not advantageous to repair such a plant, since the return after repair costs would only be 55. Notice that by this simple analysis of one possibility at the last stage we have eliminated four policies from consideration, namely BCCRR, BCRRR, BRCRR, and BRRRR. If in addition we break the tie by choosing one of the two equivalent alternatives, say that of building a new plant rather than keeping the old one, the four policies BCCRC, BCRRC, BRCRC, and BRRRC are also eliminated.

Table V gives the return $p('70, t_b, t_r)$ for 1970–71 which would result from continuing to operate the old plant. There are of course 15 entries, one for each possible combination of construction date and repair date. The net profit $p_0('70 - t_r)$ for each overhaul date is given in the left margin; the technological bonus $p_b(t_b)$, in the top margin. An entry in row t_r and column t_b, which is simply the sum of the value at the left of the row with that at the top of the column, is the actual profit $p('70, t_b, t_r)$ to be gained from the decision not to repair or build in 1970. The blank entries represent impossible combinations—plants repaired before they are built. The asterisks will be explained later.

TABLE V

Returns, $p('70, t_b, t_r)$, from *Continuing* Operations in 1970

| | | t_b | '60 | '62 | '64 | '66 | '68 |
		$p_b(t_b)$	0	20	50	80	120
t_r	$'70 - t_r$	$p_0('70 - t_r)$					
'60	10	15	15				
'62	8	20	20	40			
'64	6	30	30**	50	80		
'66	4	40	40*	60	90	120**	
'68	2	60	60*	80**	110*	140*	180*

Table VI gives the returns $r('70, t_b, t_r)$ for 1970–71 that would result from repairing the old plants in 1970. The negative repairs costs, $-c_r('70 - t_r)$ are entered at the left of each row; the technological

TABLE VI

Returns, $r('70, t_b, t_r)$, after *Repairing* the Plant in 1970

| | | t_b | '60 | '62 | '64 | '66 | '68 |
		$p_b(t_b)$	0	20	50	80	120
t_r	$'70 - t_r$	$-c_r('70 - t_r)$					
'60	10	−95	5				
'62	8	−90	10	30			
'64	6	−85	15	35	65		
'66	4	−75	25	45	75	105	
'68	2	−65	35	55	85	115	155

bonuses $p_b(t_b)$, at the top of each column. Each entry is the algebraic sum of the negative repair cost for the row, the technological bonus for the column, and 100, the net profit $p_0(0)$ for a newly overhauled plant.

Table VII gives the returns $b('70, t_b)$ after building a new plant in 1970. In all cases the net profit $p_0(0)$ will be 100 and the technological bonus $p_b('70)$ will be 180, since we have a new plant. From this must be subtracted the net construction cost $c_b('70 - t_b)$ which is a function of the age of the old plant. The negative construction costs are entered at the top of each column; and the entries in the table are obtained by adding 280 to these negative costs.

TABLE VII
Returns, $b('70, t_b)$, after *Building* a New Plant in 1970

t_b	'60	'62	'64	'66	'68
$'70 - t_b$	10	8	6	4	2
$-c_b('70 - t_b)$	−250	−200	−180	−160	−150
t_r					
'60	30*				
'62	30*	80*			
'64	30**	80*	100*		
'66	30	80*	100*	120**	
'68	30	80*	100	120	130

It is now possible to decide what to do in each of the fifteen cases simply by comparing corresponding entries in the three tables V, VI, and VII, and choosing the largest. In these tables we have placed an asterisk next to each return which is maximum for its particular combination of construction data and date of the last overhaul. Where there is a two-way tie, both of the maximum entries have been marked with two asterisks.

The optimal 1970 decisions for the fifteen possible plants are given in Table VIII, together with the 1970–71 returns associated with these decisions, which we shall denote by $f_1(t_b, t_r)$. In Table VIII, C stands for "continue operating," R stands for "repair," and B stands for "build."

Table VIII shows that in no case is repair of an old plant justified in the last two years of operation. The bottom row indicates that any plant overhauled in 1968 can be let alone in 1970. The top three rows show that any plant not repaired since 1964 should be replaced entirely. General rules of this nature give a rough indication of how many possible construction and repair policies have been eliminated by our simple

TABLE VIII
OPTIMAL DECISIONS FOR 1970–71 WITH THEIR ASSOCIATED RETURNS $f_1(t_b, t_r)$

t_r \ t_b	'60	'62	'64	'66	'68
'60	30 B				
'62	30 B	80 B			
'64	30 CB	80 B	100 B		
'66	40 C	80 B	100 B	120 CB	
'68	60 C	80 CB	110 C	140 C	180 C†

analysis of the period 1970–71. The number of policies eliminated is in fact 162, two-thirds of the total number possible, since we have chosen only one out of every three available alternatives. Thus we have already reduced the number of eligible policies from 243 to 81.

Now we are ready to analyze the period 1968–69. The number of possible combinations of construction and repair dates is now only 10. For each combination we must calculate the return for the period 1968–69 just as before. But we are not interested in maximizing this two-year return alone; we want to find how to get the highest possible return for the last *four* years of operation.

Consider for example a plant built in 1962 and last overhauled in 1964. If we continue to operate this plant without repairs, the return $p('68, '62, '64)$ for 1968–69 only, is

$$p('68, '62, '64) = p_0('68 - '64) + p_b('62)$$
$$= 40 + 20 = 60$$

But the decision not to build or repair in 1968 also affects the return in 1970–71, for it specifies that the plant in 1970 will have been built in 1962 and last overhauled in 1964. Thus the optimal 1970 decision "build," resulting from the 1968 decision "continue," will give a 1970–71 return of $f_1('62, '64)$, which from Table VIII is 80. The total return for the last two time periods (four years), which we shall denote as $p_2('62, '64)$, is therefore

$$p_2('62, '64) = p('68, '62, '64) + f_1('62, '64)$$
$$= 60 + 80 = 140$$

In general, for any plant, the decision "continue" in 1968 when followed by the optimal 1970 decision will give the following total profit for the last two stages:

$$p_2(t_b, t_r) = p('68, t_b, t_r) + f_1(t_b, t_r)$$

Table IX gives these values for all ten possibilities. The table is constructed in the same way as Table V, the only difference being that to form Table IX the numbers $f_1(t_b, t_r)$ in Table VIII corresponding to the proper row and column are also added to the usual sum of the entries $p_0('68 - t_r)$ at the left of the row and the technological bonuses $p_b(t_b)$ at the top of the column. As we have already calculated, the entry in column '62 and row '64 is 140.

TABLE IX

Two-Stage Returns, $p_2(t_b, t_r)$, from *Continuing* in 1968 and Making Optimal Decision in 1970

		t_b	'60	'62	'64	'66
		$p_b(t_b)$	0	20	50	80
t_r	'68 $-t_r$	$p_0('68 - t_r)$				
'60	8	20	50			
'62	6	30	60	130		
'64	4	40	70	140	190	
'66	2	60	100	160	210	260

Consider now the consequences of renovating in 1968 the plant built in 1962 and last overhauled in 1964 that we have been studying. The return $r('68, '62, '64)$ to be gained in 1968–69 from the newly repaired plant is

$$r('68, '62, '64) = p_0(0) + p_b('62) - c_r('68 - '64)$$
$$= 100 + 20 - 75 = 45$$

As a result of this decision, the plant entering the period 1970–71 will have $t_b = $ '62 and t_r advanced to '68. The optimal thing to do in 1970 will be either to continue operating or to build a new plant, for as Table VIII shows, either of these actions will bring a return of 80 for 1970–71. The total return $r_2('62, '64)$ obtained from a repair in 1968 and the resulting optimal decision in 1970 will be

$$r_2('62, '64) = r('68, '62, '64) + f_1('62, '68)$$
$$= 45 + 80 = 125$$

Let us consider now all possible dates of construction and previous overhaul. The two-stage returns $r_2(t_b, t_r)$ from a decision "repair" in 1968 followed by an optimal decision in 1970 are given in Table X, which is constructed in much the same way as Table VI. The only difference is that at the head of each construction year column has been added the corresponding return $f_1(t_b, '68)$, from an optimal decision in 1970. These numbers come directly from the bottom line of Table VIII. Each entry in Table X is the sum of the two numbers at the head of its column, the number at the left of its row, and 100, the value of $p_0(0)$.

TABLE X

Two-Stage Returns, $r_2(t_b, t_r)$, after *Repairing* the Plant in 1968 and Making Optimum Decision in 1970

		t_b	'60	'62	'64	'66
		$f_1(t_b, \text{'}68)$	60	80	110	140
		$p_b(t_b)$	0	20	50	80
t_r	$\text{'}68 - t_r$	$-c_r(\text{'}68 - t_r)$				
'60	8	-90	70			
'62	6	-85	75	115		
'64	4	-75	85	125	185	
'66	2	-65	95	135	195	255

It remains to analyze the decision "build" in 1968. The calculations are particularly simple in this case because the net profit, the technological bonus, and the return from an optimal decision in 1970 (Continue) are the same for all ten possible cases. That is,

$$p_0(0) = 100$$
$$p_b(\text{'}68) = 120$$

and

$$f_1(\text{'}68, \text{'}68) = 180$$

The only variable affected by the state of the old plant is the construction cost c_b. Table XI lists these two-stage returns. Each entry is simply 400 less the construction cost given at the top of each column.

TABLE XI

Two-Stage Returns, $b_2(t_b)$, after *Building* a New Plant in 1968 and Making Optimum Decision in 1970 (*Continue*)

t_b	'60	'62	'64	'66
$\text{'}68 - t_b$	8	6	4	2
$-c_b(\text{'}68 - t_b)$	-200	-180	-160	-150
t_r				
'60	200			
'62	200	220		
'64	200	220	240	
'66	200	220	240	250

In the particular case we have been studying, namely, a plant built in 1962 and repaired in 1964, the net construction cost is 180, leaving a total two-stage return of 220. This is clearly better than either repairing the old plant (r_2('62, '64) = 125) or continuing to operate with the old plant (p_2('62, '64) = 140). Hence the optimal policy for the last two stages is "build" in 1968 and "continue" in 1970. The associated maximum two-stage return f_2('62, '64) is, of course, 220.

In general we shall use $f_2(t_b, t_r)$ to denote the maximum two-stage return for a plant built in year t_b and last repaired in year t_r. The values of $f_2(t_b, t_r)$ for the ten possible combinations of t_b and t_r are given in Table XII. These maximum returns are of course obtained by comparing corresponding entries in the three preceding Tables IX, X, and XI.

TABLE XII

OPTIMAL DECISIONS FOR 1968 WITH THEIR
ASSOCIATED MAXIMUM TWO-STAGE RETURNS,
$f_2(t_b, t_r)$

t_r \ t_b	'60	'62	'64	'66
'60	200 B			
'62	200 B	220 B		
'64	200 B	220 B	240 B †	
'66	200 B	220 B	240 B	260 C

The method of dynamic programming should be clear enough by now so that we can proceed with the analysis of the remaining three time periods without further discussion. All of the pertinent calculations will be given, so that the reader can verify his own comprehension by checking some of the figures, as given in Tables XIII to XX.

TABLE XIII

THREE-STAGE RETURNS, $p_3(t_b, t_r)$, FROM *Continuing* IN 1966
AND MAKING OPTIMAL DECISIONS IN 1968 AND 1970

			t_b	'60	'62	'64
			$p_b(t_b)$	0	20	50
t_r	'66 − t_r	p_0('66 − t_r)				
'60	6	30		230		
'62	4	40		240	280	
'64	2	60		260	300	350

TABLE XIV
Three-Stage Returns, $r_3(t_b, t_r)$, after *Repairing* in 1966 and Making Optimal Decisions in 1968 and 1970

			t_b	'60	'62	'64
			$f_2(t_b, \text{'66})$	200	220	240
			$p_b(t_b)$	0	20	50
t_r	$\text{'66} - t_r$	$-c_r(\text{'66} - t_r)$				
'60	6	-85		215		
'62	4	-75		225	265	
'64	2	-65		235	275	325

TABLE XV
Three-Stage Returns, $b_3(t_b)$, after *Building* in 1966 and Making Optimal Decisions in 1968 and 1970

t_b	'60	'62	'64
$\text{'66} - t_b$	6	4	2
$-c_b(\text{'66} - t_b)$	-180	-160	-150
t_r			
'60	260		
'62	260	280	
'64	260	280	290

TABLE XVI
Optimal Decisions for 1966 with their Associated Maximum Three-Stage Returns, $f_3(t_b, t_r)$

t_r \ t_b	'60	'62	'64
'60	260 B		
'62	260 B	280 BC	
'64	260 BC	300 C	350 C†

The results for 1964–65 are given on the next page.

In the time period 1962–63, which is the fifth stage counting backwards from end to beginning, there is only one possible plant to be considered. This is the plant built two years earlier in 1960. If we continue

TABLE XVII
Continue IN 1964;
$p_4(t_b, t_r)$

t_r \ t_b	'60	'62
'60	300	
'62	320	360

TABLE XVIII
Repair IN 1964;
$r_4(t_b, t_r)$

t_r \ t_b	'60	'62
'60	285	
'62	295	335

TABLE XIX
Build IN 1964;
$b_4(t_b)$

t_r \ t_b	'60	'62
'60	340	
'62	340	350

TABLE XX
OPTIMAL DECISIONS FOR 1964,
WITH RETURNS, $f_4(t_b, t_r)$

t_r \ t_b	'60	'62
'60	340 B†	
'62	340 B	360 C

to operate this plant without repairs, and then make optimal decisions throughout the rest of the ten-year period, the return will be

$$p_5('60, '60) = p_0('62 - '60) + p_b('60) + f_4('60, '60)$$
$$= 60 + 0 + 340$$
$$= 400$$

If we repair the plant and make optimal decisions thereafter, the return will be

$$r_5('60, '60) = p_0(0) + p_b('60) - c_r('62 - '60) + f_4('60, '62)$$
$$= 100 + 0 - 65 + 340$$
$$= 375$$

If we build a new plant, the maximum possible ten-year return will be

$$b_5('60) = p_0(0) + p_b('62) - c_b('62 - '60) + f_4('62, '62)$$
$$= 100 + 20 - 150 + 360$$
$$= 330$$

We see that by continuing to operate in 1962 without repairing the plant, and by making the right decisions in the four periods following, we can realize the maximum possible return (400) over the ten-year period.

It would be interesting, of course, to know what the later decisions should be. To find out what should be done in 1964, we simply examine the optimal decisions for 1964 in Table XX. Since the optimal 1962 decision is not to repair or replace the 1960 model, the entry for $t_b = t_r = $ '62 is the only one pertinent; it says to build a new plant in 1964. Similarly, the entry for $t_b = t_r = 1964$ in Table XVI tells us to continue operating in 1966 without repairs. By continuing this process through Table XII for 1968 and Table VIII for 1970, we find that the optimal policy is CBCBC—build new plants in '64 and '68, and continue operating without repairs in '62, '66, and '70. The pertinent entries in the four optimal decisions have been marked with daggers (†) for easy reference.

IV. Feasibility Problems

A. General Remarks

This final section discusses how to find the optimum when the operating variables, or functions of them, must stay within certain predetermined limits. The functions restricted may be physical properties constrained by specifications, or quantities of materials limited by supply and demand. Any combination of the operating variables satisfying all the side conditions is said to be *feasible*. Our task is to pick out the feasible combination giving the optimum value of the criterion of effectiveness.

Optimization of a function whose variables must obey side conditions is known to the operations analysts as *mathematical programming*. In this context the word "programming" means "scheduling" or "planning," for many industrial feasibility problems involve finding optimum production schedules. Since the term "programming" is also used to indicate the writing of routines for automatic computers, the reader should guard against confusing *mathematical* programming with *computer* programming; there is no theoretical connection between the two.

The simplest kind of feasibility problem arises when the properties constrained are linear functions of the operating variables, of the following form:

$$a_1 x_1 + a_2 x_2 + \cdots + a_n x_n \leqslant b$$

$x_1, x_2, \ldots, x_n$ being the operating variables, and $a_1, a_2, \ldots a_n$, and b known constants. Fortunately, such linear inequalities describe most industrial constraints fairly well. In the chemical process industries, constraints usually involve stoichiometric relations, which are, of course, linear, or such physical properties as vapor pressure which can often be approximated by linear functions like Raoult's Law. Most of the present methods for solving feasibility problems only work when the constraints are linear. Simple as it is, the linear model has many applications, mainly to problems involving scheduling. More detailed discussions of applications are included in Sections IV, C, IV, D, and IV, E.

All of the methods discussed here involve starting with an arbitrary feasible solution, and successively generating better ones until at last the optimum is found. Section IV,B shows how to find a starting feasible solution, whenever inspection of the inequalities fails to indicate an obvious choice.

When the criterion optimized is a linear function of the operating variables, the feasibility problem is said to be one in *linear programming*. Being the simplest possible feasibility problem, it was the first one studied, and the publication in 1951 of G. Dantzig's *simplex method* for solving linear-programming problems (D2) marked the beginning of contemporary research in optimization theory. Section IV,C is devoted to this important technique.

In most linear-programming problems, the criterion is economic—production cost or net profit—the coefficients of the operating variables being the unit costs (or unit profits) in dollars per unit of variable. It is the constancy of these unit costs, independent of any changes in the operating variables, that makes the total cost linear. If, on the other hand, the unit costs increase with the production rate, we have what the economists call the "Law of Diminishing Returns." This can happen, for example, when yields or purity drop off at high production rates in a chemical process. If the unit costs themselves grow in proportion to the levels of the operating variables, the cost function is quadratic, and optimizing such a function in the face of feasibility restrictions is a problem in *quadratic programming*. Section IV,D shows that the quadratic problem is equivalent to one of finding *any* feasible solution to a certain set of linear inequalities (W2). Since the latter problem will have been solved already in Section IV,B, the only new thing in Section IV,D will be the use of the techniques of partial differentiation to derive the needed inequalities.

The concluding section describes a recent result of Dantzig and Wolfe (D1) which permits the partitioning of certain interacting linear programming problems. Their *decomposition principle* greatly extends the number of constraints that can be considered, making possible the optimization of much larger systems than before. Aside from this considerable technical advance, the method gives insight into the resolving of conflicts between local managers who must share scarce resources but who do not know how their demands affect other parts of the system. An interesting feature of the method is that it forces local managers to consider possibilities not optimal on the local level. It guides the choice of alternate possibilities by assigning penalty costs to scarce items. The decomposition principle not only is of technical interest, combining interaction problems with feasibility problems as it does, but also is of managerial interest because of its applications to the administration of decentralized decisions and the evaluation of conflicting demands.

B. Finding a Feasible Combination

1. *Inequalities*

Industrial systems which can be described entirely by equations are rare. Although, as in the case of material balances, the constraints imposed by Nature often are strict, those imposed by Man more usually allow some leeway. Product specifications fix the minimum quality permitted, but no one objects if the standards are exceeded. The amount of available raw material limits the quantity which can be produced, but usually no harm is done if some raw material is left unused. Quality specifications and material availabilities are described mathematically by inequalities rather than equations. Another type of inequality arises in industrial problems which is so obvious that it is easy to overlook. This is the fact that no material can be consumed in negative amounts. Thus, a set of *non-negative* values satisfying a set of constraints is called a *feasible solution* to the constraints.

Any inequality can be transformed into an equation by adding a new variable to the smaller side of the inequality. This new variable, which represents the difference between the two sides of the inequality, is called a *slack variable*. Like operating variables, slack variables are not allowed to take on negative values. For example, the inequality

$$10x_1 + 70x_2 \leqslant 11$$

can be converted to an equation by the addition of a slack variable u_1 to the left side, giving

$$10x_1 + 70x_2 + u_1 = 11$$

Since it is always understood that all the variables are non-negative, the inequalities

$$x_1 \geqslant 0, x_2 \geqslant 0, \quad \text{and} \quad u_1 \geqslant 0$$

need not be written explicitly.

2. *Description of Method*

When a system is tightly constrained by many inequalities, a question may arise as to whether any feasible solution to the system exists. In the next example we shall demonstrate a method which will always find a feasible solution to a set of linear inequalities if there is one. If on the other hand no feasible solution exists, the method will say so. The technique involves adding a number of *artificial variables* u' to the problem in order to create an artificial system which has an obvious feasible solution. The system variables are then manipulated to make the artificial variables vanish. If this can be done, the result is a feasible solution. If not, then no feasible solution exists.

The calculation process, which involves only simple arithmetic, is called the *simplex method*. It will be discussed more fully in the section following. This application to finding a feasible solution is an adaptation of one described by Charnes et al. (C1).

3. *Example: A Feasible Blend*

a. *Statement of problem.* Three chemicals, designated 1, 2, and 3, respectively, are to be mixed together to obtain a mixture having a maximum vapor pressure at 68°F. of 11 p.s.i.a. and a minimum mean molecular weight of 34. Physical properties and costs of the components are given below. Assume Raoult's law holds for this blend. Find a feasible blend.

Component	1	2	3	Specification
Vapor pressure at 68°F., p.s.i.a.	10	70	0	11 max.
Molecular weight	30	20	40	34 min.

b. *Solution.* Let x_1, x_2, and x_3 respectively represent the mole fractions of components 1, 2, and 3. By Raoult's Law, the vapor-pressure specification is

$$10x_1 + 70x_2 \leq 11$$

The definition of average molecular weight gives

$$30x_1 + 20x_2 + 40x_3 \geq 34$$

The material balance gives

$$x_1 + x_2 + x_3 = 1$$

This system is so simple that it doesn't take much imagination to find a feasible blend. For example, component 3 has the desired properties by itself. However, we wish to demonstrate a method which will always work, no matter how complicated the system is.

The first step is to introduce non-negative slack variables as needed to transform the inequalities into equations. In doing this, multiply any equation having a negative constant on the right side by -1, so that the right sides of all equations will be positive. This gives

$$10x_1 + 70x_2 \qquad\qquad + u_1 \qquad\quad = 11$$
$$30x_1 + 20x_2 + 40x_3 \qquad\quad - u_2 = 34$$
$$x_1 + \quad x_2 + \quad x_3 \qquad\qquad\qquad = 1$$

Next add a non-negative artificial variable to any equation not having a slack variable or having one with a negative sign.

$$10x_1 + 70x_2 \qquad\qquad + u_1 \qquad\qquad\qquad = 11$$
$$30x_1 + 20x_2 + 40x_3 \qquad\quad - u_2 + u'_2 \qquad = 34$$
$$x_1 + \quad x_2 + \quad x_3 \qquad\qquad\qquad + u'_3 = 1$$

This artificial system has an obvious feasible solution, namely $u_1 = 11$, $u'_2 = 34$, and $u'_3 = 1$, with all of the other variables zero. Such a solution, in which there is one nonzero variable for each constraint, is called a *basic solution* of the system, and the set of nonzero variables is called a *basis*. The variables not in the basis are called the *nonbasic variables*. We want to find a basic solution that is also feasible and which satisfies the side condition that all of the artificial variables must be zero. Since all variables are nonnegative, this side condition is equivalent in our case to

$$u'_2 + u'_3 = 0$$

In the first artificial basic feasible solution,

$$u'_2 + u'_3 = 35$$

Our procedure will be to manipulate the nonbasic variables one at a time in such a way that the sum $u'_2 + u'_3$ is always decreased, at the same time keeping the artificial solution feasible. Eventually this procedure will reduce the sum to zero, or else show clearly that no further reduction is possible (indicating that no real feasible solution exists).

At each stage of the calculation, the basic variables and the sum $u'_2 + u'_3$ are expressed in terms of the nonbasic variables. This immediately makes the effects of manipulations of the nonbasic variables clear. The three constraint equations already express each basic variable only in terms of nonbasic variables, since there is only one basic variable in each equation. It remains to express $(u'_2 + u'_3)$ in terms of nonbasic variables only. This is done by adding together the second and third constraint equations to obtain

$$31x_1 + 21x_2 + 41x_3 - u_2 + (u'_2 + u'_3) = 35$$

From this equation it is seen that $(u'_2 + u'_3)$ can be decreased by increasing either x_1, x_2, or x_3, or by decreasing u_2. This last possibility is ruled out because u_2 is already zero and is not allowed to become negative.

Let us choose arbitrarily to increase x_3. At first glance it might seem wise to make $x_1 = 35/11$, since this would make $(u'_2 + u'_3)$ vanish. However, substitution of this value into the second constraint would give

$$40(35/41) + u'_2 = 34$$

or $u'_2 = -0.2$, which is forbidden since u'_2 cannot be negative. We want to make x_3 just large enough to force one of the basic variables to become exactly zero and leave the basis. To do this, we set each basic variable in turn equal to zero, hold all the nonbasic variables except x_3 equal to zero, and solve for x_3. According to the first equation, u_1 is not altered by changes in x_3; according to the second, u'_2 vanishes when $x_3 = 34/40$, and according to the third, u'_3 vanishes when $x_3 = 1$. Hence we must set x_3 equal to the smaller value and displace u'_2 from the basis. The new basis is x_3, u_1, and u'_3.

Next we must express the function $(u'_2 + u'_3)$ and all of the variables in the new basis in terms of the new nonbasic variables. It is particularly easy to do this for the newest basic variable x_3, since the second equation contains only x_3 and nonbasic variables. Dividing this equation by 40, the coefficient of x_3, we obtain

$$0.75x_1 + 0.50x_2 + x_3 - 0.025u_2 + 0.025u'_2 = 0.85$$

Variable x_3 is next eliminated from the other equations and from the function $(u'_2 + u'_3)$ by substituting the expression just obtained wherever x_3 appears. For example, the third equation becomes

$$x_1 + x_2 + x_3 + u'_3 = 1$$
$$x_1 + x_2 + (-0.75x_1 - 0.50x_2 + 0.025u_2 - 0.025u'_2 + 0.85) + u'_3 = 1$$
$$0.25x_1 + 0.50x_2 + 0.025u_2 - 0.025u'_2 + u'_3 = 0.15$$

The results of these manipulations are shown in the following *tableau* of coefficients, with the function $(u'_2 + u'_3)$ written as the equation below the double line. Constants have been transposed to the right and are shown to the right of the vertical double line. For example, the equation above is the third equation in the table below.

x_1	x_2	x_3	u_1	u_2	u'_2	u'_3	
10	70	—	1	—	—	—	11
0.75	0.50	1	—	−0.025	0.025	—	0.85
0.25	0.50	—	—	0.025	−0.025	1	0.15
0.25	0.50	—	—	0.025	−1.025	—	0.15

In tableau form the basic variables are easy to locate, for they are the ones whose columns contain a single 1. The numbers in the right column are the values of the basic variables, and in the lower right corner is the current value of $(u'_2 + u'_3)$.

Since $(u'_2 + u'_3)$ has not yet been reduced to zero, we must change the basis again. Bringing in either x_1, x_2, or u_2 will decrease $(u'_2 + u'_3)$, since all of their coefficients in the expression for $(u'_2 + u'_3)$ on the fourth line are positive. For computational convenience we chose u_2. Inspection of the u_2 column shows that an increase in u_2 will not alter u_1 and it will actually increase x_3. Since only u'_3 can be reduced, it is the variable which must be forced out of the basis. The new basis is x_3, u_1, and u_2 and the resulting table is:

x_1	x_2	x_3	u_1	u_2	u'_2	u'_3	
10	70	—	1	—	—	—	11
1	1	1	—	—	—	1	1
10	20	—	—	1	−1	40	6
—	—	—	—	—	−1	−1	0

Since $(u'_2 + u'_3) = 0$, or in other words, since u'_2 and u'_3 are not in the basis, we have a solution feasible for the original problem, with $x_3 = 1$, $u_1 = 11$, and $u_2 = 6$. That is, the vapor pressure is $11 - u_1 = 0$ and the mean molecular weight is $34 + u_2 = 40$. This is the feasible solution which we remarked at the beginning was apparent from inspection of the equations. If at some stage of the calculations we had found that all of the coefficients of the function $(u'_2 + u'_3)$ were negative, indicating that no change of basis could reduce $(u'_2 + u'_3)$, and if $(u'_2 + u'_3)$ were not already zero, we would have been able to conclude that no feasible solution existed.

C. Linear Programming

1. Background and Applications

The simplest kind of mathematical programming problems arises when the inequalities and the function to be optimized are all *linear* functions of the non-negative variables. This branch of mathematical programming, the first to be studied, is called *linear programming*. It may be surprising to learn that no analysis of such problems appeared until 1951, when George Dantzig published the first practical method for solving them. His technique, which he called the *simplex algorithm,* or more simply, the *simplex method,* is still the most widely used.

The applications of linear programming have been widespread and hundreds of technical articles and many books (e.g., G2, G3) on linear programming have appeared. Many of the first uses were made by the petroleum industry, and anyone interested in scheduling interacting processes would do well to read the books by Symonds (S3) and Manne (M2), as well as the article of Charnes *et al.* (C2). The applications of linear programming go far beyond production scheduling problems. Using short illustrative examples, Vadja (V1) describes applications in such diverse domains as selection of products, transportation scheduling, network flow, investment, personnel assignment, scrap reduction, warehousing, maintenance, and shift rotation scheduling. There are also important theoretical connections between linear programming and the theory of games (M1, W1), which deals with strategy in competitive situations. We shall confine our discussion here, however, to the simple blending problem already posed.

2. Description of Method

In the previous example, the technique used to reduce the artificial variables to zero was in fact Dantzig's simplex method. The linear function "optimized" was the simple sum of the artificial variables. Any linear function may be optimized in the same manner. The process must start with a basic solution feasible with the constraints, the function to be optimized expressed only in terms of the variables not in the starting basis. From these expressions it is decided what nonbasic variable should be brought into the basis and what basic variable should be forced out. The process is iterated until no further improvement is possible.

The simplex method only examines *basic* solutions—those having exactly one nonzero variable for each constraint. It is not particularly obvious that the optimal solution must be one of the basic solutions, and one of Dantzig's contributions was to prove this fact (D2).

3. Example: Optimal Blending With Constant Unit Costs

a. Statement of problem. Find a blend of the three components in the example of Section IV,B,3 which satisfies the specifications of that example and for which the cost per pound mole is minimum. Component 1 costs $6 per pound mole; component 2, $2; and component 3, $3.

b. Solution. The cost in dollars per pound mole of blend, based on the constant unit costs given, is

$$c = 6x_1 + 2x_2 + 3x_3$$

It is desired to find non-negative values of x_1, x_2, and x_3 that minimize this function and at the same time satisfy the constraints of the previous example. We shall begin with the basis x_3, u_1, and u_2, which was shown to be feasible in the last example. At the end of the last example these variables were expressed in terms of the nonbasic variables x_1 and x_2, and so it only remains to express the new cost function in terms of these variables. Eliminating x_3 with the second line of the final table there gives

$$c = 6x_1 + 2x_2 + 3(1 - x_1 - x_2) = 3 + 3x_1 - x_2$$

The table with this new cost function is therefore

c	x_1	x_2	x_3	u_1	u_2	
—	10	70	—	1	—	11
—	1	1	1	—	—	1
—	10	20	—	—	1	6
1	−3	1	—	—	—	3

The unit cost of this blend is \$3 per pound mole, as is shown in the lower right corner. We see that c can be reduced by increasing x_2, with u_1 being forced out of the basis, since $0 < (11/70) < (6/20) < 1$. The methods already described in Section IV, 3, b give the following new table:

c	x_1	x_2	x_3	u_1	u_2	
—	0.143	1	—	0.0143	—	0.157
—	0.857	—	1	−0.0143	—	0.843
—	7.143	—	—	−0.286	1	2.86
1	−3.143	—	—	−0.0143	—	2.843

This basis is optimal, since neither of the nonbasic variables x_1 and u_1 can be brought into the basis without increasing the unit cost. Hence the optimal composition is $x_1 = 0$, $x_2 = 0.157$, and $x_3 = 0.843$. The vapor pressure is 11, just at the limit, while the mean molecular weight is 2.86 above the minimum, or 36.86. The minimum unit cost is \$2.84 per pound mole.

It is interesting to compare this cost with that of the blend just meeting all of the specifications. The composiion of such a blend would be

$$x_1 = 0.40$$
$$x_2 = 0.10$$
$$x_3 = 0.50$$

Such a blend would cost $4.10 per pound mole—44% more than the optimal blend. Hence it is actually more economical to make the product a little better than the minimum standards in this case.

D. QUADRATIC PROGRAMMING

1. Background and Applications

Quadratic programming problems are similar to those of linear programming, the difference being that in the former case the function to be optimized is a second-degree, rather than linear, function of the system variables. When unit costs are not constant, but instead are linear functions of the system variables, the cost function is quadratic rather than linear. Profit functions are approximately quadratic when market prices are depressed by increases in production (the "law of diminishing returns"). Markovitz (M4) has shown that the problem of minimizing the risk of attaining a fixed expected yield is a problem in quadratic programming. A large impetus to research in quadratic programming was provided by the fact that gasoline octane-number improvement, per unit of tetraethyl lead (T.E.L.) added, tends to decrease as the T.E.L. concentration increases. Although there do not seem to be as many applications of quadratic as of linear programming, quadratic programming deserves the attention of engineers in operations, both as a means of solving more sophisticated problems and as an introduction to the non-linear aspects of mathematical programming.

Wolfe (W2) has shown how the simplex method can be used to solve quadratic programming problems, using certain properties discovered by Kuhn and Tucker (K2) and Barankin and Dorfman (B1). Our presentation of quadratic programming is based on his work, as well on Dennis' discussion of the properties of a solution to a quadratic programming problem (D3). We have, however, revised the theoretical treatment to avoid Dennis' use of matrices and Lagrange multipliers, since these devices are not yet widely used by engineers. We have also modified Wolfe's techniques, which are strongly influenced by computer technology, to make the calculations more straightforward. The method used is intended only to illustrate the character of quadratic programming problems, and no claims as to computational efficiency are made.

2. Description of Method

It will be shown that the solution to any mathematical programming problem must satisfy not only the original constraints, but also certain side conditions. These new conditions take the form of inequalities in a new set of non-negative variables, and it will turn out that in quadratic programming problems these inequalities are linear. Thus the quadratic programming problem can be converted to one of finding a feasible solution to a set of linear inequalities. Such a feasibility problem can be solved by the simplex method, as the example demonstrated. Although the method will always find a stationary point if there is one, one cannot be certain that the point found is the minimum unless the quadratic part of the function optimized can never become negative (D3).

3. *Theoretical Development*

Suppose it is desired to minimize a cost function c when the non-negative system variables x_j are constrained by inequalities. Take any basic feasible solution and consider how c is altered by a change Δx_j in the variable x_j. There are two effects to be considered. First there is the direct cost Δc_{1j} of the material Δx_j, which is given by

$$\Delta c_{1j} = \left(\frac{\partial c}{\partial x_j}\right) \Delta x_j$$

Second there is the cost Δc_{2j} associated with adjusting all the other basic variables in order to keep the solution feasible. To make room for the perturbation Δx_j, each slack variable u_i must be adjusted by an amount

$$-\left(\frac{\partial u_i}{\partial x_j}\right) \Delta x_j$$

The total cost of this readjustment is

$$\Delta c_{2j} = -\sum_i \left(\frac{\partial c}{\partial u_i}\right)\left(\frac{\partial u_i}{\partial x_j}\right) \Delta x_j$$

The derivative $\partial c/\partial u_i$ will be abbreviated by the symbol y_i.

$$y_i \equiv \frac{\partial c}{\partial u_i}$$

In any *optimal* basis y_i cannot be negative. This may be shown by the following simple argument: either the slack variable u_i is in the optimal basis, or is not. If it is, then, because of the way the simplex method operates, y_i must be zero. On the other hand, if u_i is not in the basis, then $\partial c/\partial u_i$ cannot be negative, or else the cost could be decreased by bringing u_i into the basis. In either case, y_i is non-negative, that is

$$y_i \geqslant 0$$

Since either u_i or y_i must be zero, and since both variables are non-negative, we may say that for every constraint i,

$$u_i y_i = 0$$

In terms of the non-negative variables y_i, the total change Δc_j in cost for a change Δx_j is

$$\Delta c_j = \Delta c_{1j} + \Delta c_{2j} = \left[\frac{\partial c}{\partial x_j} - \sum_i y_i \left(\frac{\partial u_i}{\partial x_j}\right)\right] \Delta x_j$$

Let v_j designate the limit of $\Delta c_j / \Delta x_j$ as Δx_j approaches zero. Then v_j represents the rate of total change of the cost with respect to a change in x_j; note that v_j is not the same as the partial derivative $\partial c / \partial x_j$. The actual relation is

$$v_j \equiv \frac{\partial c}{\partial x_j} - \sum_i y_i \left(\frac{\partial u_i}{\partial x_j}\right)$$

In an optimal basis, v_j can also be shown to be non-negative by an argument similar to that for y_i. If x_j is in the optimal basis, then v_j is zero, since

$$\frac{\partial c}{\partial x_j} = \sum_i \left(\frac{\partial c}{\partial u_i}\right)\left(\frac{\partial u_i}{\partial x_j}\right) = \Sigma y_i \left(\frac{\partial u_i}{\partial x_j}\right)$$

On the other hand, if x_j is not in the optimal basis, then v_j must be non-negative, for otherwise the cost could be decreased by an increase in x_j. Hence, in an optimal basis,

$$v_j \geqslant 0$$

Furthermore, v_j and x_j cannot both be in the optimal basis, or, since x_j is also non-negative

$$v_j x_j = 0, \quad \text{for all } j$$

This condition and the relation above, proved

$$u_i y_i = 0, \quad \text{for all } i$$

are known as the *complementary slackness conditions*. They hold for any mathematical programming problem. The complementary slackness conditions were first established by Kuhn and Tucker (K2).

Any optimal basis for a mathematical programming problem not only must be feasible for the original constraints and satisfy the complementary slackness conditions, but also must be feasible for the equations

$$\sum_i y_i \left(\frac{\partial u_i}{\partial x_j}\right) + v_j = \frac{\partial c}{\partial x_j}$$

where y_i and v_j are non-negative. In quadratic programming problems, these equations are linear, since the $\partial u_i / \partial x_j$ are constants and $\partial c / \partial x_j$ is linear in the x_j. Hence the quadratic programming problem can be reduced to one of finding a non-negative solution for a set of linear equations, which we know can be solved by the simplex method. There is, however, no guarantee that such a feasible solution gives the lowest possible cost unless the quadratic part of the cost function cannot become negative (D3).

4. Example: Optimal Blending with Diminishing Returns

a. Statement of problem. Suppose that component 2 in the previous example comes from a process whose yield decreases as more component 2 is produced. The unit cost of component 2 is thus $(2 + 10x_2)$ per pound mole, instead of $2 as before. Find the composition of the blend having the minimum cost per pound mole, and satisfying the specifications of the preceding example.

b. Solution. The new cost per pound mole is

$$c = 6x_1 + (2 + 10x_2)x_2 + 3x_3$$
$$= 6x_1 + 2x_2 + 3x_3 + 10x_2^2$$

An equivalent problem is to find a feasible solution to the set of linear equations presented below in tableau form. The top three equations represent the original constraints, while the bottom three are the equations

$$\sum_i y_i \left(\frac{\partial u_i}{\partial x_j}\right) + v_j = \frac{\partial c}{\partial x_j}$$

for $j = 1, 2,$ and 3 respectively.

The fact that the third constraint, that on the material balance, is an equality rather than an inequality introduces a minor difficulty, since there is no slack variable u_3. Without specifically writing it down we have added a fictitious slack variable u_3 to the left side in order to have the signs of $\partial u_3/\partial x_j$ consistent. Since the fictitious variable could have been added just as easily to the right side, the signs of $\partial u_3/\partial x_j$ could all be changed. Such a change of signs would of course reverse the sign of y_3. Since the material balance restriction is always binding, y_3 must be in the optimal basis. If we have made an improper choice of signs, y_3 will merely turn out negative instead of positive. To illustrate this point, we have deliberately picked the wrong signs.

x_1	x_2	x_3	u_1	u_2	y_1	y_2	y_3	v_1	v_2	v_3	
10	70	—	1	—							11
30	20	40	—	−1							34
1	1	1	—	—							1
					−10	30	−1	1	—	—	6
	−20				−70	20	−1	—	1	—	2
					—	40	−1	—	—	1	3

Ordinarily we would introduce artificial variables and begin using the simplex method to reduce the sum of these variables to zero. However, in order to save space, as well as to demonstrate the effect of the quadratic term in the cost function, we shall start with the basis which was optimal in the linear case just solved. This basis, namely x_2, x_3, and u_2, will of course be feasible for the three original constraints. If we filled out the basis by using v_1, v_2, and v_3, the basis would be feasible and there would be no artificial variables. Although at first glance it would appear that the basis x_2, x_3, u_2, v_1, v_2 and v_3 is optimal, this is not true because of the complementary slackness condition, which prohibits having both x_2 and v_2, or both x_3 and v_3, in the same basis.

Since the complementary slackness principle does not allow the use of v_2 and v_3 in the basis, two artificial variables v'_2 and v'_3 must be introduced. With this basis, x_2, x_3, u_2, v_1, v'_2, and v'_3, the tableau of equations is as given below. Notice that the first three lines come directly from the optimal table of the preceding example. The only changes needed in the second trio of equations are in the x columns. The equation below the double line, which is the sum of the fifth and sixth equations, is simply the expression of the sum of artificial variables $(v'_2 + v'_3)$. We shall use the simplex method to minimize this sum, hopefully until it becomes zero.

x_1	x_2	x_3	u_1	u_2	y_1	y_2	y_3	v_1	v_2	v_3	v'_2	v'_3	
0.143	1	—	0.0143	—									0.157
0.857	—	1	−0.0143	—									0.843
7.143	—	—	−0.286	1									2.857
2.857			0.286		−10	30	−1	1	—	—	—	—	6
					−70	20	−1	—	1	—	1	—	5.143
					—	40	−1	—	—	1	—	1	3
2.857			0.286		−70	60	−2	1	1	1	1		8.143

Bringing any of the variables x_1, u_1, y_2, $-y_3$, v_1, or v_2 into the basis will decrease $(v'_2 + v'_3)$. However, x_1, y_2, v_2, and v_3 are ruled out by the complementary slackness principle, leaving only u_1 and y_3. We know that $-y_3$ will be in the optimal basis, since it is associated with the material balance constraint which is always binding. Hence we choose to bring in $-y_3$, displacing v'_3, since

$$|(3/-1)| < |(5.143/-1)| < |(6/-1)|$$

The new table is

x_1	x_2	x_3	u_1	u_2	y_1	y_2	y_3	v_1	v_2	v_3	v'_2	v'_3	
0.143	1	—	0.0143	—									0.157
0.857	—	1	−0.0143	—									0.843
7.143	—	—	−0.286	1									2.857
2.857			0.286		−10	−10	—	1	—	−1	—	−1	3
					−70	−20	—	—	1	−1	1	−1	2.143
					—	−40	1	—	—	−1	—	−1	−3
2.857			0.286		−70	−20	—	—	1	−1	1	−1	2.143

Of the three variables x_1, u_1, and v_2 which would reduce $(v'_2 + v'_3)$ if brought into the basis, only u_1 can enter without violating the complementary slackness principle; u_1 displaces v'_2, giving the following feasible solution:

x_1	x_2	x_3	u_1	u_2	y_1	y_2	y_3	v_1	v_2	v_3	v'_2	v'_3	
—	1	—			3.5	1	—	—	−1	1	−1	1	0.05
1	—	1			3.5	−1	—	—	0.05	−0.05	0.05	−0.05	0.95
10	—	—		1	70	−20	—	—	1	−1	1	−1	5
10				1	−10	−10	—	1	—	−1	—	−1	3
					215	−70	—	—	3.5	−3.5	3.5	−3.5	7.5
					—	−40	1	—	—	−1	—	−1	−3

The optimal composition is $x_1 = 0$, $x_2 = 0.05$, and $x_3 = 0.95$. The vapor pressure is $11 - u_1 = 3.5$; the mean molecular weight is $34 + u_2 = 39$; and the cost per pound mole is \$2.975. For comparison, the composition optimal in the linear case would cost \$3.10 per pound mole, using the quadratic cost function. The penalty v_1 associated with introducing x_1 into the blend is \$0.03 per pound mole per percent x_1 added. Since the quadratic term in the cost function cannot become negative, we know that this composition gives the lowest possible cost. It is interesting that the slight nonlinearity of the cost function gives an optimal blend that is well inside both limiting specifications.

E. Feasibility in Interacting Systems

1. *Supersystems*

Most modern chemical companies have more than one manufacturing plant. Since the plants are often widely separated geographically, each plant manager and his technical staff may have a great deal of autonomy in such matters as production planning. But the decisions of one local manager can affect the operations in plants other than his own, for within any company all plants share raw materials, markets, and capital. Thus practically every large company has a centralized planning staff for coordinating the local activities. This central staff must be familiar with every constraint involving more than one plant. At the same time, each plant manager must be thoroughly acquainted with the limitations of his own plant. Given the complexity of modern technology, it is usually as impractical to expect each plant manager to understand the company-wide interactions between all the plants as it is to require the central staff to be acquainted intimately with the workings of every plant.

We shall speak of each plant as a *system* and of the company, the assembly of systems, as a *supersystem*. Thus a supersystem is a set of interacting systems. The goal of the central planning staff is to optimize the operation of the supersystem. But since the central agency does not know the local conditions in the various systems, it cannot proceed directly with the optimization. Instead, it must give each local manager an additional set of restrictions on his system which reflect the interactions between the systems. With these artificial restrictions, each local manager can proceed to optimize his own system, ignoring the effects of his decisions on the other systems. If the artificial restrictions imposed by the central agency are well-chosen, the actions taken locally will also be optimal for the supersystem.

The central agency, of course, cannot determine the proper artificial restrictions without some information from the local systems. It would seem reasonable to have each local manager present to the central agency the plan which is optimal for his system in the absence of any artificial restrictions. By comparing the requirements of these local optimal plans with the amounts of scarce materials available, the central staff must decide how to allocate the scarce materials to the systems. The technical problem is to find the *optimum* allocation.

If the systems and their economic functions are all linear, then in principle the supersystem could be optimized directly by linear programming. This approach would eliminate not only the need for allocation quotas, but also the need for the local managers to plan their own systems. However, when one considers that the time required for the solution of a linear-programming problem on a high-speed computer is approximately proportional to the cube of the number of constraints, it immediately becomes clear that optimization of the entire supersystem at once would not be very efficient. Moreover, supersystems the size of many chemical companies could not be solved in a reasonable time even on the fastest computers available today. For these reasons the idea of optimizing the systems separately, somehow taking account of the interactions between systems at the same time, has attracted the attention of chemical engineers.

2. The Decomposition Principle (D1)

Dantzig and Wolfe recently published an exposition of a "Decomposition Principle for Linear Programs," as a practical way of solving supersystems. The decomposition principle was developed mainly as a computational device for making it possible to solve large linear-programming problems. This aspect, while certainly of great practical value, will not be discussed here because of its rather specialized character. However, Dantzig and Wolfe pointed out that the calculation process, in itself, suggests a rational method for reconciling the conflicting requirements of the various systems in a way which is best for the supersystem. It is this feature of the decomposition principle that we will discuss here. Once the best plan for the supersystem is found, it is a simple matter to construct the quotas.

3. Description of Method

The decomposition principle has the following important qualitative features. Consider first the effect on any particular system of the intersystem constraints. If the system were completely independent of all the others (that is, if the intersystem constraints were ignored) the local optimum would be a single basic feasible solution found by applying the simplex method to the local linear-programming problem. However, from the decomposition principle it can be deduced that, when the intersystem restrictions are imposed, the local plan derived from the optimum master plan will be a *weighted average* of several local basic solutions. Thus the local manager must present several local basic solutions to the central agency, which in turn transmits to him the relative weights to be placed on each proposal. For example, if a local manager proposes three alternative plans A, B, and C which are basic solutions to his local system, the central agency might say, "Use 75% of the materials required by plan A, 25% of those of plan C, and do not use plan B." From the standpoint of operations analysis, it is very interesting that some managers necessarily must propose plans which do not appear optimal at the local level. And although it is certainly not unusual in practice for a centralized administration to require alternative proposals from the local managers, it may seem strange to form a weighted average of these proposals. It is more customary to choose one of the alternatives for each system, rejecting the others, especially when the scarce commodity to be shared is money for capital investment.

If it were necessary for every local manager to submit all possible local basic solutions to the central agency, the decomposition principle would not be very efficient. Fortunately, the central agency is actually able to guide the selections made at the local levels. It does this by assigning a penalty cost to each of the scarce commodities. These penalties are found by linear programming. Each manager then finds the local basic solution which is optimal in the face of these handicaps. If this new solution is not already under consideration, the local manager submits it to the coordinating agency, which in turn readjusts the penalty costs. This exchange between the central agency and the local managers continues until all the local solutions are based upon handicaps which have already taken these particular solutions into account. At this point the central agency assigns the weight to be placed on each proposal, or, equivalently, fixes the quotas on each scarce material.

A numerical example of the decomposition principle, currently under development by the writer, has proved to be too extensive for inclusion in the present article; it is hoped to publish it elsewhere in the chemical engineering literature in the near future.

Acknowledgment

This article is based on lectures given at the Ecole Nationale Supérieure des Industries Chimiques, Nancy, France, during the academic year 1960–61 under a Fulbright professorship. Preparation of the manuscript was a joint effort of the E.N.S.I.C.; the Technion, Haifa, Israel; the Centre Interarmées de Recherche Opérationnelle, Paris; and the author's wife Jane.

References

A1. Akers, S. B., and Friedman, J., *Operations Research* 3, 429 (1955).
A2. Aris, R., *Chem. Eng. Sci.* 8, 75 (1960).
B1. Barankin, E. W., and Dorfman, R., *Univ. Calif. (Berkeley) Publs. in Statistics* 2, 285 (1958).
B2. Bellman, R., "Dynamic Programming." Princeton Univ. Press, Princeton, New Jersey, 1957.
B3. Box, G. E. P., *Biometrics* 10 (1954).
B4. Box, G. E. P., and Wilson, K. B., *J. Roy. Sta. Soc.* **B13**, 1 (1951).
C1. Charnes, A., Cooper, W. W., and Henderson, A., "An Introduction to Linear Programming." Wiley, New York, 1953.
C2. Charnes, A., Cooper, W. W., and Mellon, B., *Econometrica* 22, 193 (1954).
D1. Dantzig, G. B., and Wolfe, P., *Operations Research* 8, 101 (1960).
D2. Dantzig, G. B., *In* "Activity Analysis of Production and Allocation" (T. C.

Koopmans, ed.). Cowles Commission for Research in Economics, Monograph No. 13. Wiley, New York, 1951.
D3. Dennis, J. B., "Mathematical Programming and Electrical Networks." Technology Press and Wiley, New York, 1959.
D4. Dreyfus, S. E., *J. Soc. Indl. Appl. Math.* **8**, 425 (1960).
E1. Edie, L. C., *Operations Research* **2**, 107 (1954).
G1. Gass, S. I., "Linear Programming: Methods and Applications." McGraw-Hill, New York, 1958.
G2. Gomory, R. E., *Bull. Am. Math. Soc.* **64** (1958).
G3. Greenwald, D. U., "Linear Programming and the Simplex Algorithm." Ronald Press, New York, 1956
J1. Johnson, S. M., *Mgt. Sci.* **5**, 299 (1959).
J2. Johnson, S. M., *RAND Corp. Report* **P-856** (1956).
K1. Keifer, J., *Proc. Am. Math. Soc.* **4**, 502 (1953).
K2. Kuhn, H. W., and Tucker, A. W., *Proc. 2nd Berkeley Symposium on Math. Stat. Prob.* 481 (1951).
M1. McKinsey, J. C. C., "Introduction to the Theory of Games." McGraw-Hill, New York, 1952.
M2. Manne, A. S., "Scheduling of Petroleum Refinery Operations." Harvard Univ. Press, Cambridge, Massachusetts, 1956.
M3. Morse, P. M., and Kimball, G. E., "Methods of Operations Research." Technology Press and Wiley, New York, 1951.
M4. Markowitz, H., *Naval Res. Logistics Quart.* **3**, 111 (1956).
P1. Phillips, H. B., "Vector Analysis." Wiley, New York, 1949.
R1. Roberts, S. M., "Proceedings of a Symposium on Optimization Techniques in Chemical Engineering." New York University, New York, 1960.
S1. Schrage, R. W., *Operations Research* **6**, 498 (1958).
S2. Shapiro, E., Shapiro, S., Stillman, R. E., and Lapidus, L., "A Control Strategy for Chemical Engineering Systems." Paper presented at Am. Inst. Chem. Engrs. Mexico City Meeting, 1960.
S3. Symonds, G. H., "Linear Programming: The Solution of Refinery Problems." Esso Standard Oil (1955).
S4. Szwarc, W., *Operations Research* **8**, 782 (1960).
V1. Vadja, S., "Readings in Linear Programming." Wiley, New York, 1958.
W1. Williams, J. D., "The Compleat Strategyst." McGraw-Hill, New York, 1954.
W2. Wolfe, P., *RAND Corp. Report* **P-1205** (1957).

AUTHOR INDEX

Numbers in parentheses are reference numbers and indicate that an author's work is referred to although his name is not cited in the text. Numbers in italic show the page on which the complete reference is listed.

A

Acrivos, A., 177, *198,* 206(A1), *269*
Aerov, M. E., 229 (A1a), 231, 232(A2), 233, *269*
Aiba, S., 130, 146(A4), *198, 200*
Akers, S. B., 277(A1), *331*
Alberda, G., 212(K2), *270*
Alexander, L. G., 128, *198*
Allison, S. K., 75(C6), *115*
Altman, D., 83(A2), 85(A2), 86(C1), 96, *115*
Amsler, J., 20(A1), *58*
Amundson, N. R., 212(A4), 257, *269, 270*
Anderson, J. A., 100(A3), *115*
Argo, W. B., 216, 229, *269*
Aris, R., 212(A4), *269,* 293(A2), 296(A2), *331*
Auro, M. A., 166(A6), *198*
Autenrieth, H., 33(A2), *58*

B

Baars, G. M., 147(K8), 155(K8), *200*
Baddour, R. F., 113(B1), *115*
Badger, W. L., 35, 48(B1), *58, 59,* 125 (W10), 127(W10), 134(W10), 147 (W10), *202*
Baker, J. G., 127, *199*
Bakhurov, V. G., 212(B1), *269*
Baranaev, M. K., 168(B1), *198*
Barankin, E. W., 323, *331*
Barkelew, C. H., 258, *270*
Baron, T., 128(A5), *198,* 212(B3), *270*
Bartholomew, W. H., 159, 161, 165(K2), 167(K2), *198, 199*

Baum, S. J., 136, 137(H7), 138, 181(H4), 186, *199*
Bechtel, R. J., 121, *199*
Becker, K., 20, *58*
Beek, J., Jr., 213, 223, *270*
Beerbower, A., 151, *198*
Bellman, R., 276, 293, 297(B2), *331*
Bennett, R. C., 26(N3), 49, 50, 51, *59*
Bernard, R. A., 214(B6), 227, *270*
Bertanza, L., 18(B3), *58*
Berthoud, A., 26, *58*
Bilous, O., 257, *270*
Bissell, E. S., 125, 127(B8), *198*
Black, C. R., 127(N4), *200*
Bond, J. W., Jr., 99(B2), 112(B2), *115*
Bonin, J. H., 99(J1), *116*
Booth, A. H., 18(B5), *58*
Bosworth, R. C. L., 261, *270*
Box, G. E. P., 276, 278, 291, *331*
Brace, P. H., 98(W6), *118*
Bransom, S. H., 44, 46, *58*
Brenner, E., 135(W5), *202*
Brewer, L., 98(B3), *115*
Brinkley, S. R., Jr., 93(B4, B5), *115*
Broida, H. P., 79(B6), *115*
Brötz, W., 206, *270*
Brown, G. G., 20(P4), 21, 22, *59*
Brown, R. W., 184, *198*
Brumaqin, I. S., 126(B10), 127(B10), *198*
Bruner, L., 30(B7), *58*
Bryant, E. O., 136(W6), 178(W6), *202*
Buckley, H. E., 9, 10(B8), 13, 18(B5), 23 (B8, B9), 24(B8), 26(B9), *58*
Bundy, F. P., 107(B7), *115*
Bunnell, D. G., 229(B9), 231, *270*
Burhorn, F., 99(B8, B9), *115*

C

Cairns, E. J., 212(C1), *270*
Calderbank, P. H., 160, 161, 161(C1), *198*
Caldwell, H. B., 47(C1), *58*
Callow, D. S., 165(C4), *198*
Campbell, A. N., 4(C2), *58*

Campbell, J. M., 229(C2), 231, 232(A2), *270*
Carlson, C. J., 136, 171, *200*
Carpani, R. E., 165, *198*
Carter, J. M., 86(C1), *115*
Chaddock, R. E., 189, *198*
Chain, E. B., 165, *198*
Chambré, P. L., 177, *198*
Charnes, A., 317(C1), 321(C2), *331*
Chesnut, F. T., 98(C2), *115*
Chilton, T. H., 28, 30, *58,* 183, 184, *198*
Christiansen, J. A., 14(C4), 20(C4), *58*
Chu, Y. C., 230, *270*
Clay, P. H., 168(C6), *198*
Clouston, J. G., 110(C3), *115*
Coberly, C. A., 231, 232(C4), 233, *270*
Cobine, J. D., 68(C4), 96, *115*
Colburn, A. P., 28, 30, *58*
Comenetz, G., 98(W6), *118*
Comings, E. W., 128(A5), *198*
Compton, A. H., 75(C6), *115*
Conklin, L. H., 30(W5), *60,* 181(W7), *202*
Conn, W. M., 101(C7), *115*
Converse, A. O., 212(C5), *270*
Cooke, F., *201*
Cooper, C. M., 162(C7), 163, *198*
Cooper, W. W., 317(C1), 321(C2), *331*
Corrsin, S., 124(C8), *198*
Costrich, E. W., 134(R13), 136(R13), 139(R13), 140(R13), 141(R13), 143(R13), 145(R13), *201*
Coulson, J. M., 18, 19, 48(C5), *58*
Courant, R., 102, *115*
Couture, J. W., 180(C10), *198*
Craine, C. M., Jr., 171(W8), 174(W8), *202*
Crowell, J. H., 30(H6), *58,* 181(H8), *199*
Cummings, G. S., 184, 185, *198*

D

Dämkohler's, G., 231, 259, 260, *270*
Danckwerts, P. V., 122, *198*
Dantzig, G. B., 276, 293(D1), 315, 321, 330(D1), *331*
Davies, C. W., 18, 27(D1), *58*
Davion, M., 30(D2), 36, *58*
Davis, T. P., 98(D1), *115*
Deans, H. A., 205, 213, 257, *270*
Dennis, J. B., 323, 324(D3), 326(D3), *332*
Deterding, J. H., 131, *202*
Diamond, H. W., 32(H4), *58*

Dickerman, P. J., 99(J1), *116*
Dieke, G. H., 69(D2), 78, *115*
Dinegar, R., 14(L2), 20(L2), *59*
Diniak, A. W., 111(S2), *118*
Doering, W., 20, *58*
Dorfman, R., 323, *331*
Dorweiler, V. P., 225(D4), *270*
Douglas, J., 247, *270*
Drew, T. B., 30(G1), *58,* 183(C5), 184(C5), *198*
Dreyfus, S. E., 293, *332*
Dunkley, W. L., 154, *198, 201*
Dunlap, I. R., 184, 185, *199*
Dunning, W. J., 44(B6), 46(B6), *58*
Dust, H., 33(A2), *58*
Duvall, G. E., 101(D3), *116*
Dyer, J. A., 111(S2), *118*

E

Edie, L. C., 274(E1), *332*
Elferdink, T. H., 162(N3), *200*
Elsworth, R., 161, 165, *199*
Endoh, K., 161, 162(O11), 178, 182, 189(O12, O13), *199, 200*
Engelke, J. L., 112(E1), *116*
Ergun, S., 30(E1), *58,* 234, *270*
Everett, H. J., 125(B6), 127(B8), 128(R16), 129(R16), 130(R16), 134(R13), 136(R13), 139(R13), 140(R13), 141(R13), 143(R13), 145(R13), *198, 201*

F

Fahien, R. W., 225(D4), *270*
Ferguson, C. K., 128(F4), 154, *199*
Fernstrom, G. A., 162(C7), 163(C7), *198*
Fick, J. L., 169, 172, 176, *199*
Finkelnburg, W., 66(F1), 67(F2), 98(F1), 99(F1), *116*
Finlay, G., 98(F4), *116*
Finn, R. K., 164, 165, *199*
Flood, H., 14, 20, *60*
Flynn, A. W., 174(F3), 189, *199*
Folsom, R. G., 128(F4), 154, *199*
Forstall, W., 154(F5), *199*
Forster, E. O., 151(B3), *198*
Forsythe, W. E., 70(F5), *116*
Fossett, H., 128(F7), 151, *199*
Foust, H. C., 157, 162(F8), *199*
Fowler, R. N., 63(F6), *63*
Fox, E. A., 128(F9), 149, 153, 155, *199*
Frank, F. C., 25(F1), *58*

Frenkel, J., 16, *58*
Friedland, W. C., *199*
Friedlander, S. K., 177(F11), *199*
Friedman, A. M., 165, *199*
Friedman, J., 277(A1), *331*
Friedricks, K. O., 102, *115*
Frulla, F., 18, *60*
Fullman, R. L., 9(F4), 23(F4), 25, 26 (F4), *58*

G

Gaden, E. L., Jr., 165, 166(R3), *201*
Gaffney, B. J., 30(G1), *58*
Gallagher, J. B., 164(R14), *201*
Garrett, D. E., 24(G2), 48, *58*
Gaydon, A. G., 88, 89(G1), 90(G1), *116*
Gaylord, E. W., 154(F5), *199*
Gerdien, H., 99, *116*
Gex, V. E., 128(F9), 149, 153, 155, *199*
Giannini, G. M., 99(G3), *116*
Gibbs, J. W., 14, 16, *58*
Gilmore, F. R., 73(G4), *116*
Glass, I. I., 68(G5), 102, 104(G6), 105 (G6), 106(G5), 107(G6), 108(G6), *116*
Glasstone, S., 4(G4), *58*
Glick, H. S., 109(G7), 110(H3), *116*
Goldberg, S. A., 93(M7), *117*
Gomez, E. M., 30(P5), *59*
Gomory, R. E., 277(G2), 321(G2), *332*
Graybeal, P. E., 121, *199*
Green, S. J., 167, *199*
Greenwald, D. U., 321(G3), *332*
Gretton, A. T., 127(O6), 135(O3), 136, 150(O6), 155(O6), 180(O4), 184, 185, *200*
Griffiths, H., 48, *58*
Grosse, A. V., 84(G9), 109(G8), 113(G10), *116*
Grove, C. S., Jr., 35, 47(S10), *58, 60*
Gunness, R. C., 127, *199*

H

Hagerty, P. F., 166(S7), *201*
Hall, H. T., 107(B7), *115*
Hanratty, T. J., 30(H1), *58,* 232(H1), 233, 234, *270*
Hansen, C. F., 72(H1), *116*
Harris-Smith, R., 161(E1), 165(E1), *199*
Heath, W. S., 30(H2), *58*
Hedrich, A. L., 71(H2), *116*

Henderson, A., 317(C1), *331*
Herdan, C., 32(H3), *58*
Hertzberg, A., 107(G7), 110((H3), *116*
Herzfeld, K. F., 64(H4), *116*
Hesse, H. C., 127(B8), *198*
Hester, A. S., 32(H4), *58*
Hildebrand, F. B., 206(H4), 235(H5), 238(H4), *270*
Hinze, J. O., 168(H1), *199*
Hirano, Y., 18(H5), *58*
Hirschland, H. E., 127(O6), 150(O6), 155 (O6), *200*
Hirsekorn, F. S., 178, 189, *199*
Hixson, A. W., 27(H7), 28, 30(H6, H8), *58, 59,* 134, 136, 137, 138, 173, 177(H9), 178, 181(H4, H8, H11), 186, *199*
Hodge, H. M., 166(A6), *198*
Hodgman, C. D., 8(H9), *59*
Hoffman, A. N., 166, *199*
Hooker, T., *199*
Hohel, H. C., 92, *116*
Hougen, O. A., 10, 11(H11), 37(H11), *59,* 166(J3), *199,* 233, *270*
Houston, E. C., 42(S5), 49(S5), *60*
Huang, G. J., 182, *199*
Hughes, R. R., 123, 124, *199*
Humphrey, D. W., 182, *199*
Hunsaker, J. C., 130(H16), *199*
Huntington, R. L., 229(C2), 231, 232(C2), *270*
Hyman, M. A., 240(O1), 244(O1), *271*

I

Irvin, H. G., 229(B9), 231(B9), *270*

J

Jacobs, K. H., 99(J1), *116*
Jaffray, J., 18(M11), *59*
Jang, J. J., 37, *59*
Jebens, R. H., 183(C5), 184(C5), *198*
John, G., 161, *199*
Johnson, A. I., 182, *199*
Johnson, D. L., 166, *199*
Johnson, E. O., 71(J2), *117*
Johnson, S. M., 277(J1), 278, 282, 283, *332*
Johnstone, R. E., 187(J4), *199,* 260, *270*
Jones, A. L., 18(D1), 27(D1), *58*

K

Kalinske, A. A., 161, 162(K1), *199*
Kaplan, S., 240(O1), 244(O1), *271*
Karow, E. O., 159(B2), 161(B2), 165(B2, K2), 167(K2), *198, 199*
Karr, A. E., 174, *200*
Kauffman, H. L., 154, *200*
Kawahigasi, Z., 133, *202*
Kelley, J. C. R., 98(W6), *118*
Kells, M. C., 101(O3), *116*
Keon, J. J., 180(K5), *200*
Kiefer, J., 278, 282, 283, *332*
Kimball, G. E., 274(M3), 277(M3), *332*
Kingsley, H. A., 174(O8), *200*
Kirk, R. E., 107(K1), 109(K1), *117*
Kirschenbaum, A. D., 84(G9), *116*
Kjaer, J., 242(K1), 267, *270*
Kneule, F., 166, *200*
Knoll, W. H., 147(K8), 155(K8), *200*
Knox, K. L., 27(H7), 28, *58*
Kobe, K. A., 62(K2), *117*
Kolfenbach, J. J., 151(B3), *198*
Kolmogoroff, A. N., 168(K7), *200*
Korman, S., 111(S1, S2), *118*
Kossel, W., 25(K2), *59*
Kostin, N. M., 179(P2), *201*
Kramers, H., 147, 155, *200,* 212(K2), 232, 233, *270, 271*
Kraussold, H., 184, 185, *198*
Krekels, J. T. C., 30(V4), *60*
Kroll, A. E., 136, 145, *200*
Krzywoblocki, M., 128, *200*
Kuhn, H. W., 323, 325, *332*
Kunii, D., 214(Y1), *271*
Kwong, S. S., 230, *270*

L

Laity, D. S., 172, *200*
LaMer, V. K., 14(L1, L2), 15, 16(L1), 20(L2), 21, *59*
Lamont, A. G. W., 180(L2), *200*
Landenburg, R. W., 71(L1), *117*
Langlois, G. E., 159(V2), 168(V2), 172 (V2), *202*
Lapidus, L., 205, 213, 257, *270,* 278(S2), 292(S2), 293(S2), *332*
Lapple, C. E., 73, 74(L2), *117*
Latinen, G. A., 214(L1), 228, *270*
Lewis, B., 93(B5), *115*
Lichtman, R. S., 184(R15), 185(R15), *201*

Lightfoot, E. N., Jr., 165, *199*
Linton, W. H., 30(L3), *59*
Lochte-Holtgreven, W., 68(L4), 78, 80 (L3, L4), 81(L3), 99(L5), 101(L3), *117*
Longwell, P. A., 102(L6), *117*
Lonsdale, K., 2(L4), *59*
Lotz, A., 99, *116*
Lovell, C. L., 134, 136, 139, *201*
Lowan, A. N., 240, *270*
Lucks, C. F., 70(R1), *118*
Luedecke, V. D., 134, 136, 177(H9), *199*
Lukens, B. E., 136(W6), 178(W6), *202*
Lyons, E. J., 125, 126(L3), 127(L3), 192, *200*

M

McCabe, W. L., 14(M3), 23(M1, M2), 24 (M2), 28, 29, 35(T5), 35(M1), 36, 47 (M2), 48(M2, M3), *59, 60*
McCamy, I. W., 42(S5), 49(S5), *60*
McCune, L. K., 30(M5), *59*
McGinn, J. H., 99(M5), *117*
Mack, D. E., 128(R16), 129(R16), 130 (R16), 136, 144, 145, 157(F8), 162(F8), *199, 200, 201*
Mack, E. M., 177, 182, *200*
McKinsey, J. C. C., 321(M1), *332*
MacLean, G., 125, *200*
MacPherson, H. G., 68(M1), 98(M1), *117*
Maeda, S., 229(M1), 231, *270*
Maeker, H., 99(B9), *115*
Mahoney, L. H., 130(R11), 184(R15), 185(R15), *201*
Malter, L., 71(T2), *117*
Mann, C. A., 171, 173, 189, *200*
Manne, A. S., 321, *332*
Mare, R., 13, *59*
Margoshes, M., 113, *117*
Margrave, J. L., 97, *117*
Markovitz, H., 323, *332*
Marriner, R. E., 177, 182, *200*
Marshall, W. R., Jr., 231, 232(C4), 233, *270*
Martelli, G., 18(B3), *58*
Martin, J. J., *200*
Martin, W. A., 68(G5), 102(G5), 104 (G6), 105(G6), 106(G5), 107(G6), 108 (G6), *116*
Martinek, F., 76(M4), *117*
Mason, W. D., 162(N3), *200*

Matveen, S. F., 179(P2), *201*
Matz, G., 48(M7), *59*
Mellon, B., 321(C2), *331*
Meny, R. B., 154, *200*
Metzner, A. B., 145(M8), *200*
Mickley, H. S., 235(M2), *271*
Miers, H., 12, *59*
Millard, B., 44(B6), 46(B6), *58*
Miller, F. D., 127, *198, 200*
Miller, P., 42(M9), 44(M9), 49(M9), *59*
Miller, R. S., 213, 223, *270*
Miller, S. A., 162(C7), 163(C7), 171, 173, 178, 189, *198, 200*
Milne, W. E., 235(M3), *271*
Möhle, W., 151, *200*
Mohler, F. L., 70(M6), 71(M6), 78, *117*
Molstad, M. C., 166(S7), *201*
Montgomery, J. B., 166(H12), *199*
Montillon, G. H., 35, *59*
Montmory, R., 18(M11), *59*
Moore, J. K., 166(H12), *199*
Morales, M., 225(M4), *271*
Morrison, M. S., 135(W5), *202*
Morse, P. M., 274(M3), 277(M3), *332*
Morton, A. A., *200*
Murphree, E. V., 30(M12), *59*, 180, *200*
Myers, J. W., 93, *117*

N

Nagata, S., 136, 151, *200*
Nancollas, G. H., 18(D1), 27(D1), *58*
Nelson, H. A., 162(N3), *200*
Newitt, D. M., 127, *200*
Newkirk, J. B., 22, *59*
Newman, H. H., 26(N3), 49, 50, 51, *59*
Nielsen, A.E., 14(C4), 18(N4), 20(C4), 26(N4), 27(N5), 37, 38, 39, *58, 59*
Nikolskii, B. P., 30(N6), *59*

O

O'Brian, G. G., 240, 244(O1), *271*
O'Connell, F. R., 136, 144, *200*
Okress, E. K., 98(O1, W6), *117, 118*
Oldshue, J. Y., 125, 126(R12), 127(R12), 129(R12), 135(O3), 136, 150, 155, 164 (R14), 174, 180(O4), 184, 185, 194, 195, *200, 201*
Olney, R. B., 136, 171, 174(O8), *200*
Olson, R. W., 229(B9), 231(B9), *270*
Ornstein, L. S., 79(O2), *117*
Ostwald, W., 12, 14, *59*

Othmer, D. F., 107(K1), 109(K1), *117*
Overcashier, R. H., 174, *200*
Oyama, Y., 130, 161, 162(O11), 178, 182, 189(O12, 13), *199, 200*

P

Pai, S., 128, *200*
Paladino, S., 165(C4), *198*
Palermo, J. A., 36, 47(S10), *59, 60*
Pardue, D. Q., 71(H2), *116*
Parker, N. H., 192, *200*
Patterson, G. N., 68(G5), 102(G5), 106 (G5), *116*
Pavlushenko, I. S., 179, *201*
Peck, A. C., 109(D1), *117*
Penner, S. S., 82(P2, P3), 86, 94, 109, *117*
Perry, J. H., 8(P2), *59*, 126(P3), 127(P3), 137(P3), 177(P3), 181(P3), *201*
Perry, R. L., 154, *198, 201*
Peters, T., 99(B9, P5), *115, 117*
Peterson, M. H., *199*
Phillips, G. A., 135(W5), *202*
Phillips, H. B., 290(P1), *332*
Pierce, D. E., 183(P6), *201*
Pigford, R. L., 157, 160(S5), 168, 181(S5), *201*
Polack, J. A., 230, *271*
Polejes, J., 166(J3), *199*
Poritsky, H., 71(P6), *117*
Pound, G. M., 20, *59*
Pramuk, F. S., 186, *201*
Prausnitz, J. M., 212(C1), *270*
Preckshot, G. W., 20(P4), 21, 22, *59*
Preining, O., 99(P7), *118*
Prelat, C. E., 30(P5), *59*
Prosser, L. E., 128(F7), 151, *199*
Pursell, H., 184, 186, *201*

Q

Quinton, J. H., 229(Q1), 231, 232(Q1), 233, *271*

R

Ranz, W. E., 212(R1), *271*
Rea, H. E., 169(F1), 172(F1), 176(F1), *199*
Read, W. T., 26(R2), *59*
Reamer, H. H., 102(L6), *117*
Reavell, B. N., 127, *201*
Reed, C. E., 235(M2), *271*
Reike, H., 33(A2), *58*

Rhodes, F. H., 183(R2), *201*
Rhodes, R. P., 166(R3), *201*
Ricci, J. E., 4(R3), 30(R3), *59*
Richardson, J. F., 18, 19, 48(C5), *58*
Richardson, L. F., 240, *271*
Riegel, E. R., 134(R4), *201*
Rightmire, B. G., 130(H16), *199*
Roberts, A. G., 131, *201*
Roberts, S. M., 293(R1), 296(R1), *332*
Rodger, W. H., 169, *201, 202*
Rosenbaum, G. P., 24(G2), 48, *58*
Roth, N. G., 166(A6), *198*
Roxburgh, J. M., 165, *198*
Rushton, J. H., 125(B6), 126(R12), 127 (B8, S2), 128, 129, 130, 131, 134(R13), 136, 139, 140, 141, 143, 145, 151(R10), 157(F8), 162(F8), 164, 169(R6), 174, 184, 185, 189, 194, 195, *198, 199, 200, 201*
Russell, H. W., 70(R1), *118*

S

Sachs, J. P., 125, 127(S2), 131, 132, *201*
Saeman, W. C., 42(M9, S1, S4, S5), 43, 44(M9, S1), 45, 47, 49(M9, S1, S2, S3, S4, S5), 52, 54, 55, *59, 60*
Sage, B. H., 102(L6), *117*
Saito, H., 166(J3), *199*
Satterfield, C. N., 92(H5), *116*
Sauer, T. C., 30(W5), *60*, 181(W7), *202*
Schaefer, V., 14(S6), 18(S6), *60*
Scheibel, E. G., 174, *200*
Schierholtz, O. J., 39, 40, *60*
Schlichtkrull, J., 40, 41, 42, *60*
Schoen, H. M., 47(S10), 48(S9), *60*
Schrage, R. W., 292(S1), *332*
Schultz, J. S., 165, *201*
Schultze, W., 25, *60*
Schwartz, C. E., 225(S2), 227, *271*
Scott, M. A., 184(B9), *198*
Scribner, B. F., 113, *117*
Sehr, R. A., 230(S3), *271*
Seitz, F., 3(S11), *60*
Sfat, M. R., 159(B2), 161(B2), 165(B2, K2), 167(K2), *198, 199*
Shanahan, C. E. A., *201*
Shapiro, E., 278, 292(S2), 293(S2), *332*
Shapiro, S., 278(S2), 292(S2), 293(S2), *332*
Sheer, C., 111(S1, S2), *118*
Sherwood, T. K., 30(L3), *59*, 157, 160(S5), 168, 181(S5), *201*, 235(M2), *271*

Shipp, G. C., 127(N4), *200*
Shu, P., 166, *201*
Shurter, R. A., 161, 162(W1), *202*
Singer, E., 212(S4), 230, *271*
Sinke, G. C., 97(S6), *118*
Smith, J. C., 14(M3), 29, 48(M3), *59*
Smith, J. M., 216, 225, 227, 228, 229(B9), 230, 231(B9), 242(S5), *269, 270, 271*
Smith, N. O., 4(C2), *58*
Smith, R. W., 93(M7), *117*
Smith, S., 100(A3), *115*
Smit-Miessen, M. M., 69(S3), 79(S3), 80(S3), *118*
Snyder, J. R., 166, *201*
Spier, J. L., 69(S3), 79(S3), 80(S3), *118*
Spinn, C. W., 225(M4), *271*
Squire, W., 109(G7), 110(H3), *116*
Steffans, C., 68(S4), 88(S4), *118*
Stevens, R. P., 28, *59*
Stillman, R. E., 278(S2), 292(S2), 293 (S2), *332*
Stillwell, C. W., 6(S12), *60*
Stoops, C. E., 134, 136, 139, *201*
Storrow, J. A., 230, 231, 232(Q1), 233, *270, 271*
Stranski, I. N., 25(S13), *60*
Strong, H. M., 107(B7), *115*
Stull, D. R., 97(S6), *118*
Suits, C. G., 71(P6, S7), 74, *117, 118*
Summerford, S. D., 136(W6), 178, *202*
Sylvester, J. C., *199*
Symonds, G. H., 321, *332*
Szware, W., 277(S4), *332*

T

Tammann, G., 18, *60*
Taylor, J. S., 125, 132, 146(T1), *201*
Telkes, M., 18(T2), 22, *60*
Tennant, B. W., 132, *201*
Tenney, A. H., 137(H10), 173, 178, *199*
Tereschkevitz, W., 160, *201*
Terry, P. B., 183(P6), *201*
Teverovskii, E. N., 168(B1), *198*
Thoenes, D., Jr., 232, 233, *271*
Thompson, A. R., 16(T3), 48(T3), *60*
Thring, M. W., 187(J4), *199, 260, 270*
Ting, H. H., 32(T5), *60*
Tolloczko, S. T., 30(B7), *58*
Toyne, C., 184(B9), *198*
Tregubova, E. L., 168(B1), *198*

Treybal, R. E., 168, 171, 172, 174(F3), 176, 189, *199, 200, 201, 202*
Trice, V. G., 169(R6, T6), *201, 202*
Tucker, A. W., 323, 325, *332*
Turnbull, D., 18(T6), 22, *60*
Turnbull, L. G., 70(R1), *118*
Tyler, W. S., 32(T7), *60*

U

Ugolini, F., 165(C4), *198*
Uhl, V. W., 136, 184, 186, *200, 202*
Umnik, N. N., 229 (A1a), 231, 232(A2), 233, *269*
Unsöld, A., 69(U1), *118*

V

Vadja, S., 321, *332*
Valeton, J. J. P., 26, *60*
Van der Sluis, J., 165(C4), *198*
van de Vusse, J. G., 135(V1), 136, 148, 149, 155, *202*
Van Hook, A., 18, 21, 27(V2), *60*
Van Krevelen, D. W., 30(V4), *60*
Van Ness, H. C., 182, *199*
Velykis, R. B., 154, *200*
Verma, A. R., 26(V5), 30(V5), *60*
Vermeulen, T., 159, 168, 169(F1), 172 F1), 176(F2), *199, 202*
Vesterdal, H. G., 151(B3), *198*
Vinyard, M. N., 171(W8), 174(W8), *202*
Volmer, M., 14, 20, 21, 25, *60*
Vonnegut, B., 18(T6), 22, *60*
Von Weimarn, P., 20(V10), *60*

W

Wachman, H., 99(M5), *117*
Waeser, B., 151, *200*
Wagner, C., 30(W1), *60*
Wakao, 233, *271*
Walas, S. M., 242(W1), *271*
Wall, F. T., 7(W2), *60*
Watson, K. M., 10, 11(H11), 37(H11), *59*, 233, *270*
Wegrich, O. G., 161, 162(W1), *202*
Wehner, J. F., 213, *271*
Weidenbaum, S. S., 121, 123, *202*
Weinberg, F. J., 93, *118*

Wells, A. F., 23(W3), *60*
Wentorff, R. H., 107(B7), *115*
West, A. S., 184, 185, *198*
Westwater, J. W., 186, *201*
Whetstone, J., 24(W4), *60*
White, A. M., 135, 136, 178, *202*
Whittemore, E. R., 125(W10), 127(W10), 134(W10), 147(W10), *202*
Wilbur, D. A., 68(C4), 96, *115*
Wilburn, N. P., 102(L6), *117*
Wildhack, W. A., 73(W2), *118*
Wilhelm, R. H., 30(M5, W5), *59, 60,* 159(B2), 161(B2), 165(B2), 181(W7), *198, 202,* 212(S4), 213, 214(B6), 227, 230, *270, 271*
Wilkens, G. A., 30(H8), *59,* 181(H11), *199*
Williams, C. M., 159(V2), 168(V2), 172(V2), *202*
Williams, G. C., 92(H5), *116*
Williams, J. D., 321(W1), *332*
Williams, V., 161(E1), 165(E1), *199*
Wilson, K. B., 258, *271,* 276, 278, 291, *331*
Wingrad, R. E., 171, 174, *202*
Winter, E. F., 131, *202*
Winternitz, P. F., 93, *118*
Wolfe, P., 276, 293(D1), 315(W2), 323, *331, 332*
Wolfhard, H. G., 88(G1), 89(G1), 90(G1), *116*
Wood, J. C., 125(W10), 127, 134(W10), 147, *202*
Woolley, H. W., 97, *118*
Wroughton, D. M., 98(O1, W6), *117, 118*

Y

Yagi, S., 214(Y1), 233, *271*
Yamamoto, K., 133, *202*
Yanagimoto, M., 151(N2), *200*
Yokoyama, T., 136, 151(N2), *200*
Yoon, C. Y., 214(Y3), *271*

Z

Zdanovskii, A. B., 30(Z1), *60*
Zwietering, T. N., 180, *202*

SUBJECT INDEX

A

Absorption of oxygen in air oxidation, 163–164, 165
Additive property, calculation of, 209
Agitation, 119–202
   of liquid-liquid systems, interfacial area, 168 ff
   solid-liquid, critical impeller speed, 178 ff
   variables, 193–195
Agitators, rotating, 147–151
Algorithm, simplex, 321 ff
Arc, Gerdian, 99
Arcs, 98–99
Argo-Smith theory, 229–230

B

Baffles, effect of, 126–127
Barium sulfate precipitation, 38
Barkelew's criterion, 258–259
Becker and Doering equation, 20
Berthoud-Valeton crystal growth model, 26–27
Black body radiation, 82
   intensity, 62
Blade length effect on impeller power, 144
Blending
   of liquids, gas jet, 154
   optimal, with diminishing returns, 326–328

C

Calcium sulfate dihydrate growth rate, 39–40
Carnot cycle, 62
Catalyst
   activity, 260, 264
   pellets, bed of, 211
Cells, unit, 5
Chemical equations, reduction to an independent set, 205–211

Combustion
   of ethylene with oxygen, 87–92
   products, equilibrium composition of, 85–87
   rocket propellant, 85
   temperature, 85–87
Composition in terms of concentration of key components, 208
Conductivity, effective thermal, 229–232
Contact time, gas, 158
Contours, 288
Contour tangent, 288–290
Conversions
   relation between concentrations and, 207–209
   virtual, 209–210
      of selected key reactions, 218–219
Cooling medium mass rate of flow, 236
Crystal, 2
   growth, 22–31
      spiral, 26
      theories, 26–28
   habit, 24
   size determination, 31–47
   surface developement, 24–26
   types, 2–3
Crystallization
   continuous, 42–47
   equipment, 47–56
   from solution, 1–60
Crystallizers, 47–56
   circulating magma, 50–52
   design criteria, 52–53
   operation, 54–56
   perfectly-mixed, 44–47
   vacuum, 50–52, 55

D

Dantzig's simplex method, 276
Decision
   optimal, 294, 301–302
   problems, multistage, 293

SUBJECT INDEX

Decomposition principle, 315
  for linear programming, 330–331
Density, effective mean, 169
Derivative, central-difference form approximation, 238, 242
Difference equations, first-order, solution of, 237
Diffusion
  eddy, 212
    material balance in, 216
    nature of transverse, 214–216
    operator, 215–216, 221
  molecular, 121 ff
  processes, axial, 212–214
Diffusivity, eddy, 227–229
Direction of steepest ascent, 290
Discharge, rate of, 128–132
Dislocations, screw, 25–26
Dissolution, 4, 30–31
Doppler broadening effect, 77–78
Dynamics, bubble, 157 ff

E

Electron
  free energy function, 97
  temperature, 70–71
Embryo, crystal rate-determining, 14
Energies
  bond, for diatomic molecules, 84–85
  statistical, 65–66
Energy
  conservation, 102
  free, of nucleation, 16–17
  modes, 64–65
Enthalpy balance, 216–219
  relation to temperature and composition, 216–217
  expansions in terms of measured quanties, 219
Entropy, particle, 97
Error
  order of, 239–240
  truncation, 237, 239
Ethynol, in ethynol-water system, molecular diffusion time, 124
Excitation temperatures, electronic, 78–79
Expansion of gas through De Laval nozzle, 93–96

Extraction rates in liquid-liquid systems, 174–176

F

Faces, parallel displacement of, principle of, 23
Feasibility
  in interacting systems, 328–331
  problems, 275–276, 314–331
Fibonacci
  numbers, 282
  search, 282–283, 285
Fines removal, 52–54
Flow
  isentropic, criteria for, 95
  patterns, 125
  visualization, 131
Fluid induction, volumetric rate, 154
Flux
  radiant, 78
  superficial diffusive, 215
Forces, interatomic, 6
Frenkel formula for nucleogenesis, 16–17
Friction factor, 234–235
Fronde group, 139, 141
Fugacity relationships at saturation, 6–7

G

Gas
  density measurement of temperature, 74–77
  "holdup," 158 ff
  -liquid systems, 157–167
  temperatures, 71–78
Gassing, effect of on power requirements, 161–162
Gradient direction, 290
Growth
  constant, crystallization, 36
  crystal, 22–31
  of potassium alum crystals, 36
  of potassium chloride, 36
  rate, linear, concentration dependence, 39

H

Heat
  balance equations, 236
  of reaction, 87, 92
  transfer

in agitated systems, 183–187
  coefficient, 262
    over-all, 222–223
    to tube wall, 232–234
    at wall of reactor, 229–230
  radial, in packed reactor, 265
  radiative, special treatment, 249
  transport in packed beds, 229–230
Heating
  induction, 98
  resistive, 98

I

Impeller, 126, 127
  diameter, influence of, 194–195
  discharge flow, 194
  Reynold's number, 189–190
  rotating, power input to, 134
  speed, "critical," in solid-liquid agitation, 178 ff
Impulse, specific, 94
Inequalities, 316
Insulin crystal growth, 40–42
Intensity
  blackbody radiation, 62
  of segregation, 122
Interaction problems, 275–276, 292–314
Interval effectiveness ratio, 282
Ionization temperature, 69

J

Jet mixing, 128
Jets, free, 151–154

K

Kelvin-Gibbs equation for vapor pressure of droplet, 16
Kinetics, nucleation, 17–22
Kossel site, 25

L

La Mer formula for free energy change in nucleation, 16
Lapple equation for mass rate of gas flow, 73
Length, "cut-off," 123
Liquid-liquid systems, 167–176

M

Mach number, 103
Mass
  conservation, 102
  rate of flow, 244
  transfer
    coefficient, 28–29
    from bubbles, liquid shear effect, 160–161
    in solid-liquid agitation, 180–183
Material balance, use of to introduce rates, 217–218
Matrix, stochiometric
  complete, 205–206
  reduced, 206–207
Maxwell-Boltzmann Law, 63
McCabes "ΔL law," 23, 36–37
Minimax plan, 281
Mixedness, 122
Mixing
  index
    in liquid-liquid systems, 173–174
    in sand-water suspensions, 178
  processes, 119–202
    general characteristics, 121–133
    systems, performance parameters of, 191–192
  time, 146–157
    effect of impeller type on, 148–149
Mole fractions, calculation of, 208–209
Momentum conservation, 102

N

Nitrogen, atmospheric, fixation of, 108–109
Nucleation, 13–22
  energy relationships, 15–17
  temperature effect, 18–19
Nucleogenesis, Frenkel equation for, 16–17
Nusselt number, 234

O

Operations analysis in chemical engineering, 274–275
Optimization
  methods, 273–332
  simple, 293–295

Oxidation, air, of sodium sulfite solution, 163
Oxygen transfer in anti-biotic production, 165

P

Particle size distribution function, 35
Peclet number, 213
 change of, 224
 relation to Reynolds number, 228
Planck's law, 62, 82
Plasma jet, 99–100
Potassium
 alum growth, 36
 chloride growth, 36
Power
 consumption in oil-water system agitation, 171
 input
  to mixing systems, 194
  per unit volume of material to be mixed, 189
 number, 137, 139, 145
 requirements
  effect of gassing, 161–162
  gas-liquid systems, 161–162
  liquid-liquid systems, 171–173
  one liquid-phase systems, 134–146
  solid-liquid agitated systems, 177
Precipitation of barium sulfate, 38
Pressure
 change across shock front, 103, 104
 drop in catalytic reactor, 246–247
 fictitious partial, 88
 of reactor cooling medium, 243
 within catalytic reactor, 223–224
Problem
 basic solution of, 318
 feasible blend, 317–320
Product removal, crystal size distribution, 43, 47
Program, long-range construction and repair, 302–314
Programming
 dynamic, 293, 297–314
  example, 302–314
  generalization, 300–301
  last stage, 297–298
  last two stages, 298–300
  optimal decisions, 301–302
 linear, 315, 320–323
  decomposition principle for, 330–331
 mathematical, 276, 314
 quadratic, 315, 323–328
  example, 326–328
  theoretical, 324–326
Propeller pitch effect on impeller power, 144

R

Rate
 of crystal growth, 24–29
 equations, reduction to an independent set, 205–211
 relationship, nucleation, 19
 of solution in solid-liquid systems, 180–183
Rates
 use of material balance to introduce, 217–218
 virtual, of selected key reactions, 218–219
Reaction and transport processes, simultaneous, 211–224
Reactions, highly exothermic, flame temperatures for, 83–84
Reactors
 adiabatic catalytic, 222, 249–250
 catalytic, calculation of pressure within, 223–224
 packed catalytic, 203–271
 packed tubular, stability of, 257–259
 tubular catalytic
  boundary conditions, 221–222
  equations for, 219–221
  hot spot in, 284–286
  numerical calculation of equations, 235–257
   explicit partial difference equations, 241–244
   implicit partial difference equations, 244–249
   introduction of r^2 as radial variable, 240–241
   one-dimensional approximation, 235–236
   sources of error, 236–240

stability in partial difference equations, 240
Recombination
  atom-atom, 96
  ion-atom, 96-98

S

Sackur-Tetrode equation, 97
Saha equation, 69
Scale effects, 290–292
Scale-up
  of heterogeneous agitated systems, 187–190
  of mixing vessels, 154–156
Scaling of packed tubular reactors, 259–268
Search
  Fibonacci, 282–283, 285
  lines of, 287
  problems, 275–276, 277–292
    multivariable, 286–292
    one variable, 279–286
    sequential dichotomous, 281–282
    two-variable dichotomous, 292
    uniform, 280–281
    with a variable number of experiments, 284
Segregation scale, 122
Sensitivity, parametric, 257
Series-optimization problems, 293, 295–296
Sheer-Korman high intensity arc, 111
Shock
  front
    pressure change across, 103, 104
    temperature change across, 102
  tube, temperature rise in, 105, 106
  waves, 101–107
Similarity, geometrical, 188, 192–193
Simplex method, 276, 321 ff
Size distribution, continuous crystallization, 45–46
Slackness condition, complementary, 325
Solid-liquid agitated systems, 176–183
Solubility, 6
  crystal size dependence, 9
  principle, 4–13
Solution, 4
  crystallization from, 1–60
  seeded, crystal growth in, 32–42

Sound velocity measurement of gas temperature, 71–73
Spargers, effect on system performance, 166
Stagnation region, 112
Stark effect, 80
Stefan-Boltzmann Law, 82
Substance, crystalline, 2
Supersaturation, 6, 11–13
  nucleation and crystal growth rate dependence, 32–33
  profile for unagitated bath operation, 55–56
Supersolubility, 12–13
Supersystems, 328–330
Surface area distribution, differential, 34
Surfaces, response, 286–287
Suspension
  sand-water, 178, 179
  solids, 178–180
System model, 274
Systems
  crystallographic, 3–4
  gas-liquid, 157–167
    basic studies, 157–161
    performance studies, 162–166
    power requirements, 161–162
  liquid-liquid, 167–176
    basic studies, 168–171
    performance studies, 173–176
    power requirements, 171–173
  one-liquid-phase, 133–157
    performance studies, 146–157
    power requirements, 134–146
  solid-liquid agitated, 176–183
    basic studies, 177
    performance studies, 178–183
    power requirements, 177

T

Tail flame, 98, 99
Tank diameter effect on impeller power, 144
Taylor expansion, 239, 289
Temperature
  colorimetric, 83
  combustion, 85–87
  definition, 62, 63
  effect on nucleation, 18–19

empirical, 69, 81–83
high, 61–118
   electrical sources, 96–101
   equilibrium factors, 67–68
   means for attaining, 83–107
   measurement, 70–83
   mechanical sources, 101–107
  ionization, 69, 79–81
  line reversal, 82
  of reactor cooling medium, 243
  rotational, 79
  scale, international, 63
  statistical-mechanical, 63–68, 70–79
  thermodynamic, 62, 70
  total brightness, 82
  translational, 70–78
  vibrational, 79
Thrust, rocket engine, 93–94
Time
  -lag in nucleation, 20–21
  mixing, 146–157
  total recirculation, 152
Torque measurement, 134
Transport
  of heat in packed beds, 229–230
  radiative, 230–231
  and reaction processes, simultaneous, 211–224
Truncation error, 237, 239
Turbulence
  parameters, 132–133
  production, 127

U
Unimodality, 279–280, 287–288

V
Vapor pressure, droplet, 16
Variable
  nonbasic, 318
  slack, 316
Velocity
  of fluid discharge, 128–132
  jet, minimum, 152
  profile in packed tube, 225–227
  superficial air flow, 158
  translation, 23
Vessels, agitated, 121–133
Viscosity, effective mean, in oil-water mixtures, 172

W
Weber group, 170
Weight
  distribution
    crystal, 34
    differential, in continuous crystallization, 44–45
  statistical, 64
Wilson's criterion, 258
Wire, exploding, 100–101
Work of forming a new phase, Gibb's formula for, 16

Y
Yield, optimal, 294, 301

DOES NOT CIRCULATE